Lecture Notes in Physics

Edited by J. Ehlers, München, K. Hepp, Zürich and
H. A. Weidenmüller, Heidelberg
Managing Editor: W. Beiglböck, Heidelberg

19

Proceedings of the Third International Conference on Numerical Methods in Fluid Mechanics

Vol. II
Problems of Fluid Mechanics

July 3-7, 1972
Universities of Paris VI and XI
Edited by Henri Cabannes and Roger Temam

Springer-Verlag
Berlin · Heidelberg · New York 1973

ISBN 3-540-06171-1 Springer-Verlag Berlin · Heidelberg · New York
ISBN 0-387-06171-1 Springer-Verlag New York · Heidelberg · Berlin

Offsetprinting and bookbinding: Julius Beltz, Hemsbach/Bergstr.

Editors' Preface

This issue of Lecture Notes in Physics is the second of two volumes constituting the Proceedings of the Third International Conference on Numerical Methods in Fluid Mechanics, which was held at the University of Paris VI, from July 3 to 7, 1972. Three general lectures and forty eight short individual communications were presented at this conference; the complete proceedings are published here. The general lectures were given by Professor A. DORODNICYN, Director of the Computing Center of the Academy of Sciences of the Soviet Union, who presented the Soviet works dealing with the solution of Navier-Stokes equations; by P. MOREL, professor at the University of Paris VI and Director at the Laboratory of Dynamical Meteorology of the National Center of scientific research (C.N.R.S.), who presented the Problems of numerical simulation of geophysical flows; by Professor R.D. RICHTMYER of the University of Colorado, U.S.A., who spoke on Methods for (generally unsteady) Flows with Shocks.

The individual communications have been separated into two groups: Fundamental Numerical Techniques and Problems of Fluid Mechanics; in each group they are published in the alphabetic order of the author, or of the first of the authors.

Volume I contains the three general lectures and the thirteen communications on Fundamental Numerical Techniques. Volume II contains the thirty five communications on Problems of Fluid Mechanics.

This Conference follows the conferences with the same topic hold at Novossibirsk, U.S.S.S. in 1969, and at Berkeley, U.S.A. in 1970 (the proceedings of which appeared in Lecture Notes in Physics, Vol. 8). The French Organizing Committee was sponsored by Commissariat à l'Energie Atomique, Electricité de France, Union des Chambres Syndicales des Industries du Pétrole, in France, and also by the Office of Naval Research and Air Force Office of Scientific Research, in the U.S.A. The Universities of Paris VI and Paris XI, and the Centre National de la Recherche Scientifique also helped the Committee in a much appreciated manner.

We wish to thank all the persons who contributed to the success of the Conference, the participants for their scientific contributions, our colleagues and younger researchers for their help in the organization and Mrs. M.T. CARTIER and Miss S. DELABEYE for their excellent secretarial work.

Finally we wish to express our appreciation to Dr. W. BEIGLBÖCK and the Springer-Verlag Company for the rapid publication of these proceedings in the series of Lecture Notes in Physics.

January 25, 1973

HENRI CABANNES
ROGER TEMAM

Contents

Volume I

General Lectures

Fundamental Numerical Techniques

Volume II

A NUMERICAL METHOD FOR HIGHLY ACCELERATED LAMINAR BOUNDARY-LAYER FLOWS

R.C. Ackerberg and J.H. Phillips

Polytechnic Institute of Brooklyn Graduate Center
Farmingdale, New York, 11735, U.S.A.

A second-order-accurate implicit finite difference method is developed to study the boundary-layer flows that occur just upstream of a trailing edge which is attached to a free streamline. An important feature of this technique is the use of an asymptotic expansion to satisfy the boundary condition at the edge of the boundary layer while retaining a rapid algorithm for inverting the system of linear equations for each Newton iteration. The method is applied to the Kirchhoff-Rayleigh flow past a finite flat plate set perpendicular to a uniform stream. Computed velocity profiles are found to be in excellent agreement with those obtained from an asymptotic solution (Ackerberg (1970), (1971a), (1971b)) with pointwise differences being less than 1.2% over two-thirds of the profile. A detailed description of the method is given in Ackerberg and Phillips (1973).

This work was supported by the U.S. Army Research Office — Durham under Grant No. DA-ARO-D-31-124-71-G68.

References

1. Ackerberg, R. C. "Boundary-Layer Separation at a Free Streamline. Part 1. Two-dimensional Flow". J. Fluid Mech. 44, p. 211, (1970).

2. Ackerberg, R. C. "Boundary-Layer Separation at a Free Streamline. Part 2. Numerical Results". J. Fluid Mech. 46, p. 727, (1971a).

3. Ackerberg, R. C. "Boundary-Layer Separation at a Free Streamline - Finite Difference Calculations". Proceedings of the Second International Conference on Numerical Methods in Fluid Dynamics edited by M. Holt. Published as Lecture Notes in Physics No. 8, Springer-Verlag, p. 170, (1971b).

4. Ackerberg, R.C. and Phillips, J.H. "A Numerical Method for Highly Accelerated Laminar Boundary-Layer Flows". SIAM Journal on Numerical Analysis 10, part 1, (1973).

RELAXATION METHODS FOR TRANSONIC FLOW ABOUT WING-CYLINDER COMBINATIONS AND LIFTING SWEPT WINGS

Frank R. Bailey

Ames Research Center, NASA

Moffett Field, California 94035

and

William F. Ballhaus

U. S Army Air Mobility Research and Development Laboratory

Moffett Field, California 94035

INTRODUCTION

It has recently been demonstrated that relaxation methods are a powerful numerical tool for obtaining steady-state solutions to the two-dimensional transonic potential equations. The basic numerical procedure, first introduced by Murman and Cole (1971), accounts for the mixed elliptic-hyperbolic character of the governing equations by using a mixed finite-difference scheme. The general procedure is to employ centered differences when the flow is locally subsonic and one-sided differences when it is locally supersonic. In this paper we extend the mixed elliptic-hyperbolic relaxation method to the transonic small disturbance equation in three dimensions. In particular, we consider transonic flow over thin lifting wings with sweep and taper and about nonlifting wing-cylinder combinations. We restrict our treatment to freestream Mach numbers less than one and to wings with subsonic trailing edges.

BASIC EQUATION AND BOUNDARY CONDITIONS

The governing equation for small disturbance transonic perturbation potential (Spreiter (1953) and the corresponding pressure coefficient can be written

$$\left(1 - M_\infty^2 - (\gamma + 1)\, M_\infty^2\, \phi_x\right) \phi_{xx} + \phi_{yy} + \phi_{zz} = 0 \tag{1}$$

$$C_p = -2\, \phi_x \tag{2}$$

respectively, where M_∞ is the free-stream Mach number and the ϕ is the perturbation potential divided by the free-stream velocity. In small disturbance theory the flow tangency condition at the wing surface is linearized and applied on the wing mean plane ($z = 0$) giving

$$\phi_z\Big|_{u,\ell} = \left.\frac{df}{dx}\right)_{u,\ell} \tag{3}$$

where $\left.\frac{df}{dx}\right|_u$ and $\left.\frac{df}{dx}\right|_\ell$ are the slopes of the upper and lower surfaces and include the effect of thickness, camber and angle of attack. In the case of a lifting wing the Kutta condition is applied, thus forcing the flow to leave all subsonic trailing edges smoothly. In the small disturbance theory the Kutta condition is satisfied by requiring that ϕ_x (pressure) be continuous across the trailing edge. In addition, provision must be made for a trailing vortex sheet downstream of the wing trailing edge. The vortex sheet is assumed to be straight and lie in the wing mean plane $z = 0$ with the conditions that ϕ_x and ϕ_z be continuous and ϕ be discontinuous through it. Due to the continuity of pressure through the vortex sheet, the jump in potential at any span station, $y = y_o$, is independent of x and is equal to the circulation about the wing section defined by

$$\Gamma(y_o) = -\oint d\phi(x,y_o,z) \tag{4}$$

for any path enclosing the wing section.

The outer flow boundary conditions for a nonlifting wing are that the perturbation velocities tend to zero with increasing distance from the wing. In the numerical method this is approximated by specifying free-stream conditions far from the wing. In the case of a lifting wing the pertrubation velocities, ϕ_y and ϕ_z, far downstream do not vanish due to the presence of the vortex sheet. At an infinite distance downstream the motion due to the vortex sheet becomes two-dimensional in the (y,z) Trefftz plane and this motion is described by the two-dimensional Laplace equation.

BASIC NUMERICAL PROCEDURE

The basic feature of the numerical method is to account for the mixed elliptic-hyperbolic nature of the governing transonic equation by central differencing the streamwise derivatives when the coefficient of ϕ_{xx} is positive and backward differencing when the coefficient is negative. Consider a three-dimensional rectangular domain and let the mesh be evenly spaced with the streamwise coordinate, $x = j\Delta x$, the spamwise coordinate, $y = k\Delta y$, and the vertical coordinate, $z = \ell\Delta z$. At each mesh point the equation type (i.e., elliptic or hyperbolic), is determined by the sign of the expression

$$V = 1 - M_\infty^2 - (\gamma + 1) M_\infty^2 \frac{\phi_{j+1} - \phi_{j-1}}{2\Delta x} \tag{5}$$

If $V > 0$ the flow is subsonic, and the x derivatives are approximated by the centered difference

$$\left[1 - M_\infty^2 - (\gamma+1) M_\infty^2 \phi_x\right]\phi_{xx} = \left[1 - M_\infty^2 - (\gamma+1) M_\infty^2 \frac{(\phi_{j+1} - \phi_{j-1})}{2\Delta x}\right] \frac{\phi_{j+1} - 2\phi_j + \phi_{j-1}}{(\Delta x)^2} \tag{6}$$

If $V > 0$ the flow is supersonic, and the x derivatives are approximated by the backward difference

$$\left[1 - M_\infty^2 - (\gamma+1) M_\infty^2 \phi_x\right]\phi_{xx} = \left[1 - M_\infty^2 - (\gamma+1) M_\infty^2 \frac{\phi_j - \phi_{j-1}}{2\Delta x}\right] \frac{\phi_{j+1} - 2\phi_j + \phi_{j-2}}{(\Delta x)^2} \tag{7}$$

Notice that the derivative ϕ_x is also backward differenced. The y and z derivatives are replaced everywhere by the usual centered formula except at the wing root, $k = 0$, where the symmetry condition gives

$$\phi_{yy} = 2\frac{(\phi_1 - \phi_2)}{(\Delta y)^2} \tag{8}$$

and at the boundary $\ell = 1$ which is placed half a mesh spacing off the $z = 0$ plane. At these points the wing boundary condition is incorporated by writing

$$\phi_{zz} = \frac{1}{\Delta z}\left[\frac{\phi_2 - \phi_1}{\Delta z} - (\phi_z)_{z=o}\right] \tag{9}$$

Note that applying the wing boundary condition in this manner requires that the values of ϕ on the wing mean plane itself must be found by some procedure such as extrapolation. Studies made by Krupp (1971) on solutions for blunt nosed lifting airfoils have shown that in the region of the nose the best results are obtained by linear extrapolation.

The set of nonlinear algebraic equations obtained from the difference formulas are solved iteratively by a line-relaxation algorithm. Each vertical line is successively relaxed by marching toward the increasing y direction in an x = constant plane; the process is repeated for each x = constant plane in the increasing x-direction.

NONLIFTING WINGS AND WING-BODY COMBINATIONS

The numerical method can be applied to rectangular nonlifting wings in a straightforward manner. For swept and tapered wings, however, complications can occur, since in general the boundary points defining the wing shape do not fit naturally in a Cartesian grid network. A special case, which can be easily described in a Cartesian grid is that of an untapered swept wing. In this case, an equally spaced mesh may be used with $\Delta y = \Delta x/\tan\Lambda$, where Λ is the sweep angle, thus permitting the same number of chordwise mesh points on each wing section. An illustrative example is shown in Fig. 1 for a 30° sweptback wing of aspect ratio 4 and with a 6 percent (streamwise) parabolic arc section. The results calculated for $M_\infty = 0.908$ show an embedded shock wave at the wing root that weakens and becomes oblique as it proceeds outboard from the root. The calculations for this example were carried out on a 70 X 31 X 21 (xyz) grid which was evenly spaced in the (xy) plane (Δx = 5% chord and Δy = 8.66% chord) and required 140 iterations corresponding to 40 minutes of computer time on an IBM 360/67.

The transonic relaxation method has also been applied to wing-body combinations that can be represented by boundary conditions applied on combined mean planar and cylindrical control surfaces. The problem is recast into cylindrical coordinates for which Eq. (1) becomes

$$\left[1 - M_\infty^2 - (\gamma+1)M_\infty^2\,\phi_x\right]\phi_{xx} + \frac{1}{r}(r\phi_r)_r + \frac{1}{r^2}\phi_{\theta\theta} = 0 \tag{10}$$

The finite difference approximations for the derivatives in Eq. (10) are essentially the same as those that were applied to Eq. (1). The body boundary condition($r\phi_r = R\,\mathrm{d}R/\mathrm{d}x$ where R is the body radius) is applied on the cylindrical control surface $r_{k=1}$ and can be written

$$\left.\frac{1}{r}(r\phi_r)_r\right|_{k=1} = \frac{2}{r_1\Delta r}\left[\left(\frac{r_1+r_2}{2}\right)\left(\frac{\phi_2-\phi_1}{\Delta r}\right) - R\,\frac{\mathrm{d}R}{\mathrm{d}x}\right] \tag{11}$$

while the wing boundary condition is given by Eq. (9) with Δz replaced by $r\Delta\theta$.

The results for the 30° swept wing on a straight cylinder and on a symmetrically indented cylinder based on Mach-one area-ruling are shown in Fig. 2. Note that the area-ruling eliminates the embedded shock waves on the wing. These calculations were carried out on a 77 X 30 X 23 $(xr\theta)$ grid and required 200 iterations and one hour computation time on the IBM 360/67.

NUMERICAL TREATMENT OF LIFTING WINGS

In the numerical procedure for treating lifting wings, the wing is placed in the finite difference grid as shown in Fig. 3. The circulation at each span station (defined in Eq. (5)) is determined by the jump in potential at the trailing edge from the relation

$$\Gamma^{n+1} = \Gamma_k^{\,n} - \omega\left[\Gamma_k^{\,n} - \left(\phi^n_{JTE,k,o^+} - \phi^n_{JTE,k,o^-}\right)\right] \tag{12}$$

where n is the iteration count and ω is a relaxation parameter. New values of Γ are obtained at each iteration with the values of ϕ being obtained by extrapolating the values above and below the trailing edge. The continuity of pressure through the vortex sheet is maintained by holding the value of Γ, given by Eq. (12), fixed in x along the entire length of the vortex sheet and by setting $\phi_{j,k,o^+} = \phi_{j,k,o^-} + \Gamma_k$.

Difference formulas for ϕ_{zz} at the vortex sheet may be derived by noting that jumps in ϕ occur only at the vortex sheet and only odd functions may jump. Since the jump is independent of x, the solution at the vortex sheet decouples into even, ϕ^e, and odd, ϕ^o, solutions with ϕ^e satisfying Eq. (1) and ϕ^o satisfying

$$\phi^o_{yy} + \phi^o_{zz} = 0 \tag{13}$$

At the sheet itself the odd solution is given by

$$\phi^o(x,y,o\pm) = \pm 1/2\Gamma(y) \tag{14}$$

Therefore, ϕ_{zz} at the sheet can be written

$$\phi_{zz}\Big|_{o\pm} = \phi^e_{zz}\Big|_o \mp 1/2\Gamma_{yy} \tag{15}$$

The difference approximation for Eq. (15) is applied at points $(j,k,o-)$ (see Fig. 3) and is given by

$$\phi_{zz}\Big|_{j,k,o-} = \frac{4}{(\Delta z)^2}\left[\left(\phi_{j,k,1} - \Gamma_k\right) - 2\phi_{j,k,o-} + \phi_{j,k,-1}\right] + \frac{1}{2(\Delta y)^2}\left(\Gamma_{k+1} - 2\Gamma_k + \Gamma_{k-1}\right) \tag{16}$$

The difference formula for points $(j,k,1)$ is the usual centered difference with $\phi_{j,k,o+}$ replaced by $\phi_{j,k,o-} + \Gamma_k$. The wing tip and edge of the vortex sheet are placed midway between grid points, thus avoiding differencing at the tip singularity. The required value of potential just outboard of the tip is found by interpolation.

The infinity boundary conditions far from the wing and vortex sheet are given at some finite distance by an approximate analytical expression for the far field solution (see Klunker (1971)). The dominant term in the expression is due to lift and is proportional to the circulation integrated over the wing. The conditions at the downstream boundary, i.e., Trefftz plane, are found by relaxing Eq. (13) with boundary condition, Eq. (14), along with the rest of the flow field.

DIFFERENCING SCHEMES FOR SWEPT AND TAPERED WINGS

We now consider the application of the relaxation method to lifting wings with swept and tapered planforms. Experience with calculations about two-dimensional lifting airfoils has shown that very small mesh spacing (less than 1% chord) is required in the nose region, particularly for blunt leading edges. Satisfying this requirement with an even spaced mesh would require a prohibitive number of mesh points. An alternate approach is to use a coordinate transformation and map any swept or tapered planform into a rectangle. Such a transformation, valid for wings with finite tip chords, is given by

$$\xi(x,y) = \frac{x - x_{\mathrm{LE}}(y)}{c(y)} \qquad \eta = y \qquad z = z \tag{17}$$

where $x_{\mathrm{LE}}(y)$ is the value of x at the leading edge and $c(y)$ is the ratio of the local chord to the root chord. The governing small disturbance equation can be rewritten in terms of the new independent varialbes ξ,η,z in the form

$$\left\{\left[1 - M_\infty^2 - (\gamma+1)\frac{M_\infty^2}{U_\infty}\,\phi_\xi\,\frac{1}{c}\right]\frac{1}{c^2} + {\xi_y}^2\right\}\phi_{\xi\xi} + 2\xi_y\,\phi_{\xi\eta} + \xi_{yy}\phi_\xi + \phi_{\eta\eta} + \phi_{zz} = 0 \tag{18}$$

and the pressure coefficient becomes

$$C_p = -\frac{2}{c}\,\phi_\xi \tag{19}$$

The transformation given by Eq. (17) shears x to remove the sweep and stretches x to remove the taper. The effects of sweep and taper on the boundary conditions are thereby removed from the boundary conditions themselves and incorporated into the governing Eq. (18). In the region outboard of the tip the same rate of stretching and shearing is used unless $c(y)$ becomes much less than one. In this case $c(y)$ is set equal to a constant, $c(y_p)$, for values of y between some point $y_p>1$ and the far field boundary.

Treatment of the wing root boundary condition at $\eta = 0$ in the transformed coordinate system requires special consideration since the derivatives of ϕ with respect to η are discontinuous there. The condition of symmetry leads to the relation

$$\phi_{xy})_{y=o} = \phi_{\xi\eta} + \xi_{xy}\phi_{\xi} + \xi_y\phi_{\xi\xi} = 0 \tag{20}$$

which can be used to eliminate $\phi_{\xi\eta}$ from Eq. (18). Since $c(y)=1$ at the root section, the governing equation then reduces to

$$\left[1 - M_\infty^2 - (\gamma+1)M_\infty^2\phi_\xi - \xi_y^2\right]\phi_{\xi\xi} + (\xi_{yy} - 2\xi_y\xi_{xy})\phi_\xi + \phi_{\eta\eta} + \phi_{zz} = 0 \tag{21}$$

The $\phi_{\eta\eta}$ term in Eq. (21) is replaced by expressing ϕ at point 2 (see Fig. 4) in a Taylor series about point 1 and using the symmetry condition. The final form of the equation to be relaxed at the root boundary becomes

$$\left[1 - M_\infty^2 - (\gamma+1)M_\infty^2\phi_\xi - \zeta^2\right]\phi_{\xi\xi} + \left[\xi_{yy} + \left(\frac{2}{\Delta\eta} - 2\xi_{xy}\right)\zeta\right]\phi_\xi + \frac{2(\phi_2-\phi_1)}{(\Delta\eta)^2} + \phi_{zz} = 0 \tag{22}$$

with the ξ and z derivatives replaced by the already mentioned difference formulas and where

$$\zeta = \frac{n\Delta\xi}{\Delta\eta} + \xi_y$$

At each point η is picked such that $|\zeta|$ is made as small as the mesh will allow. For a wing with no taper ($\xi_{yy} = \xi_{xy}=0$) a zero value of ζ would reduce Eq. (22) to the untransformed equation.

In subsonic regions centered difference formulas applied to Eq. (18) give essentially the same solution as that found by applying centered formulas to Eq. (1). Unfortunately, however, a difficulty arises in differencing Eq. (18) in supersonic regions. It occurs if the initiation of backward differencing in the ξ direction commences when M, the local Mach number, becomes supersonic; that is when

$$M \equiv M_\infty\left[1 + \frac{(\gamma+1)}{c}\phi_\xi\right]^{1/2} > 1 \tag{23a}$$

In such a case the coefficient of $\phi_{\xi\xi}$ is still positive since it contains the term ξ_y^2, and the calculations do not converge. Furthermore, in this case the numerical domain of dependence can not include the analytical domain of dependence traced out by the local characteristics. This is illustrated in Fig. 5.

This difficulty can be overcome by substituting for the condition of backward differencing, the requirement that

$$M > (1 + c^2\xi_y^2)^{1/2} \tag{23b}$$

which amounts to the condition that the coefficient of $\phi_{\xi\xi}$ changes sign, or alternatively, that the component of local Mach number normal to the local sweep angle becomes supersonic. It should be emphasized that the criterion given by Eq. (23b) is successful for supercritical flow fields only if local Mach numbers are sufficiently large to ensure backward differencing at shock waves. For example, application of this method at $M_\infty = 0.908$ to the 30° swept wing shown in Fig. 1 produced no detectable shock wave because at no point did the local Mach numbers satisfy condition (23b), although they did, of course, satisfy condition (23a). The method appears to give satisfactory results, however, for wings with moderate sweep angles in flows with sufficiently high local Mach numbers, examples of which are given below.

It should be pointed out that the above difficulty can be alleviated, and the ability to capture weak oblique shocks by means of Eq. (18) can be improved if a skewing technique, similar to that constructed for the root section in Eq. (22), is used in the supersonic region. The object is to find a computational molecule in the supersonic region which is aligned as closely as possible to the (xy) coordinates. Such a scheme applied to the 30° swept wing at $M_\infty = 0.908$ with the angle of the skewed computational molecule differing from the (xy) molecule by less than four degrees gave the same results as those shown if Fig. 1.

Subcritical ($M_\infty = 0.752$) and supercritical ($M_\infty = 0.853$) results obtained using the transformation method (with root skewing only) are shown in Figs. 6 and 7 for flow about a lifting swept wing at two degrees angle of attack. The constant chord, 23.75° sweptback wing with a Lockheed C141 airfoil section (11.4% thick streamwise) was tested in the NASA Ames 11-Foot Transonic Wind Tunnel by Cahill and Stanewsky (1969). The results for $M_\infty = 0.752$ are compared in Fig. 6 with both the experimental results and those obtained by the subsonic panel method of Saaris and Rubbert (1972). The present results agree will with those obtained by the panel method but both numerical methods show more lift than the experiment. The present method also shows more lift than the experiment at $M_\infty = 0.853$ (see Fig. 7), as well as a shock location aft of the experimental one. It should be mentioned that inviscid solutions generally give more lift than the experiment when compared at the same geometric angle of attack. The principal cause is that viscous effects at the trailing edge (apparently a separation and formulation of a thin turbulent wake) decrease the circulation, thereby causing the loss in lift. The associated decrease in expansion also causes the experimental shock to occur further upstream. This is not to be confused with shock induced separation which, it is believed, does not occur in the experimental data shown.

These numerical solutions were obtained using an unevenly space $(\xi\eta,z)$ grid of 68 X30 X 49 points and 7 hours of computation time for both solutions on the IBM 360/67 computer. Convergence was established when the lift changed less than 0.02 percent per iteration. The three relaxation parameters required in the method were set at 1.4 in subsonic regions, 0.7 in supersonic regions, and 1.0 for the circulation equation. Experimentation with the circulation relaxation parameter indicated that the value of one was the best choice. The use of higher values caused oscillations to occur.

CONCLUSIONS

A mixed elliptic-hyperbolic relaxation method has been applied to the study of a nonlinear small perturbation equation modeling steady, three-dimensional transonic flow. Certain nonlifting wing-body combinations were computed without difficulty, and a numerical procedure for treating lifting wings without bodies was presented. In an effort to simplify the treatment of swept and tapered wings with blunt leading edges, a coordinate transformation has been introduced to map the wing planform into a rectangle. Certain difficulties introduced by this transformation were explored, and results found from its use under valid circumstances were presented and compared with experiment.

REFERENCES

Cahill, J. F. and Stanewsky, E., Air Force Flight Dynamics Lab, AFFDL-TR-69-78 (1969).
Klunker, E. B., NASA TN D-6530 (1971).
Krupp, J. A., The Boeing Company, Rep. D180-12958-1 (1971).
Murman, E. M. and Cole, J. D., AIAA J, 9, 114-121 (1971).
Saaris, G. R. and Rubbert, P. E., AIAA Paper 72-188 (1972).
Spreiter, J. R. NACA Rep. 1153 (1953).

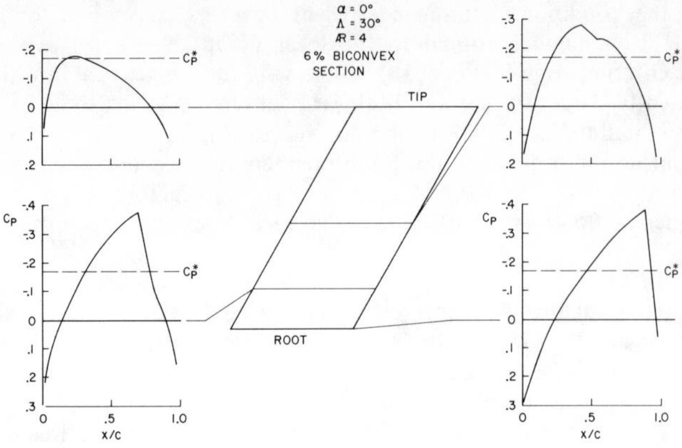

Fig. 1. C_p distribution on 30° swept wing at M = 0.908

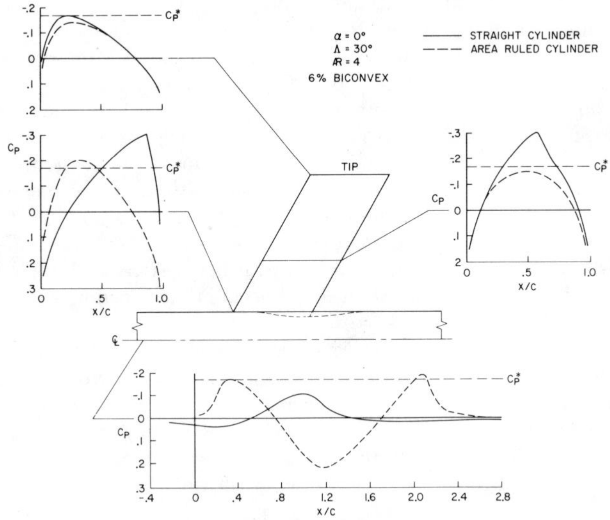

Fig. 2. C_p distribution on cylinder-wing combinations at M = 0.908

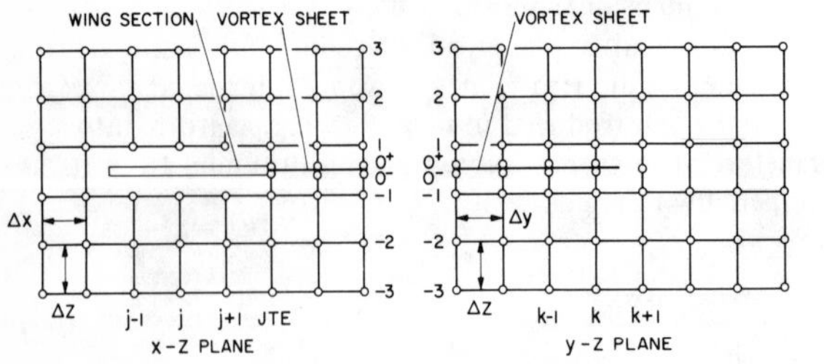

Fig. 3. Finite difference grid

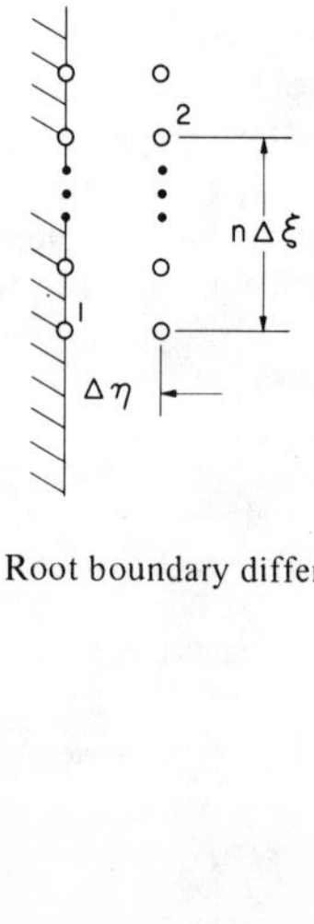

Fig. 4. Root boundary differencing

TAN Λ = cξy
TAN θ = $\sqrt{M^2-1}$

$\theta < \Lambda$ NUMERICAL DOMAIN DOES NOT INCLUDE
ANALYTICAL DOMAIN – NONCONVERGENT

$\theta \geq \Lambda$ NUMERICAL DOMAIN INCLUDES ANALYTICAL
DOMAIN – CONVERGENT

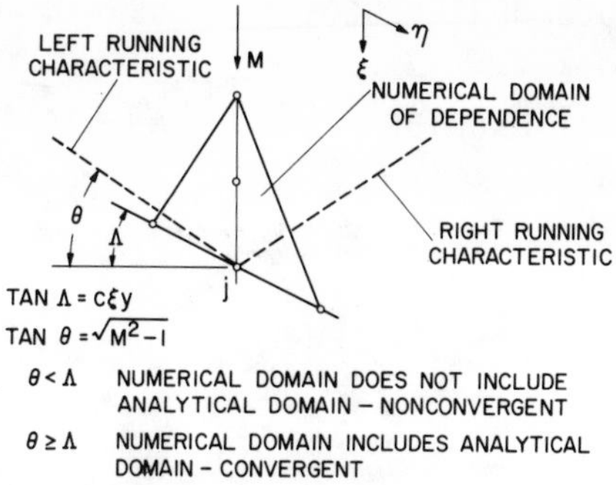

Fig. 5. Numerical and analytical domains of dependence for point j

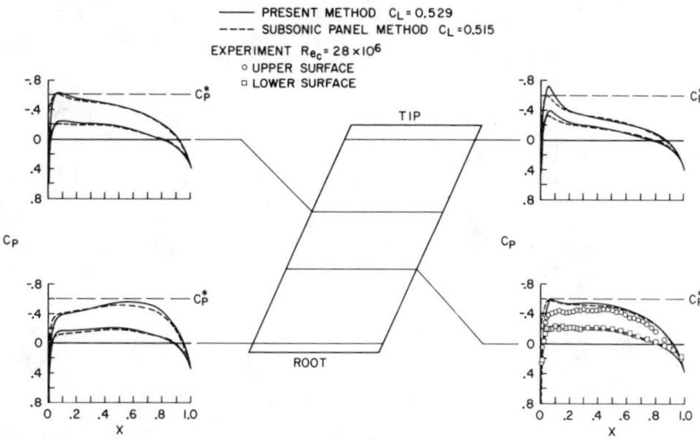

Fig. 6. C_p distribution on C141 swept panel model, M_∞ = 0.752, α = 2°

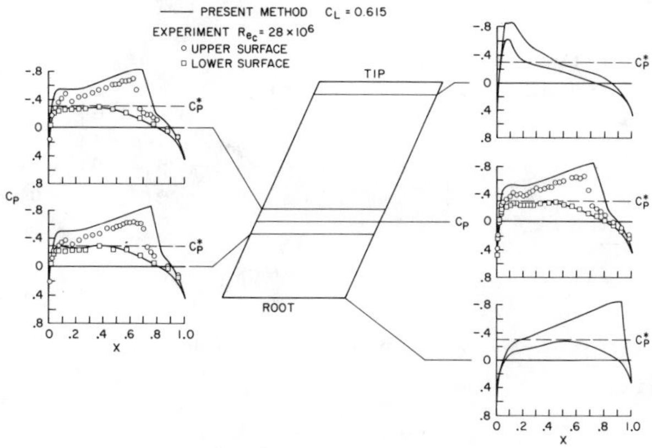

Fig. 7. C_p distribution on C141 swept panel model, M_∞ = 0.853, α = 2°

CALCUL D'UN ECOULEMENT VISCOELASTIQUE DANS UNE CAVITE CARREE

F. BAUDIER* et P. AVENAS**

I. INTRODUCTION

Les lignes d'écoulement d'un fluide viscoélastique, le plus souvent un polymère fondu ou en solution, diffèrent de celles d'un écoulement newtonien dans la même géométrie et avec le même nombre de Reynolds. Le but de notre étude est d'évaluer numériquement cette différence en fonction de l'élasticité du fluide. Nous avons choisi l'étude de l'écoulement plan dans une cavité rectangulaire, déjà faite numériquement et expérimentalement dans le cas newtonien, parce qu'elle peut contribuer à la compréhension de certains problèmes technologiques :

- modèle de tourbillon engendré par un écoulement dans l'anfractuosité d'une paroi, et dont la forme peut être modifiée par la présence d'un polymère en solution dans le fluide (problème des agents de réduction de perte de charge)
- rotation du polymère fondu dans un filet de vis d'extrudeuse...

II. EQUATIONS DE L'ECOULEMENT

a) Lois de comportement

Nous avons envisagé une loi de comportement du type de Rivlin-Ericksen d'ordre 2 :

$$\Sigma = -pI + \omega_1 B_1 + \omega_2 B_1^2 + \omega_3 B_2 \qquad (\underline{1})$$

où : $B_1 = \frac{1}{2}(\nabla\vec{u} + \vec{u}\nabla)$; $B_2 = \frac{\delta B_1}{\delta t} = \frac{dB_1}{dt} - \nabla\vec{u}.B_1 - B_1.\vec{u}\nabla$

$\delta/\delta t$ est la dérivation convective introduite par Oldroyd (1),
$\omega_1 = 2\mu$ où μ est la viscosité usuelle
ω_2 et ω_3 sont des paramètres caractérisant l'élasticité du fluide.

Les liquides viscoélastiques réels ont a priori des lois de comportement plus compliquées. On a montré récemment (2) qu'une loi ne

---

* Centre des Matériaux, Ecole des Mines de Paris, 60 bd Saint-Michel PARIS 6e

** Centre des Matériaux, Groupe Commun Mines-ENSTA, 32 Bd Victor PARIS 15e permettait de prévoir le comportement des fluides polymériques à forte vitesse de déformation que si elle comportait un terme de dérivation des contraintes : c'est le cas de la loi, type loi de Maxwell, introduite par White et Metzner (3) :

$$\Sigma' + \tau \frac{\delta \Sigma'}{\delta t} = 2 \mu B_1 \qquad (\underline{2})$$

où τ est un temps de relaxation et Σ' l'excès de contrainte par rapport à une pression hydrostatique.

Cependant le calcul à partir de la loi ($\underline{2}$) est rendu très difficile dans la mesure où le tenseur des contraintes n'est pas connu explicitement.

Nous avons donc traité dans un premier temps le cas de la loi ($\underline{1}$) qui peut être considérée comme l'approximation d'ordre 2 à vitesses de déformation faibles de toute loi de comportement plus compliquée.

b) Equilibre dynamique

En l'absence de forces de masse, ρ étant la densité du fluide, l'équation équivalente à celle de Navier-Stokes est la suivante :

$$\rho \frac{d\vec{u}}{dt} = - \nabla p + \mu \Delta \vec{u} + \omega_2 \nabla B_1^2 + \omega_3 \nabla B_2 \qquad (\underline{3})$$

Ces équations sont compliquées. Dans le cas plan, l'équation projetée sur Ox est la suivante :

$$\rho \frac{du}{dt} = - \frac{\partial p}{\partial x} + \mu \Delta u + \omega_2 \frac{\partial}{\partial x}\left[\left(\frac{\partial u}{\partial x}\right)^2 + \frac{1}{4}\left(\frac{\partial u}{\partial y} + \frac{\partial v}{\partial x}\right)^2\right]$$
$$+ \frac{\omega_3}{2}\left[\frac{d}{dt}\Delta u - \frac{\partial u}{\partial x}\left(3\frac{\partial^2 u}{\partial x^2} - \frac{\partial^2 u}{\partial y^2}\right) - \frac{\partial u}{\partial y}\left(\frac{\partial^2 v}{\partial x^2} - \frac{\partial^2 v}{\partial y^2}\right) - 2\frac{\partial v}{\partial x}\frac{\partial^2 u}{\partial x \partial y}\right]$$

Ces équations prennent une forme beaucoup plus simple si on élimine la pression en introduisant le tourbillon de la vitesse.

Soit $\Omega = \frac{1}{2} \operatorname{rot} \vec{u}$; ($\underline{3}$) devient, dans le cas plan :

$$\rho \frac{d\Omega}{dt} = \mu \Delta \Omega + \frac{\omega_3}{2} \frac{d\Delta\Omega}{dt} \qquad (\underline{4})$$

ω_2 n'intervenant plus car le terme B_1^2 est équivalent à une pression dans le cas plan.

En introduisant la vitesse U et la longueur L de la paroi mobile, on peut mettre ($\underline{4}$) sous une forme adimensionnelle :

$$\rho \frac{UL}{\mu} \frac{d\Omega}{dt} = \Delta \Omega + \frac{\omega_3 U}{2\mu L} \frac{d\Delta\Omega}{dt}$$

On fait apparaître ainsi le nombre de Reynolds et un nombre caractérisant l'élasticité du fluide, et appelé nombre de Weissenberg par White et Metzner (4). Le sens physique de ces nombres est le suivant :

$$R = \rho \frac{UL}{\mu} = \frac{\text{forces d'inertie}}{\text{forces de viscosité}} = \text{nombre de Reynolds}$$

$$W = \frac{-\omega_3 U}{2 \mu L} = \frac{\text{forces élastiques}}{\text{forces de viscosité}} = \text{nombre de Weissenberg.}$$

c) Système d'équations étudié

La forme des équations nous a donc incité à appliquer une

méthode en tourbillon et fonction de courant. Nous avons étudié le système stationnaire suivant :

$$\Delta\Omega = R\left[\frac{\partial\psi}{\partial y}\frac{\partial\Omega}{\partial x} - \frac{\partial\psi}{\partial x}\frac{\partial\Omega}{\partial y}\right] + W\left[\frac{\partial\psi}{\partial y}\frac{\partial\Delta\Omega}{\partial x} - \frac{\partial\psi}{\partial x}\frac{\partial\Delta\Omega}{\partial y}\right] \qquad (\underline{5})$$

$$\Omega = \frac{1}{2}\Delta\psi \qquad (\underline{5'})$$

III. PROBLEME DES CONDITIONS AUX LIMITES

a) Position du problème

L'équation (5) contient des dérivées d'ordre 3 en Ω, soit d'ordre 5 en ψ. Les conditions aux limites habituelles, à savoir ψ et $\nabla\psi$ données sur les parois, semblent donc a priori insuffisantes pour que le problème soit bien posé.

La difficulté essentielle que l'on rencontre ici est qu'aucune propriété physique des liquides viscoélastiques ne suggère d'imposer dans le calcul une condition aux limites autre que celle de non glissement aux parois. Des études expérimentales (5) n'ont pas mis en évidence un comportement anormal des solutions polymériques au voisinage des parois permettant par exemple de poser une condition sur le tourbillon. Le comportement d'un fluide viscoélastique ne peut donc être décrit que par des équations d'écoulement d'ordre supérieur à celles de l'écoulement newtonien, mais aucune condition aux limites supplémentaire n'est imposée par l'expérience. Cette situation a priori défavorable est cependant particulière dans la mesure où le calcul montre que les termes d'ordre supérieur dans l'équation (5) ne jouent aucun rôle au voisinage des parois.

b) Couche limite viscoélastique

En effet, le calcul de l'écoulement près d'une paroi, avec les approximations usuelles de la couche limite loin d'un point de stagnation, montre que les termes d'élasticité n'entraînent aucune modification par rapport à l'écoulement newtonien. Ceci a été montré (6) de même dans un cas plus général (fluides d'Oldroyd et fluides simples de Noll).

On peut donc penser que l'équation (5) est d'un ordre 5 dégénéré en ψ et de ce fait aucun résultat théorique ne permet d'affirmer qu'une condition aux limites supplémentaire est nécessaire pour que le problème soit bien posé.

Nous avons donc cherché, du moins dans un premier temps, à résoudre notre problème avec les seules conditions aux limites habituelles de non glissement aux parois.

IV. METHODE NUMERIQUE

a) Schéma aux différence finies

Nous avons appliqué au système (5) une méthode aux différences finies du type Gauss-Seidel, analogue à celle appliquée par Burggraf (7) au cas newtonien, en considérant les deux termes non linéaires comme un second membre.

Les équations (5') et (5) dans un maillage de pas D, en différences finies d'ordre D^2, s'écrivent, à l'intérieur du domaine :

$$\psi_{ij} = \mathcal{F}(\psi_{ij},\Omega_{ij}) = \frac{1}{4}\left(\psi_{i-1,j}+\psi_{i+1,j}+\psi_{i,j-1}+\psi_{i,j+1}\right) - \frac{D^2}{4}\Omega_{ij}$$

$$\begin{aligned}\Omega_{ij} = \mathcal{G}(\psi_{ij},\Omega_{ij}) = {} & \frac{1}{4}\left(\Omega_{i-1,j}+\Omega_{i+1,j}+\Omega_{i,j-1}+\Omega_{i,j+1}\right) \\ & - \frac{R}{16}\Big[(\psi_{i+1,j}-\psi_{i-1,j})(\Omega_{i,j+1}-\Omega_{i,j-1}) - (\psi_{i,j+1}-\psi_{i,j-1})(\Omega_{i+1,j}-\Omega_{i-1,j})\Big] \\ & - \frac{W}{16D^2}\Big[(\psi_{i+1,j}-\psi_{i-1,j})(\Omega_{i,j+2}-4\Omega_{i,j+1}+4\Omega_{i,j-1}-\Omega_{i,j-2}+\Omega_{i+1,j+1}+\Omega_{i-1,j+1}-\Omega_{i+1,j-1}-\Omega_{i-1,j-1}) \\ & \quad - (\psi_{i,j+1}-\psi_{i,j-1})(\Omega_{i+2,j}-4\Omega_{i+1,j}+4\Omega_{i-1,j}-\Omega_{i-2,j}+\Omega_{i+1,j+1}+\Omega_{i+1,j-1}-\Omega_{i-1,j+1}-\Omega_{i-1,j-1})\Big]\end{aligned}$$

Des schémas asymétriques pour les dérivations ont été utilisés aux points situés à une distance D à l'intérieur des frontières. Une rangée de points fictifs, situés à une distance D à l'extérieur des frontières, a été introduite pour écrire les conditions aux limites sur $\nabla\psi$:

$\psi = 0 \quad ; \quad \frac{\partial\psi}{\partial y} = -1 \quad ; \quad \frac{\partial\psi}{\partial x} = 0$ sur la paroi mobile

$\psi = 0 \quad ; \quad \frac{\partial\psi}{\partial x} = \frac{\partial\psi}{\partial y} = 0$ sur les parois fixes.

La méthode du type Gauss-Seidel consiste à calculer les valeurs à l' itération n+1 par :

$$\psi_{ij}^{n+1} = \mathcal{F}(\psi_{ij}^{n'}, \Omega_{ij}^{n'}) \quad ; \quad \Omega_{ij}^{n+1} = \mathcal{G}(\psi_{ij}^{n'}, \Omega_{ij}^{n'})$$

en utilisant les nouvelles valeurs de ψ_{ij} et Ω_{ij} dès qu'elles sont connues, (n' égale n ou n+1).

Lorsque les valeurs de R et de W augmentent, il est nécessaire d'introduire une sous-relaxation dans la méthode pour assurer la convergence. D'où le schéma :

$$\psi_{ij}^{n+1} = (1-K)\,\psi_{ij}^{n} + K\,\mathcal{F}(\psi_{ij}^{n'}, \Omega_{ij}^{n'})$$

$$\Omega_{ij}^{n+1} = (1-K)\,\Omega_{ij}^{n} + K\,\mathcal{G}(\psi_{ij}^{n'}, \Omega_{ij}^{n'})$$

A chaque itération, les résidus des équations sont définis par :

$$\mathcal{R}_n(\psi) = \max\left|\psi_{ij}^{n} - \mathcal{F}(\psi_{ij}^{n'},\Omega_{ij}^{n'})\right| = \max\frac{|\psi_{ij}^{n+1}-\psi_{ij}^{n}|}{K}$$

$$\mathcal{R}_n(\Omega) = \max\left|\Omega_{ij}^{n} - \mathcal{G}(\psi_{ij}^{n'},\Omega_{ij}^{n'})\right| = \max\frac{|\Omega_{ij}^{n+1}-\Omega_{ij}^{n}|}{K}$$

La convergence de la méthode se traduit par la décroissance de $\mathcal{R}_n(\psi)$ et $\mathcal{R}_n(\Omega)$.

Nous avons adopté le plus souvent les conditions de départ assurant le repos du fluide dans la cavité et la vitesse 1 de la paroi mobile.

b) Sous-relaxation et convergence

Burggraf (7) a remarqué que le calcul pouvait conduire à des résultats erronés concernant la forme des écoulements si le maillage adopté était trop grossier. Une convergence jusqu'à la 4e décimale sur les valeurs de ψ semble toutefois obtenue lorsque le pas D décroît jusqu'à 0,02. Nous avons donc adopté dans nos calculs un maillage 50x50.

Dans ces conditions, lorsque W=0, la méthode est convergente (jusqu'à la précision de la machine), sans relaxation dans les cas où R égale 0 et 100 ; la sous-relaxation est nécessaire lorsque R égale 400.

La sous-relaxation est nécessaire beaucoup plus rapidement lorsque W prend des valeurs croissantes. Les valeurs du paramètre de sous-relaxation nécessaires pour assurer la convergence sont sensiblement indépendantes de R. Dans les trois cas R=0, 100, 400, pour W=0,1, la sous-relaxation avec K=0,4 est nécessaire pour assurer une convergence (Tableau I).

TABLEAU I : cas où W = 0,1

		$\mathcal{R}(\psi)$	$\mathcal{R}(\Omega)$	Nombre d'itérations
R = 0	K = 0,5	$7{,}34.10^{-5}$	$1{,}06.10^{-1}$	4000 → oscillations
	K = 0,4	$2{,}47.10^{-6}$	$2{,}64.10^{-5}$	4000 convergence
R=100	K = 0,5			moins de 1000 divergence
	K = 0,4	$8{,}99.10^{-6}$	$1{,}68.10^{-4}$	3500 convergence
R=400	K = 0,4	$6{,}58.10^{-6}$	$3{,}07.10^{-4}$	3500 convergence

Nous avons utilisé des méthodes consistant à faire varier le paramètre de relaxation en fonction de l'évolution des résidus au cours des itérations. Ceci nous a permis, dans quelques cas, d'éviter la recherche d'un paramètre fixe optimal par tâtonnement.

En particulier, dans un assez grand nombre de cas, nous avons constaté que la loi de variation de K suivante était efficace :

$$\text{si} \left| \frac{\mathcal{R}_n(\psi)}{\mathcal{R}_{n-1}(\psi)} - \frac{\mathcal{R}_{n-1}(\psi)}{\mathcal{R}_{n-2}(\psi)} \right| > 2.10^{-3} \qquad K_{n+1} = 0{,}99\ K_n$$

$$\text{si} \left| \frac{\mathcal{R}_n(\psi)}{\mathcal{R}_{n-1}(\psi)} - \frac{\mathcal{R}_{n-1}(\psi)}{\mathcal{R}_{n-2}(\psi)} \right| < 5.10^{-4} \qquad K_{n+1} = K_n + 0{,}1(1-K_n)$$

Par exemple, dans le cas R=100, W = 0,05, partant de K=1, on obtient une valeur de $\mathcal{R}(\psi)$ inférieure à 10^{-6} en 2021 itérations, K évoluant le plus souvent entre 0,65 et 0,85. Par une méthode de ce type, il semble possible d'obtenir une convergence pour des valeurs de W supérieures à

0,1, la limite actuelle (le calcul diverge pour W=0,2 lorsque K est fixé à 0,1).

V. RESULTATS

En définitive, quelle que soit la méthode de relaxation, nous avons admis comme critère de convergence le fait que $\mathcal{R}(\psi)$ soit inférieur à 10^{-6}, les calculs étant arrêtés à 4000 itérations lorsque cette limite n'était pas atteinte.

Nous calculons dans chaque cas les coordonnées x, y du centre du vortex, ainsi que les valeurs ψ_c et Ω_c en ce point. Les résultats ainsi obtenus s'interprètent différemment selon que R=0 ou $R \neq 0$.

a) Cas où R = 0

Physiquement, il s'agit du cas où les forces d'inertie sont négligeables devant les forces de viscosité, ces dernières pouvant être du même ordre que les forces élastiques. Mathématiquement, l'équation (5) se réduit à :

$$\Delta\Omega = W \frac{d}{dt} \Delta\Omega \tag{6}$$

La solution (dont on démontre qu'elle est unique) de $\Delta\Omega = 0$ reste donc solution de (6).

Dans ces conditions, ou bien l'équation (6) admet d'autres solutions et le problème est mal posé, ou bien elle n'en admet pas d'autre et ceci signifie que, pour les fluides envisagés, les forces élastiques ne peuvent modifier l'écoulement lorsque les forces d'inertie sont très petites.

Numériquement, on constate que, dans le domaine de variation possible pour W, les solutions obtenues pour ψ et Ω sont identiques au quatrième chiffre significatif près (Tableau II).

TABLEAU II

W	x	y	ψ_c	Ω_c
0	0,50000	0,764568	0,099982	1,6012
0,05	0,500006	0,764772	0,099889	1,5998
0,1	0,499754	0,764558	0,099936	1,5964

Cette constatation semble indiquer que lorsque R=0, la méthode utilisée, dans l'intervalle de variation de W, converge vers la solution de $\Delta\Omega = 0$.

b) Cas où $R \neq 0$

Les raisonnements précédents ne s'appliquent plus dans ce cas, où l'existence même de la solution n'est pas assurée théoriquement. Numériquement, des modifications sensibles de la forme d'écoulement ont effectivement été obtenues pour R = 100 et R = 400 (Tableau III).

Les modifications des lignes d'écoulement sont visibles sur les figures.

On constate les points suivants :

- le centre du vortex est déplacé horizontalement dans le sens du mouvement de la paroi et verticalement vers le haut,

- le débit du vortex principal diminue mais le tourbillon au centre augmente lorsque W augmente,

- la surface des tourbillons contrarotatifs des coins inférieurs augmente lorsque W augmente.

Tous ces effets, nuls lorsque R=0, sont d'autant plus accusés que R est grand. Lorsque R = 400 et W = 0,1, le débit du tourbillon est diminué de 10 % environ par rapport au cas newtonien.

TABLEAU III

R	W	x	y	ψ_c	Ω_c	$\mathcal{R}(\psi)$	$\mathcal{R}(\Omega)$
R = 0	$\forall$W	0,500	0,765	0,09998	1,601	$2,3.10^{-10}$	$3,9.10^{-7}$
R = 100	0	0,382	0,739	0,1023	1,580	$2,3.10^{-10}$	$3,4.10^{-7}$
	0,01	0,380	0,742	0,1014	1,590	10^{-6}	$3,6.10^{-5}$
	0,05	0,376	0,752	0,0984	1,622	10^{-6}	$3,3.10^{-5}$
	0,1	0,370	0,759	0,0952	1,621	9.10^{-6}	$1,7.10^{-4}$
R = 400	0	0,439	0,614	0,1058	1,103	$1,9.10^{-7}$	$7,1.10^{-6}$
	0,01	0,437	0,615	0,1048	1,103	10^{-6}	$3,7.10^{-5}$
	0,05	0,433	0,625	0,1005	1,110	10^{-6}	$3,2.10^{-5}$
	0,1	0,428	0,644	0,0951	1,129	$6,6.10^{-6}$	$3,1.10^{-4}$

VI. CONCLUSIONS

Une méthode de résolution numérique de l'écoulement newtonien a pu être appliquée à l'écoulement d'un fluide Rivlin-Ericksen d'ordre 2, caractérisé par deux nombres sans dimension, de Reynolds et de Weissenberg.

La convergence de la méthode n'a pas pu être obtenue jusqu'à présent pour une valeur de W supérieure à 0,1. Il faut noter que ce domaine de variation n'est pas sans intérêt car les valeurs de W réalisées dans des cas technologiques sont beaucoup plus petites que celles de R (W est plus souvent très inférieur à 10).

Les résultats obtenus semblent montrer que le débit du vortex diminue lorsque l'élasticité du fluide augmente, cette diminution étant d'autant plus nette que le nombre de Reynolds est élevé.

Gilligan et Jones (8), à partir d'hypothèses rhéologiques semblables, ont obtenu, pour les vortex à l'arrière d'un cylindre, un résultat opposé, mais non contradictoire puisque la configuration est différente. Des relations entre études numériques et expérimentales sont encore nécessaires pour relier clairement ces résultats à des propriétés physiques des fluides viscoélastiques.

Enfin, lorsque R=0, la méthode converge vers la même solution, à l'approximation numérique près, quel que soit W. Ceci constitue un "indice d'unicité" de la solution du problème numérique mais met en cause la validité de la loi de comportement pour les fluides viscoélastiques réels. En effet, Ultman et Denn (9) ont obtenu expérimentalement une distorsion importante des lignes d'écoulement viscoélastique par rapport à l'écoulement newtonien dans un cas $R = 2.10^{-4}$ et W = 3,2 ; cette distorsion a été interprétée dans l'étude en attribuant au fluide une loi de comportement du type (2).

En conclusion, notre étude apporte une contribution au domaine encore récent du calcul numérique des écoulements viscoélastiques mais des problèmes restent non résolus au niveau de la loi de comportement du fluide et au niveau de l'analyse numérique des problèmes aux limites non linéaires sans propriété d'èllipticité.

BIBLIOGRAPHIE

(1) J.G. OLDROYD - Proc. Royal Soc. A200, 523 (1950)

(2) B.J. MEISTER, R.D. BIGGS - AI ch E J. 15, 643 (1969)

(3) J.L. WHITE, A.B. METZNER - J. Appl. Polym. Sci. 7, 1867 (1963)

(4) J.L. WHITE, A.B. METZNER - AI Ch E J. 11, 324 (1965)

(5) G.J. REUSSWIG, F.F. LING - Appl. Sci. Res. 21, 260 (1969)

(6) K.R. FRATER - Z.A.M.P. 20(5), 712 (1969)

(7) O.R. BURGGRAF - J. Fluid Mech. 24, 113 (1966)

(8) S.A. GILLIGAN, R.S. JONES - Z.A.M.P. 21, 786 (1970)

(9) J.S. ULTMAN, M.M. DENN - Chem. Eng. J. 2, 81 (1971)

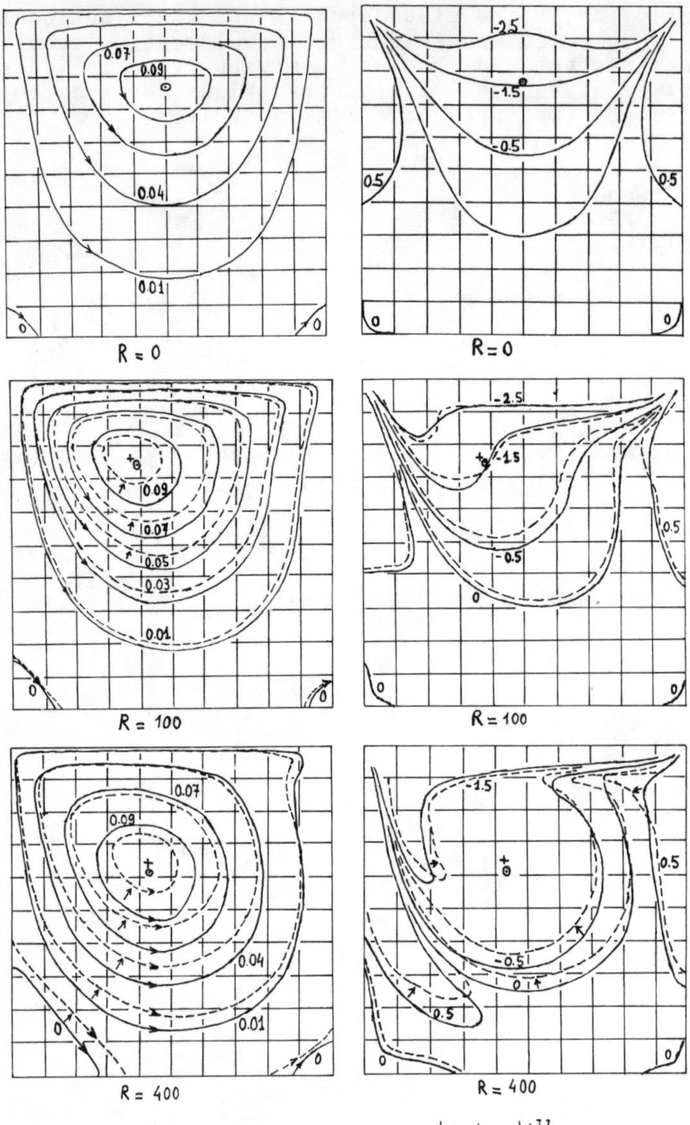

(a) lignes de courant

——————— : W = 0

– – – – – – : W = 0.1

(b) tourbillon

⊙ : Centre du vortex (W = 0)

+ : " " (W = 0.1)

OSCILLATIONS LIBRES D'UN BASSIN EN ROTATION

Claude Bellevaux* et Michel Maillé**

Université Paris VI France

I - Position du problème

On cherche les modes d'oscillation libres d'un fluide parfait, pesant dans un bassin en rotation.

Ce problème relativement ancien a été posé par LAMB (1) et n'a jamais été résolu complètement ni du point de vue théorique ni du point de vue numérique.

A partir des équations de l'Hydrodynamique et des hypothèses habituelles en Océanographie, le problème posé se ramène à la recherche des solutions périodiques en t du système d'équations :

$$
I \quad \left\{ \begin{array}{l} \dfrac{\partial \bar{u}}{\partial \bar{t}} - \bar{\omega}\,\bar{v} = -g\dfrac{\partial \bar{e}}{\partial \bar{x}} \\[2mm] \dfrac{\partial \bar{v}}{\partial \bar{t}} + \bar{\omega}\,\bar{u} = -g\dfrac{\partial \bar{e}}{\partial \bar{y}} \\[2mm] \dfrac{\partial \bar{e}}{\partial \bar{t}} + \dfrac{\partial}{\partial \bar{x}}(\bar{h}\,\bar{u}) + \dfrac{\partial}{\partial \bar{y}}(\bar{h}\,\bar{v}) = 0 \end{array} \right.
$$

dans un domaine Ω de frontière $\partial\Omega$ représentant le bassin étudié.

Dans ces équations, on a utilisé les symboles :

$\bar{u}$, $\bar{v}$: vitesses moyennes horizontales pour une tranche verticale dans un système d'axes $O\bar{x}, O\bar{y}$ orienté Est-Nord.

$\bar{e}$: dénivellation de la surface du bassin par rapport au repos.

g : accélération de la pesanteur.

$\bar{h}$: profondeur supposée variable du bassin $\left(\bar{h}(\bar{x},\bar{y}) \geq \underline{h} > 0\right)$.

$\bar{\omega}$: paramètre de Coriolis $(\bar{\omega} = 2\Omega \sin\varphi$, φ latitude moyenne, $\Omega = 2\pi/24\,h)$

$\bar{t}$: temps.

---

* L.I.M.S.I. BP 30 - 91 ORSAY .

** Université Paris VI - 9, Quai St Bernard - 75 PARIS

On introduit les grandeurs de référence :

L : diamètre du bassin.

H : profondeur maximale.

Et on définit les grandeurs adimensionnelles :

$$x = \bar{x}/L \ , \ y = \bar{y}/L \ , \ h = \bar{h}/H \ , \ \omega = L\bar{\omega}/\sqrt{gH}$$

On cherche les solutions de la forme :

$$\text{II} \quad \begin{cases} \bar{U}(x,y,\bar{t}) = \sqrt{gH} & e^{i\frac{\sqrt{gH}}{L}k\bar{t}} & u(x,y) \\ \bar{V}(x,y,\bar{t}) = \sqrt{gH} & e^{i\frac{\sqrt{gH}}{L}k\bar{t}} & v(x,y) \\ \bar{e}(x,y,\bar{t}) = H & e^{i\frac{\sqrt{gH}}{L}k\bar{t}} & e(x,y) \end{cases}$$

En portant (II) dans le système (I) et en éliminant U et V on obtient l'<u>équation aux dérivées partielles</u> :

$$\frac{\partial}{\partial x}\left(h\frac{\partial e}{\partial x}\right) + \frac{\partial}{\partial y}\left(h\frac{\partial e}{\partial y}\right) + ic\left(\frac{\partial}{\partial y}\left(h\frac{\partial e}{\partial x}\right) - \frac{\partial}{\partial x}\left(h\frac{\partial e}{\partial y}\right)\right) - (\omega^2 - k^2)e = 0$$

Où :

$$c = \omega / k$$

Le long des côtes $\partial\Omega$ on écrit que le flux normal moyen est nul, ce qui donne la <u>condition aux limites</u> :

$$\frac{\partial e}{\partial n} + ic\frac{\partial e}{\partial s} = 0$$

e et k sont solutions d'un problème aux valeurs propres qui admet la solution triviale $k = \omega$. La difficulté essentielle pour la recherche des autres modes propres provient du fait que k intervient non linéairement dans le paramètre

II - <u>Résultats théoriques</u>

On procède en deux étapes :

<u>1ère étape</u> - Le paramètre c est fixé indépendamment de k et on cherche les valeurs propres $k_n(c)$.

<u>2ème étape</u> - On cherche pour chaque n les solutions de l'équation : $k_n(c) = \frac{\omega}{c}$.

1°. <u>Problème à c fixé</u>

On introduit la forme quadratique :

$$a_c(u,u) = \iint_\Omega \left[h\left\{\left|\frac{\partial u}{\partial x}\right|^2 + \left|\frac{\partial u}{\partial y}\right|^2 + ic\left(\frac{\partial u}{\partial x}\frac{\partial \bar{u}}{\partial y} - \frac{\partial u}{\partial y}\frac{\partial \bar{u}}{\partial x}\right)\right\} + \omega^2|u|^2\right]dx\,dy$$

Si $|c| < 1$ on démontre la coercivité de la forme $a_c(\cdot\,,\,\cdot)$:

$$a_c(u,u) \geq (h/H)\,\mathrm{Min}\,(1-|c|,\omega^2)\,\|u\|^2_{H^1(\Omega)}$$

On en déduit la formulation variationnelle :

$$k_i^2(c) = \underset{H_i \subset H^1(\Omega),\, \dim H_i = i}{\mathrm{Min}} \quad \underset{u \in H_i,\, |u|_{L^2(\Omega)}=1}{\mathrm{Max}} \quad a_c(u,u).$$

Et on démontre (2) les théorèmes :

<u>Théorème 1</u> - Il existe une suite infinie de valeurs propres $k_n(c) \geq \omega$ et tendant vers $+\infty$ pour $n \to +\infty$. Les valeurs propres $k_n(c)$ dépendent continûment du paramètre $c\,(|c| < 1)$ et pour $c \to 1-0$ toutes les valeurs propres $k_n(c)$ tendent vers ω.

<u>Théorème 2</u> - Soient $\{\mu_n\}_{n\geq 1}$ les valeurs propres toutes positives du problème "sans rotation" $(\omega = 0)$. On a les inégalités suivantes :

$$(1-c)\,\mu_n^2 + \omega^2 \leq k_n^2(c) \leq (1+c)\,\mu_n^2 + \omega^2$$

2°. <u>Problème général</u>

On considère le schéma itératif suivant :

$$\text{III}\quad \begin{cases} c^{(0)} = 0 & , \quad k_n^{(0)} = \sqrt{\mu_n + \omega^2} \\ c^{(i)} = \omega / k_n^{(i-1)} & , \quad k_n^{(i)} = k_n(c^{(i)}) \end{cases}$$

On a les théorèmes :

<u>Théorème 3</u> - Le schéma (III) est convergent et $k_n^{(i)}$ tend pour $i \to +\infty$ vers ω ou vers la solution, si elle existe, de l'équation :

$$k_n(c) = \frac{\omega}{c}$$

<u>Théorème 4</u> - Si ω est "assez petit" ou si n est "assez grand" pour chaque valeur propre μ_n du problème "sans rotation" $(\omega = 0)$ il existe une valeur propre $k_n > \omega$ du problème avec rotation $(\omega \neq 0)$ qui dépend continûment du paramètre ω.

III - <u>Résolution numérique</u>

1°. <u>Choix et construction d'un maillage curviligne</u>

La géométrie des domaines étudiés étant compliquée, il est nécessaire pour la commodité du traitement numérique, d'effectuer une représentation R du domaine Ω sur un domaine Ω' de formes rectangulaires et ayant les mêmes caractéristiques topologiques. Dans le plan de Ω' on choisit un maillage rectangulaire régulier $x' = c^{te}$, $y' = c^{te}$ et l'on choisit la représentation R de façon à ce que l'image des lignes $x' = c^{te}$ et $y' = c^{te}$ dans le plan de Ω soient les lignes de niveau de deux fonctions harmoniques (figure 1).

2°. Choix des éléments finis

Les éléments finis φ_i choisis sont définis dans le plan de Ω' comme indiqué sur la figure 2. On a choisi les éléments finis de support minimal permettant l'introduction des dérivées mixtes .

3°. Equations discrètes

On cherche une solution approchée $U = \sum_{i=1}^{N} U_i \varphi_i$

Par la méthode de Galerkine et à partir de la forme bilinéaire $a_c(\cdot,\cdot)$ on obtient pour les inconnues $(U_i)_{i=1,\dots,N}$ et $\lambda = k^2 - \omega^2$ le système d'équations :

$$\sum_{j=1}^{N} a_{ij} U_j = \lambda \sum_{j=1}^{N} c_{ij} U_j$$

Avec :

$$a_{ij} = \iint_{S_i \cap S_j \cap \Omega} \left[\alpha \frac{\partial \varphi_i}{\partial x'} \frac{\partial \varphi_j}{\partial x'} + \beta \left(\frac{\partial \varphi_i}{\partial x'} \frac{\partial \varphi_j}{\partial y'} + \frac{\partial \varphi_i}{\partial y'} \frac{\partial \varphi_j}{\partial x'} \right) + \gamma \frac{\partial \varphi_i}{\partial y'} \frac{\partial \varphi_j}{\partial y'} + ic \left(\frac{\partial \varphi_i}{\partial x'} \frac{\partial \varphi_j}{\partial y'} - \frac{\partial \varphi_i}{\partial y'} \frac{\partial \varphi_j}{\partial x'} \right) \right] h\, dx'dy'$$

$$c_{ij} = \iint_{S_i \cap S_j \cap \Omega} J\, \varphi_i\, \varphi_j\, dx'dy'$$

Où :

$$J = \frac{\partial x}{\partial x'} \frac{\partial y}{\partial y'} - \frac{\partial x}{\partial y'} \frac{\partial y}{\partial x'} \quad , \quad \alpha = \frac{\left(\frac{\partial x}{\partial y'}\right)^2 + \left(\frac{\partial y}{\partial y'}\right)^2}{J}$$

$$\beta = -\frac{\frac{\partial x}{\partial x'} \frac{\partial x}{\partial y'} + \frac{\partial y}{\partial x'} \frac{\partial y}{\partial y'}}{J} \quad , \quad \gamma = \frac{\left(\frac{\partial x}{\partial y'}\right)^2 + \left(\frac{\partial y}{\partial x'}\right)^2}{J}$$

La matrice $A = \{a_{ij}\}$ est "creuse" (sept diagonales non nulles), hermitienne et vérifie :

$$\sum_{j=1}^{N} a_{ij} = 0 \qquad \forall i \ .$$

4°. Calcul des valeurs propres à fixé

On cherche les solutions du système précédent correspondant aux plus petites valeurs propres (pratiquement $p = 3$) : on utilise une méthode d'itérations inverses simultanées. (Méthode de BAUER (3).

On part de p vecteurs $U_1^{(0)}, U_2^{(0)}, \dots, U_p^{(0)}$ orthogonaux; après résolution par itérations successives $(S.O.R)$ des systèmes linéaires :

$$A\, \tilde{U}_k^{(i+1)} = U_k^{(i)} \quad , \; k = 1, 2, \dots, p \ .$$

On obtient $\left\{ \mathcal{U}_k^{(i+1)}, k=1,2,\ldots,p \right\}$ par orthogonalisation et normalisation de $\left\{ \tilde{\mathcal{U}}_k^{(i+1)}, k=1,2,\ldots,p \right\}$.

5°. Calcul des modes d'oscillation

On utilise conjointement le schéma itératif III et la méthode précédemment décrite (§ 4).

IV - Résultats

Dans le cas étudié du Bassin Méditerranéen, la période pendulaire est de 17h 56mn. Curieusement, il se trouve que le premier harmonique μ_1 du problème sans rotation a une période de 18h 47mn proche de la période pendulaire. Les courbes $\lambda_1(c)$ et ω/c n'ont pas de points communs pour $|c| < 1$: le mode fondamental ne peut donc avoir, s'il existe, qu'une période supérieure à la période pendulaire. Le premier mode pour $|c| < 1$ est un mode d'ordre 2 (deux points amphidromiques) et a pour période 7h 59 mn. (Voir figure 3 et 4 la phase et l'amplitude de ce mode).

Références

(1) H. LAMB - Hydrodynamics Cambridge University Press 1932.

(2) C. BELLEVAUX, M. MAILLE - C.R.A.S. - Série A (1970) t.270, p 1622-1625.

(3) H. RUTISHAUSER. Num Math, 16 (1970), 205-223.

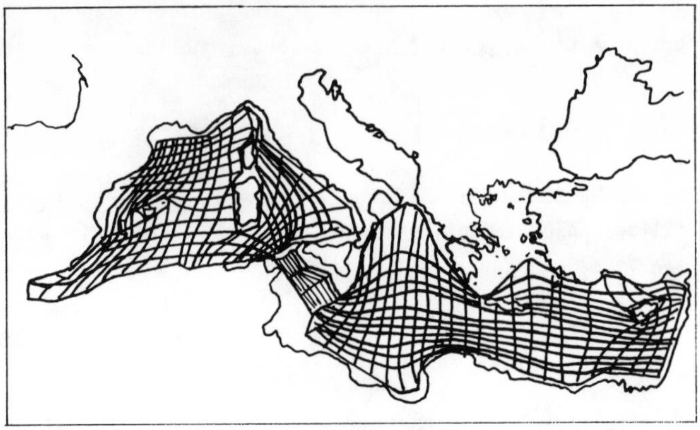

Fig. 1

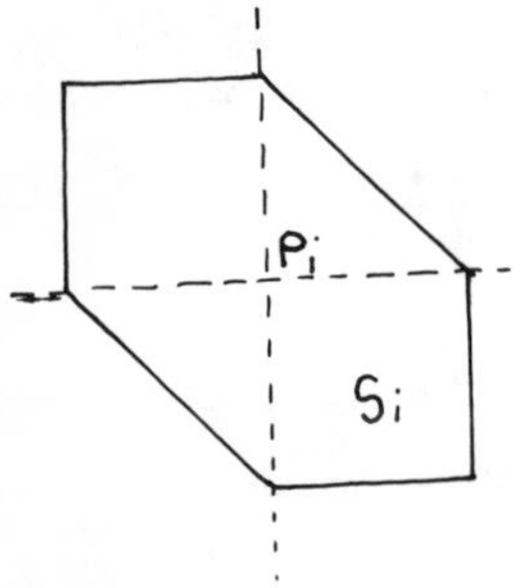

- supp $\varphi_i = S_i$
- $\varphi_i(P_i) = 1$
- φ linéaire par morceaux

Fig. 2

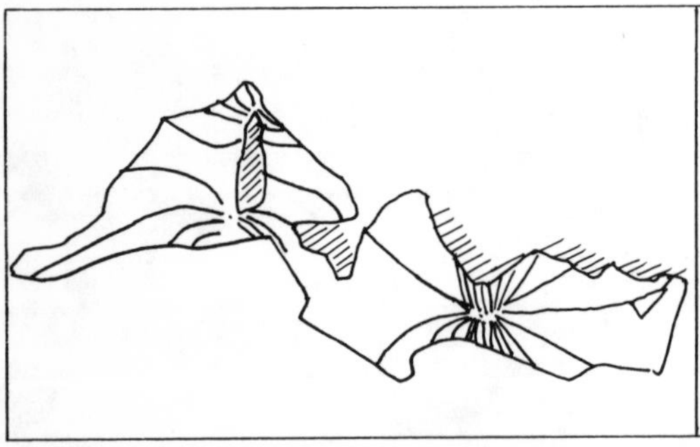

Fig. 3

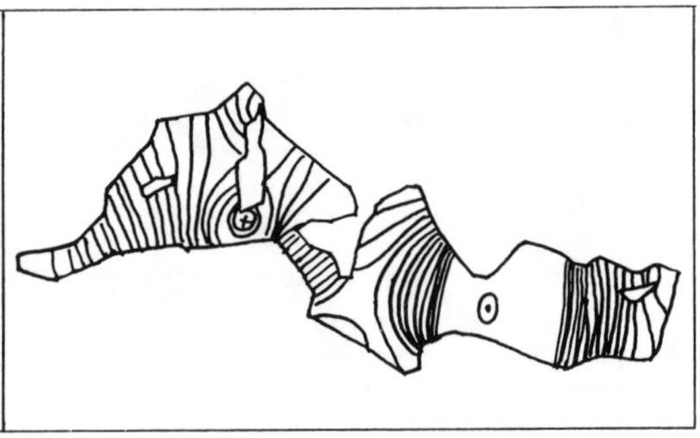

Fig. 4

NUMERICAL APPROACH FOR INVESTIGATING SOME TRANSSONIC FLOWS

O. M. Belotserkovskii, Yu. M. Davidov

Computing Center, Academy of Sciences
Moscow, U.S.S.R.

As is known, even at present the investigation of transsonic problems of aerodynamics including vorticity (especially supercritical and transsonic regimes) is fraught with many difficulties. As a matter of fact, in a number of cases it is these regimes which are the most complicated from the point of view of design, the study of problems of stability and guidance, etc. At the same time, analytical, numerical and experimental methods have not been developed here sufficiently, therefore there are only individual instances of the study of a complete flow pattern.

Recently, in the Computing Center of the Academy of Sciences of the U.S.S.R., systematic calculations of super-critical and transsonic flows including vorticity have been carried out. These apply to plane and axisymmetrical bodies and are based on the method of "large particles". In this paper, without discussing the details of the method, we shall only be concerned with the formulation of the boundary conditions of the problem for this class of flows and present some calculated results.

1. The development of the nonstationary "large particles" method [1,2] was stimulated by the works of F. Harlow, M. Rich and C. Hirt [3,5]. We adhere to the main principles of the organization of the calculation procedure of the Harlow "particle-in-cell" method. Nevertheless, it seems reasonable that, for gas dynamics problems, we should not confine ourselves to the discrete model of a continuous medium comprising a combination of particles of a fixed mass in a cell. Instead, we consider continuous flows of "large particles" whose mass coincides with the mass of an Euler cell at a given instant of time. As is known, in such an approach each time cycle is divided into three stages ("Eulerian", "Lagrangian" and "Final"), while greater attention is paid to the development of a numerical algorithm which might be used for a wide class of problems concerned with motion of a compressible gas.

As a result, we have obtained some divergent-conservative and dissipative-steady difference schemes which allow us to consider a wide class of transsonic problems as well. The development in this direction has been carried out since 1965. Some results were reported at the 3d All-Union Conference on Theoretical and Applied Mechanics [6] (Moscow, January, 1968), as well as at the 1st (Novosibirsk, August, 1969) [7] and the 2nd (Berkeley, U.S.A., October, 1970) [8] Conferences on Numerical Methods in Gas Dynamics.

Let us briefly describe the main principles of the "large particles" method. The region of integration is covered by a fixed (over space) Euler net composed of rectangular cells with sides Δx, Δy (ΔZ, Δz are along a cylindrical coordinate system).

In the first ("Eulerian") stage of calculations only those quantities change which are related to a cell as a whole, and the fluid is supposed to be instantaneously decelerated. Hence, the convective terms of the form $\mathrm{div}(\phi\rho\vec{w})$ where $\phi = (1,u,v,E)$, corresponding to displacement effects, are eliminated from eq. (1). Then it follows from the equation of continuity, in particular, that the density field will be "frozen" and the initial system of equations will be of the form

$$\rho \frac{\partial u}{\partial t} + \frac{\partial p}{\partial x} = 0 \quad , \quad \rho \frac{\partial v}{\partial t} + \frac{\partial p}{\partial y} = 0 \quad , \quad \rho \frac{\partial E}{\partial t} + \mathrm{div}(p\vec{w}) = 0 \tag{1}$$

Here we have used both the simplest finite-difference approximations and, to improve the calculation stability, the schemes of the method of integral relations [9], in which "sweeping-through" approximations of the integrands with respect to rays $(N = 3,4,5)$ are used.

In the second ("Lagrangian") stage we find mass flows ΔM^n actoss the cell boundaries at $t^n + \Delta t$. In this case we assume the total mass to be transferred only by a velocity component normal to the boundary. Thus, for instance.

$$\Delta M^n_{i+\frac{1}{2},j} = \langle\rho^n_{i+\frac{1}{2},j}\rangle\langle U^n_{i+\frac{1}{2},j}\rangle \; \Delta y \; \Delta t \qquad , \text{etc.} \tag{2}$$

The subscript $< >$ denotes the values of ρ and U across the cell boundary. The choice of these values is of great importance since they substantially influence the stability and accuracy of calculations. The consideration of the flow direction is characteristic of all possible ways of writing down ΔM^n.

Here different kinds of representation for ΔM^n are considered of first and second orders of accuracy. These are based on central differences, without account being taken of the flow direction and so on, as well as by means of the discrete model of a continuous medium comprising a combination of particles of a fixed mass in a cell.

Lastly, in the third ("Final") stage we estimate the final fields of the Euler flow parameters at the instant of time $t^{n+1} = t^n + \Delta t$ (all the errors in the solution of equations are "removed"). As was pointed out, the equations at this stage are laws of conservation of mass M, impulse $\vec{P}$ and total energy E written down for a particular cell in the difference form

$$F^{n+1} = F^n + \Sigma\Delta \, F^n_{rp} \qquad \text{where} \quad F = (M, P, E) \tag{3}$$

According to these equations, inside the flow field there are no sources or sinks of M, P and E and their variation in time Δt is caused by the interaction across the external boundary of the flow region.

It follows from the very character of the construction of the calculation scheme that a complete system of nonstationary gas dynamics equations is essentially solved here, while each calculation cycle represents a completed process in calculating a given time interval. Besides all initial nonstationary equations, the boundary conditions of the problem are satisfied and the real fluid flow at the time in question is determined.

Thus, the "large particles" method allows us to obtain the characteristics of nonstationary gas flows and by means of the stability process their steady magnitudes as well. Such an approach is especially applicable to problems in which a complete or partial development of physical phenomena with respect to time takes place. For example, in studying transsonic gas flows, flows around finite bodies, flow in local supersonic zones, separation regions, and so on develop comparatively slowly while the major part of the field develops rather rapidly. In contrast to the FLIC - method [10] our investigation is wholly devoted to systematic calculations of a wide class of compressible flows in gas dynamics problems (transsonic regimes; discontinuity, separation and "injected" flows, etc.).

The divergent forms of the initial and difference equations are considered in the "large particles" method; the energy relation for total energy E is used; different kinds of approximations are used in the 1st and 2nd stages; additional density calculations are introduced in the final stage, which helps us to remove fluctuations and makes it possible to obtain satisfactory results with a relatively small network (usually 1 - 2.5 thousand cells are used), and so on. All this results in completely conservative schemes, i.e., laws of conservation for the whole net region are an algebraic consequence of difference equations. Fractional cells are introduced for the calculation of bodies with a curvature in the slope of the contour [11].

The investigation of the schemes obtained (approximation problems, viscosity, stability, etc.) was carried out successively by considering the zero, the first and the second differential approximations [5,1,2]. These investigations show that the "large particles" method yields divergent - conservative and dissipative-steady schemes for "sweeping-through" calculations. These enable us to carry out stable calculations for a wide class of gas dynamics problems without introducing explicit terms with artificial viscosity. It may be of particular significance in studying flows around bodies with a curvature in the slope of the contour since the ways of introducing explicit terms with artificial viscosity are different for whole and fractional cells. Moreover, by varying only the second stage of the calculation procedure we can arrive at the conservative "particle-in-cell" method so that the calculational algorithm is of general use.

As for discontinuities the stability of calculations is provided here by the presence of approximate viscosity in the schemes (dissipative terms in difference equations), which results in "smearing" shock waves into several calculating cells, the formation of a wide boundary layer near the body, and so on. It should be stressed that the magnitude of the approximate viscosity is proportional to a local flow velocity and to the dimension of the difference net, therefore its effect is practically evident only in zones with high gradients.

2. The boundary conditions of the problem were realized by introducing layers of "fictitious" cells along the region boundaries [4,1,2,10,11].

The conditions on the body are close to those of "attachment". It can be seen in Fig. 1 where the density profile is given for the conditions of "attachment" (dashed line) and "non-flow" (solid line). It is seen that even in the vicinity of the body the difference between these two cases is insignificant and at some distance from the body it disappears entirely.

As it turned out, the right "open" boundary of the region introduces the greatest disturbances in calculating transsonic gas flows. For the evaluation of its influence and for the choice of the optimum dimensions of the net the calculations were carried out with the help of nets of different sizes; the "matching" of the flow fields took place (when one of the internal columns was used as the initial one for a new field); a comparison between the asymptotic form and experiment was made as well [1,2].

Figure 2 shows the results of the calculations of an "overcritical" flow ($M_\infty = 0.9$) around a semi-infinite cylinder of various lengths $\ell/R \sim 2 \div 3$ in the region considered (as if the "moving-in" of the body took place). If the flow field ahead of the body is established rather quickly, then the parameters to the right from a corner point become steady only at $\ell/R \sim 2 \div 3$. The data of the calculations with the help of a large net are given as an example in Fig. 2d; the commonly used region (Figs. 2a - 2c, about 2.5 thousand nodes) is shown here by a dashed line.

In Fig. 3 a comparison is given between the results of the calculations (solid lines) and the analytical data obtained from the asymptotic (dashed line) for a sonic flow around the same body [1,2]. Here "1" is the sonic line; "2" is the boundary characteristic; "3" is the line showing the departure of the velocity vector from the horizontal, and "4" is the shock wave. It is seen that already at a distance of 2 - 3 radii from the body good agreement is observed.

3. Let us now give some of the results of the calculations of transsonic and "overcritical" flows around profiles, plane and axisymmetrical bodies obtained by

the "large particles" method.

It is reasonable to characterize the overcritical regimes of transsonic flows around bodies by the value of the critical Mach number of the oncoming flow M_∞^* (when a sonic point develops on the body) as well as by the extent of a local supersonic zone (as compared to a characteristic dimension of the body) and by its intensity (maximum supersonic velocity relaized in the zone).

Figure 4 (series 4.1 - 4.8) presents flow field patterns (lines M = const.) for a 24% circular arc profile ($\nu = 0$) extending from purely subsonic ($M_\infty = 0.6$) to supersonic regimes ($M_\infty = 1.5$). Dynamics of the formation and development of a local supersonic zone, transitions through the critical Mach number (here $M_\infty^* = 0.65$), sound velocity, and so on are shown.

Figures 4.2 - 4.7 illustrate a supercritical flow around a profile ($0.65 < M_\infty < 1$). One can distinctly see the position of the shock in the region of crowded lines M = const which bounds the local supersonic line together with the sonic line (M = 1). The region of low velocities is located behind the shock wave. When the velocity of the flow increases, it reaches the parameters of an undisturbed flow at a large distance from the body. With $M_\infty \geq 0.9$ the zone becomes considerable both in size and in intensity (supersonic velocities are attainable up to M = 1.7 - 1.8) and in case of a sonic flow (Fig. 4.7) lines of the level M = 1 end at infinity.

The asymmetry of the whole flow pattern is noticeable (even at purely subsonic velocities - Fig. 4.1) which results from non-potentiality of the flow (supercritical regimes) and from the presence of viscous effects as well (subsonic regimes; formation of a wake behind the body, etc.)

In the case of a supersonic flow around a profile (Fig. 4.8 $M_\infty = 1.5$) a shock wave ahead of the body develops which bounds the disturbed region. Behind the wave, in the vicinity of the axis of symmetry a region of subsonic velocities is realized, afterwards the flow velocity along the contour of the body increases and, as a result, an "ending" shock occurs near the stern of the body.

For comparison the results of the calculations by the above method of a flow around a 24% axisymmetrical "spindlelike" body ($\nu = 1$) ($0.8 \leq M_\infty \leq 2.5$) are given in Fig. 5. Here a critical regime occurs already at $M_\infty^* = 0.\overline{8}6$; local supersonic zones as compared to the plane case are less developed and of a weaker intensity (for example, values of $M \sim 1.3 - 1.4$ are realized) and so on, although, naturally, the main singularities of a transsonic flow are seen here too.

In Fig. 6 a comparison is given between the flow fields calculated by the above method (solid line) and those of the Wood and Gooderum experiment (dashed line) [12] for subcritical (Fig. 6a $M_\infty = 0.725$) and supercritical (Fig. 6b $M_\infty^* = 0.761$) flows around a 12% profile (in accordance with the calculations and the experiment $M_\infty^* = 0.74$).

The analysis of the internal reference tests as well as the results of the comparisons reveal that the error in the calculations carried out with the help of the "large particles" method does not usually exceed several per cent.

The calculations were carried out using a BESM-6 computer; the time of the calculation in this case did not exceed an hour.

REFERENCES

1. Belotserkovskii, O. M., Davidov, Yu. M. Inf. Bull. SO AN SSSR Chys. Met. Splosh. Sred. 1, N3, 3-23 (1970).
2. Belotserkovskii, O. M., Davidov, Yu. M. J. Vych. Matem. Phys. 11, N1, 182-207 (1971).
3. Evans, M. W., Harlow, F. H. Los Alamos Scientific Laboratory, Rept. N. LA-2139 (1957).

4. Rich, M. Los Alamos Scientific Laboratory, Rept. N LAMS-2826 (1963).
5. Hirt, C. W. J. Comp. Phys. 2, N4, 339-355 (1968).
6. Belotserkovskii, O. M., Popov, F. D., Tolstykh, A. I., Fomin, V. N., Kholodov, A. S. J. Vych. Matem. Matem. Phys. 10, N2, 401-416 (1970).
7. Belotserkovskii, O. M. J. Comp. Phys. 5, N3, 587-611 (1970).
8. Belotserkovskii, O. M. Lect. Notes Phys. 8, 255-263 (1971).
9. Belotserkovskii, O. M., Chuskin, P. I. J. Vych. Matem. Matem. Phys. 2, N5, 731-759 (1962).
10. Gentry, R. A., Martin, R. E., Daly, B. J. J. Comp. Phys. 1, 87-118 (1966).
11. Davidov, Yu. M. J. Vych. Matem. Matem. Phys. 11, N4, 1056-1063 (1971).
12. Ferrari, C., Tricomi, F. G. Academic Press, New York and London (1968).

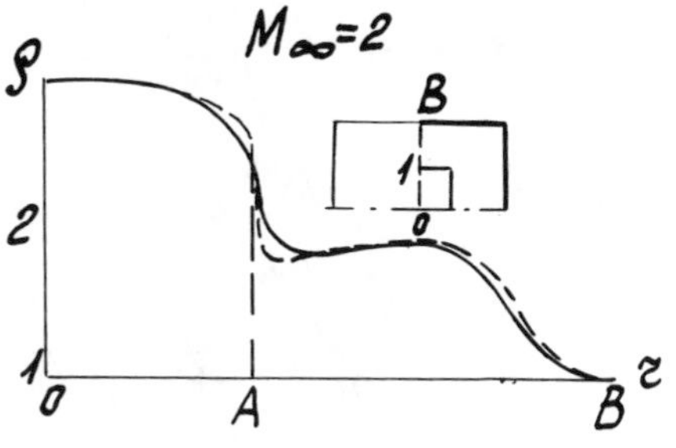

Fig. 1

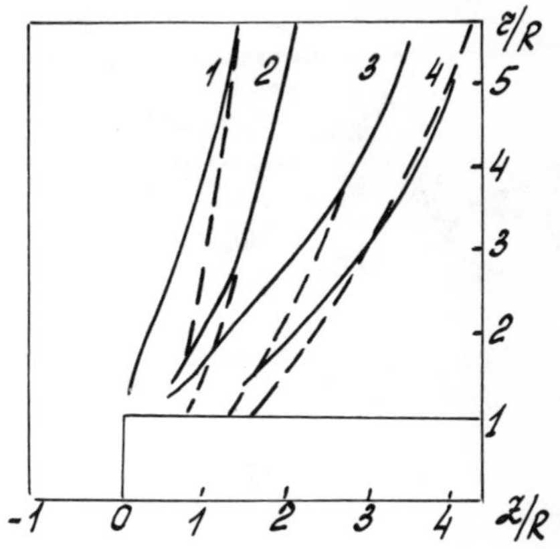

Fig. 3

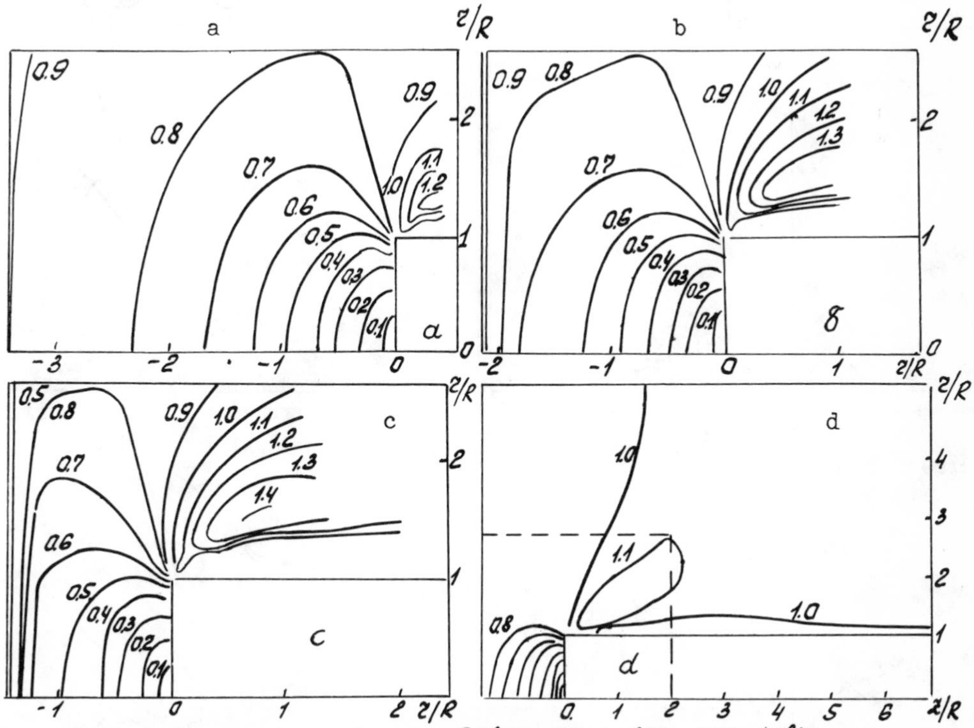

Fig. 2. $M_\infty=0.9$ a-$l/R=0.56$ б-$l/R=2.0$; c-$l/R=2.72$ d-$l/R=7$

Plane case ($\nu = 0$) $\delta = 24\%$ $M^*_\infty = 0.65$

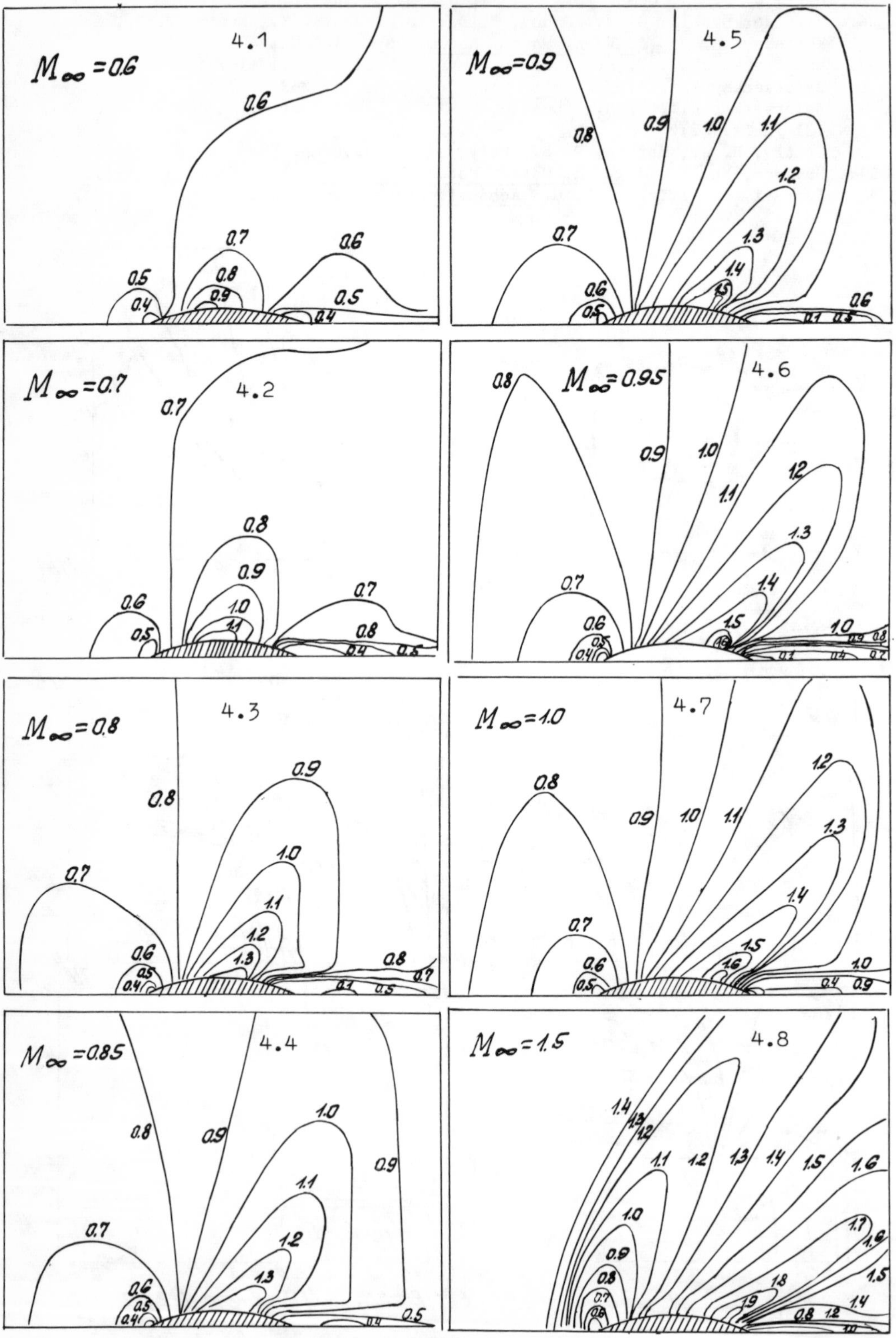

Fig. 4

Axisymmetric case (ν = 1) $\delta = 24\%$ $M^{*}_{\infty} = 0.86$

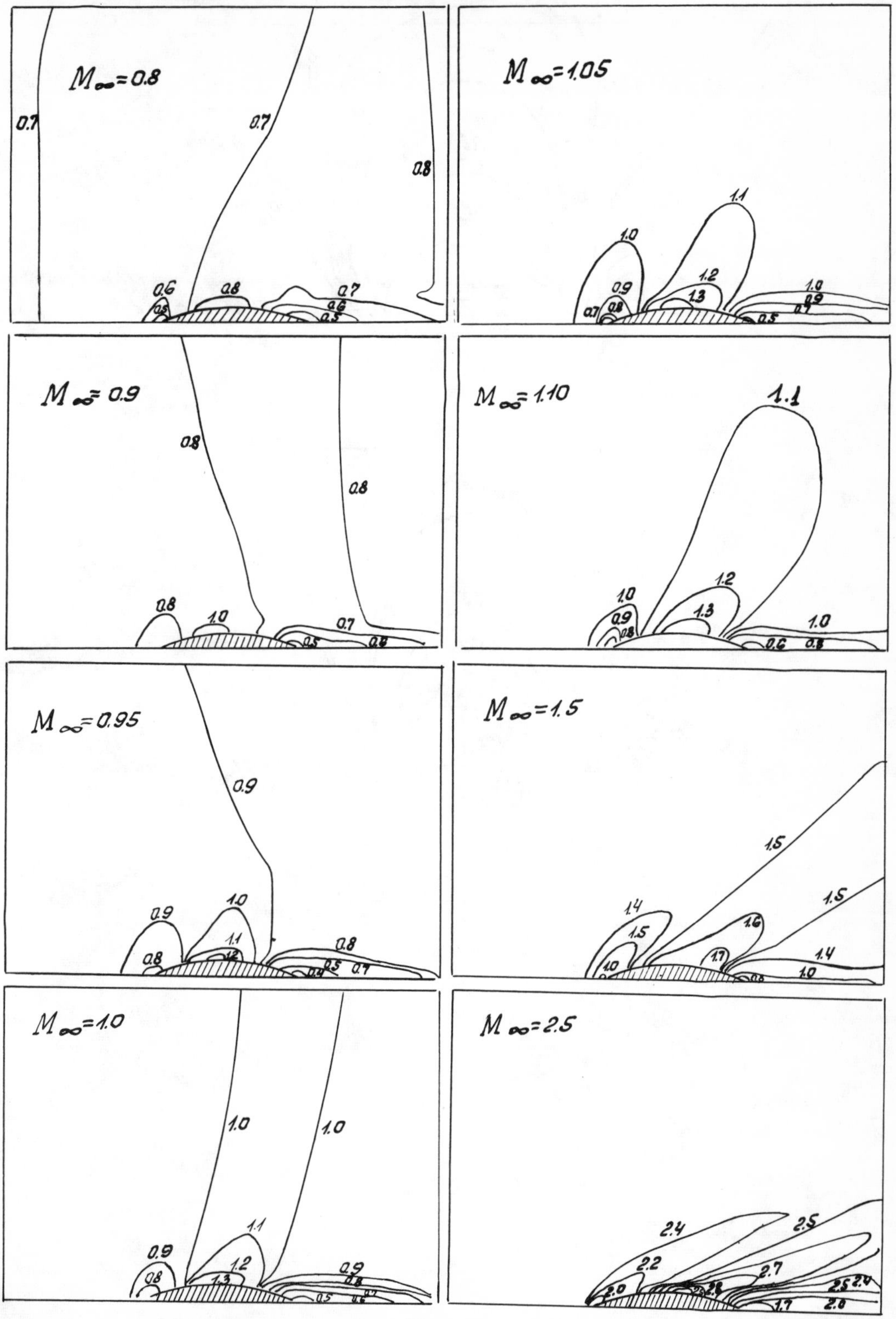

Fig. 5

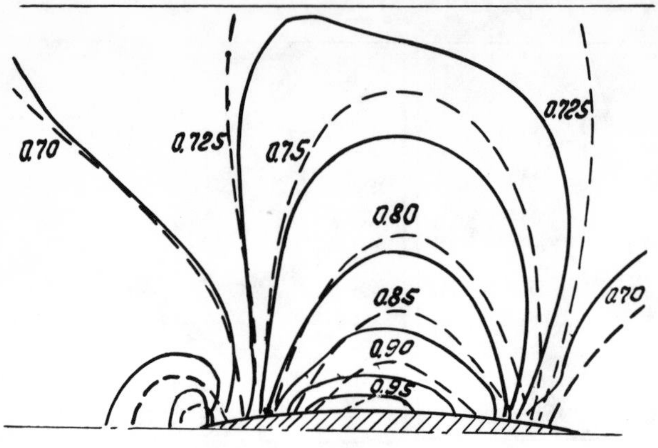

$M_\infty = 0.725$

Fig. 6 a

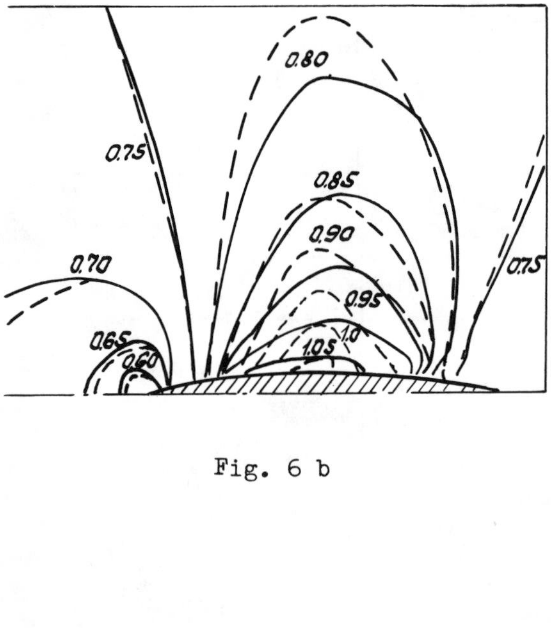

$M_\infty = 0.761$

Fig. 6 b

THE COMPUTATION OF THREE-DIMENSIONAL VISCOUS INTERNAL FLOWS

W. R. Briley
United Aircraft Research Laboratories
East Hartford, Connecticut USA

A procedure for computing three-dimensional, incompressible, viscous internal flows is outlined and the results of an application to the flow in straight rectangular ducts is summarized. The procedure is reasonably general and can be used to compute high Reynolds number flows which, when viewed from an appropriate coordinate system, have strong convection in one predominate direction, called the primary flow direction, with a secondary flow in the transverse coordinate plane. The procedure utilizes approximate governing equations which require a coordinate system as input to define the primary and secondary flow directions, and an (inviscid) first approximation to the static pressure gradients arising from curved flow geometries. The approximate equations are parabolic and are solved by stepwise integration in the primary flow direction from prescribed upstream initial conditions, the entire flow field being obtained by a sequence of two-dimensional numerical calculations. Use of the present procedure results in a substantial savings of computer time and storage compared to that required for solution of the three-dimensional Navier-Stokes equations. Furthermore, the method accounts automatically for interaction between viscous and inviscid flow regions and, consequently, is free from the problems of patching boundary-layer and inviscid flow solutions. A similar approach was recently taken by Patankar and Spalding (Ref. 4), who also computed the flow in the entrance region of a straight duct. Their governing equations are equivalent to those used here; however, the two methods employ different numerical techniques and differ somewhat in rationale. Although this initial application of the present procedure is restricted, in the interest of simplicity, to flow in straight rectangular ducts, extensions of the procedure to treat more complicated problems are possible and seem warranted by the favorable results obtained for rectangular ducts.

FORMULATION OF GOVERNING EQUATIONS

The governing equations utilized in the present procedure are derived from the Navier-Stokes equations using what are referred to here as "parabolic flow approximations". For rectangular ducts, cartesian coordinates (x,y,z) are used, with primary flow in the z-direction and secondary flow in the x-y plane (see Fig. 1). Using subscripts to denote partial derivatives, the Navier-Stokes equations can be written as

$$ww_z = -\frac{1}{\rho}(P+p)_z - uw_x - vw_y + \nu\left[w_{xx} + w_{yy} + w_{zz}\right] \tag{1}$$

$$wu_z = -\frac{1}{\rho}(P+p)_x - uu_x - vu_y + \nu\left[u_{xx} + u_{yy} + u_{zz}\right] \tag{2}$$

$$wv_z = -\frac{1}{\rho}(P + p)_y - uv_x - vv_y + \nu\left[v_{xx} + v_{yy} + v_{zz}\right] \tag{3}$$

$$u_x + v_y + w_z = 0 \tag{4}$$

In these equations, u, v, and w are velocity components in the x, y, and z directions, respectively; ν is the kinematic viscosity; ρ is the (constant) density, and the static pressure has been written as the sum of an inviscid pressure, P, and a viscous pressure correction, p, to be explained subsequently.

To create a more amenable system of equations, it is assumed firstly, that streamwise viscous diffusion for all three velocity components can be neglected by dropping the last of the bracketed terms in each of Eqs. (1) through (3), i.e., all second derivatives with respect to the primary flow or z-direction are discarded. Secondly, the inviscid pressure, P, is assumed known from a potential-flow solution and the components of its gradient are treated as source terms. For the straight constant-area duct problem, a uniform axial velocity with zero secondary flow is assumed at the duct entrance, and the potential flow with constant axial velocity and pressure is used; in this special case, P is constant and its gradient is zero. Finally, the viscous pressure correction, p, in the primary flow equation, Eq. (1), is treated separately from that in the secondary flow equations, Eqs. (2) and (3). In Eq. (1), the p_z term is redefined to be a mean viscous pressure drop which is a function of z only (i.e., constant in the x-y plane) and which is computed as part of the solution from the requirement that the integral mass flow in the axial direction be conserved. Thus, p_z is replaced by $dp_m(z)/dz$. No restrictions are placed on p in the secondary flow equations; in effect, p is allowed to vary in the x-y plane in such a way as to ensure that the continuity equation, Eq. (4), is satisfied at every point in the flow field. With these assumptions, the governing equations for the straight constant-area duct become

$$ww_z = -\frac{1}{\rho}\frac{d}{dz}p_m(z) - uw_x - vw_y + \nu\left[w_{xx} + w_{yy}\right] \tag{5}$$

$$wu_z = -\frac{1}{\rho}p_x - uu_x - vu_y + \nu\left[u_{xx} + u_{yy}\right] \tag{6}$$

$$wv_z = -\frac{1}{\rho}p_y - uv_x - vv_y + \nu\left[v_{xx} + v_{yy}\right] \tag{7}$$

$$u_x + v_y + w_z = 0 \tag{8}$$

together with the integral constraint on mass flow through the duct, which for impermeable walls can be written as

$$\int_{-H/2}^{H/2} \int_{-L/2}^{L/2} w \; dx \; dy = \dot{m}/\rho \tag{9}$$

where $\dot{m}$ is the mass flow rate through the duct. It can be shown that the set of equations, Eqs. (5) through (9), constitute a parabolic system and can be solved by stepwise integration in the axial or z- direction from a specified set of upstream initial conditions.

One important restriction arises from the parabolic character of the governing equations: the axial or primary flow velocity component must remain positive. In other words, there can be no separation of the primary flow; secondary flow separation can occur, however, and thus the equations possess the necessary generality to describe the formation of streamwise vortices. Clearly, however, the analysis is not intended for application to geometries having abrupt changes in cross-sectional area or to flow through sharp elbows, since the primary flow separates in these cases.

The application to flow in a straight rectangular duct provides a test of all the essential features of the method except the important one in which an inviscid static pressure field, P, is input to account _a priori_ for the elliptic influence of a curved geometry, in much the same way that a pressure distribution is "imposed" in boundary layer solutions. To test this feature, the present method can be employed to compute the flow in a curved duct. In this case, the inviscid pressure, P, would be obtained from a potential flow solution for flow through the duct, and a new set of governing equations would be derived after writing the Navier-Stokes equations in a coordinate system appropriate for the duct geometry under consideration. The coordinate system would be curvilinear and would consist of an axial coordinate which defines the primary flow direction, and two transverse coordinates which define the secondary flow planes; the secondary flow planes must be perpendicular to the walls. The approximate governing equations would be obtained, as in the straight duct case, by neglecting viscous diffusion in the axial direction, and by computing the mean viscous pressure drop so as to satisfy an integral formula for mass flow through the duct. Based on the experience that potential flow analysis yields a good prediction of the pressure field in many cases of practical importance, it is anticipated that this treatment of the pressure terms would be adequate.

COMPUTED RESULTS AND COMPARISON WITH EXPERIMENT

A finite-difference technique, based on an alternating-direction-implicit (ADI) scheme was used to compute solutions of Eqs. (5) through (9). A description of the numerical method is given by Briley in Ref. 2, together with a more detailed explanation of the analysis. Solutions are presented here for two ducts having aspect ratios of 1:1 and 2:1. These solutions are presented in nondimensional form and are compared with experimental velocity profile and pressure drop measurements from three sources (Beavers, et al., Ref. 1; Goldstein and Kreid, Ref. 3; and Sparrow, et al., Ref. 5). The following definitions are relevant: D is the hydraulic diameter, equal to four times the cross-sectional area of the duct divided by its perimeter; w_0 and p_0 are the axial velocity and static pressure at the duct entrance; $w_{\mathrm{c\!\!\!l}}$ is the centerline axial velocity; and Re is the Reynolds number, $w_0 D/\nu$. The solutions were computed for Reynolds numbers of 1000 and 1333, respectively, for the 1:1 and 2:1 ducts.

The computed mean pressure drop and centerline velocity development are shown in Figs. 2 and 3 for the two ducts, and these are found to be in good agreement with the experimental measurements. The effect on centerline velocity development of neglecting the secondary flow is shown in Fig. 3. The secondary flow has only a mild influence on the primary flow development because the transverse velocity components are relatively small; the magnitude of u and v for these solutions being of the order of a few tenths of one percent of w_0. A more detailed comparison between the computed and measured axial velocity development is given in Figs. 4 and 5, where velocity profiles at selected axial locations are shown; again, the computed and experimental results are in good agreement. Computed secondary-flow velocity profiles at one axial location are shown in Figs. 6 and 7 for each of the two ducts. The secondary flow is away from the walls and toward the center of the duct. It is perhaps worth noting that the secondary flow does not have a structure consisting of streamwise vortices.

In each of the present solutions, advantage was taken of the symmetry of the flow field about the $y = 0$ plane. The solutions were computed using 21 x 11 and 21 x 21 grids for the 1:1 and 2:1 ducts, respectively; required about 75 axial steps to reach the fully-developed region, and required 2 to 5 minutes of UNIVAC 1108 computer time.

REFERENCES

1. Beavers, G. S., Sparrow, E. M., and Magnuson, R. A. International Journal of Heat and Mass Transfer. 13, 689-701 (1970).

2. Briley, W. R. United Aircraft Research Laboratories Report L110888-1. East Hartford, Connecticut (1972).

3. Goldstein, R. J., and Kreid, D. K. Journal of Applied Mechanics. 34, 813-818 (1967).

4. Patankar, S. V., and Spalding, D. B. Report BL/TN/A/45. Imperial College of Science and Technology, London, England (1971).

5. Sparrow, E. M., Hixon, C. W., and Shavit, G. Journal of Basic Engineering. 89, 116-124 (1967).

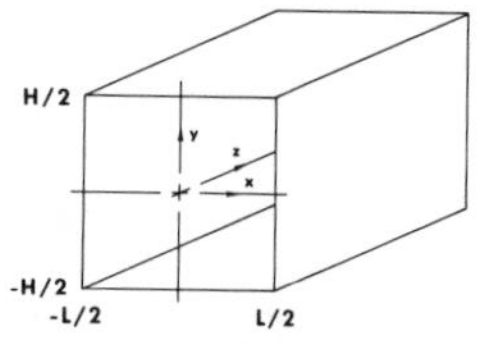

Fig. 1. Duct geometry and coordinate system

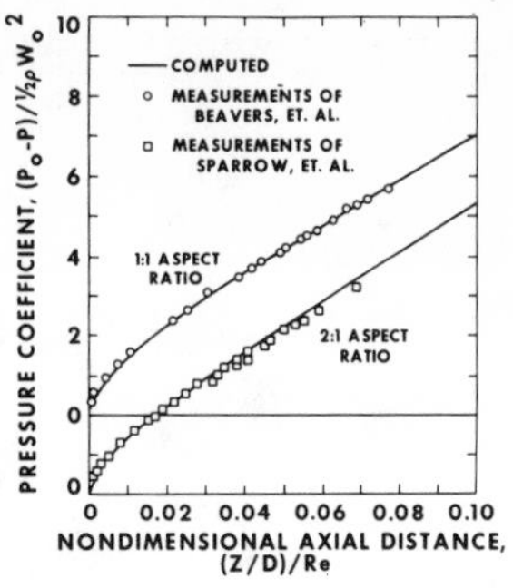

Fig. 2. Comparison between computed and measured pressure drop for rectangular ducts

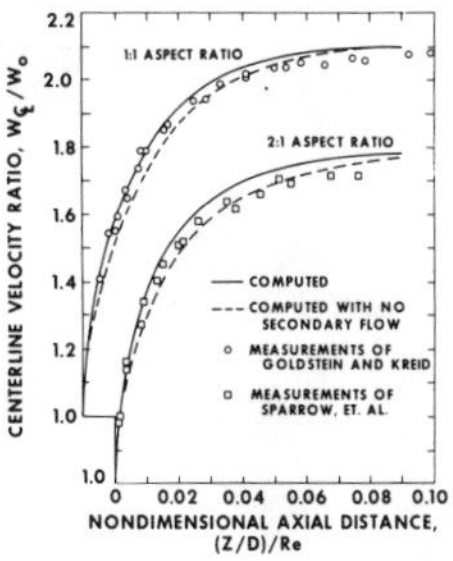

Fig. 3. Comparison between computed and measured development of center-line velocity for rectangular ducts

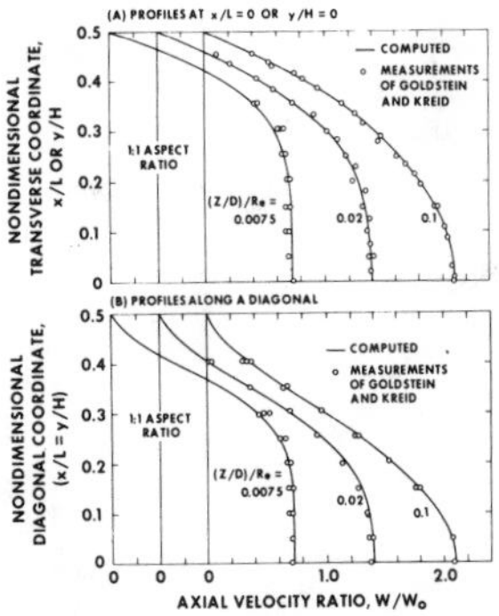

Fig. 4. Comparison between computed and measured axial velocity profiles for square duct

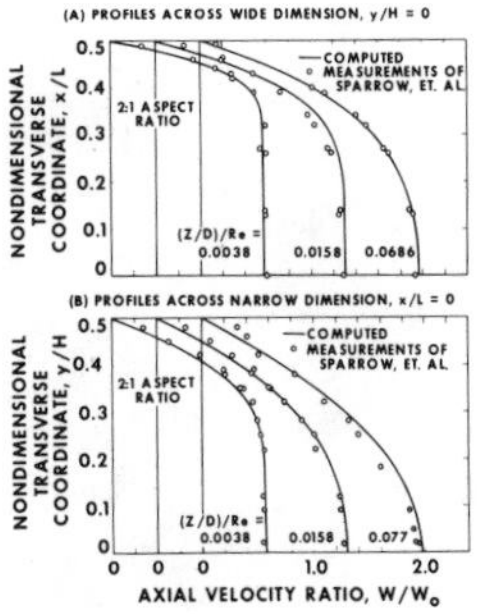

Fig. 5. Comparison between computed and measured axial velocity profiles for rectangular duct

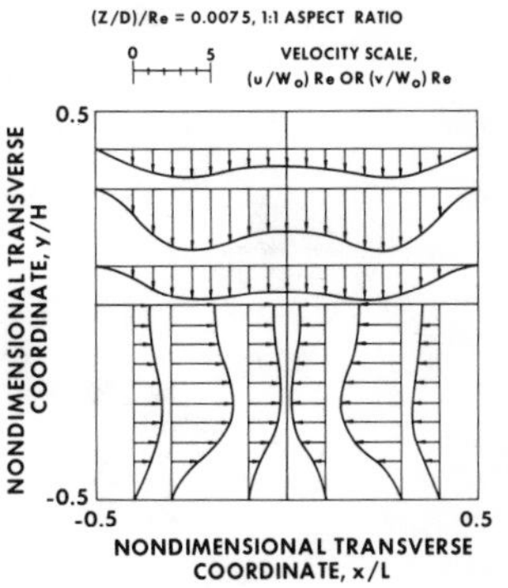

Fig. 6. Secondary flow velocity Profiles for square duct

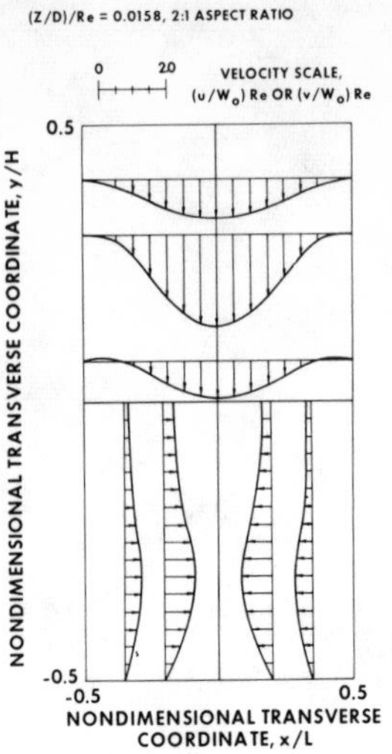

Fig. 7. Secondary flow velocity profiles for rectangular duct

INVISCID REATTACHMENT OF A SEPARATED SHEAR LAYER

Odus R. Burggraf*
Department of Mathematics, University College London

1. INTRODUCTION

Several theoretical models for calculating the reattachment process of a separated shear flow are available at present; however, they are based on contradictory assumptions as to the dominant physical processes involved. The well-known model of Chapman (1958) assumes the reattachment to be predominantly inviscid and for convenience takes the reattachment pressure to be the static pressure far downstream. Another well-known model is that of Lees and Reeves (1964) who treat the reattachment zone by a variant of the momentum-integral method of boundary-layer theory, requiring that the viscous terms remain important throughout reattachment and that the transverse pressure variation be negligible.

These seemingly contradictory theories, in this writer's opinion, may each be applicable to distinct classes of separated flows. In particular, Chapman's model may be applicable to cases in which the scale of the separated region is established by a characteristic body length (Class 1). This allows the relatively thin separated shear layer to develop independent of the surface conditions. On the other hand, the Lees-Reeves model would then apply when the extent of the separated region does not scale with body length and the shear layer remains relatively close to the surface (Class 2). In assessing the relevance of these models to various types of separated flows, it is desirable to compare details of the model-flow field with experiment. This is possible for the Lees-Reeves model, since calculations of detailed flow properties have been made for a variety of flows, including shock-boundary layer interaction (Class 2), with very good results, and base flows (Class 1) with rather less spectacular agreement with experiment. For Chapman's model, no such detailed calculations are available since the theory has been used to compute only overall flow properties, such as base pressure and mean heat flux. It is generally felt that this model, though qualitatively correct, is grossly oversimplified. In the present work, a detailed analysis of the reattaching flow is made on the basis of the assumption that the reattachment process is predominantly inviscid. The results of this analysis will be seen to substantiate Chapman's model of reattachment.

The assumption of an inviscid reattachment process can be motivated by an order-of-magnitude argument. If the length of the separated region remains finite

---

*Permanent address: Department of Aeronautical and Astronautical Engineering, The Ohio State University, Columbus, Ohio, U.S.A.

as the Reynolds number grows without bound (as appears to be the case for a supersonic free stream), then for a flow such as that behind a backward-facing step, the shear layer approaches the wall at a finite angle. Thus it seems reasonable to assume that the length of the reattachment zone scales like the thickness of the reattaching shear layer; i.e. as $R_*^{-\frac{1}{2}}$, where R_* is the characteristic Reynolds number. If the coordinates are rescaled as

$$x = (\bar{x}/L)R_*^{1/2}, \quad y = (\bar{y}/L)R_*^{1/2}$$

it is found that the viscous terms in the Navier-Stokes equations become negligible for $R_* \to \infty$. However, for reattachment on a solid wall, a viscous sublayer must form to satisfy the no-slip condition.

Two cases must be considered: (1) reattachment takes place asymptotically far downstream and the reattachment pressure coincides with the downstream pressure, as vizualized by Chapman, or (2) reattachment occurs locally and the dividing streamline intersects the wall with finite angle, corresponding to a peak in the pressure distribution. In the first case, the viscous sublayer is analogous to a member of the Falkner-Skan family of boundary layers with thickness

$$\delta_v \sim (\nu \bar{x}/\bar{u})^{1/2}$$

where $\bar{u}$ is the velocity just outside the sublayer. In the second case, the sublayer at reattachment is analogous to the conventional Hiemenz stagnation-region boundary layer; thus the thickness of the viscous sublayer scales as

$$\delta_v \sim \left(\nu / \frac{d\bar{u}}{d\bar{x}}\right)^{1/2}$$

where $\frac{d\bar{u}}{d\bar{x}}$ is the longitudinal velocity gradient just outside the sublayer. In either case, $(\bar{u}/\bar{x})$ or $\left(\frac{d\bar{u}}{d\bar{x}}\right)$ scales like $\bar{u}/\delta_s$ in the reattachment zone, where δ_s is the thickness of the separated shear layer. Hence

$$\delta_v \sim (\nu \delta_s/\bar{u})^{1/2} \sim \left(\frac{\nu L}{\bar{u}} R_*^{-\frac{1}{2}}\right)^{\frac{1}{2}}$$

or

$$\frac{\delta_v}{L} \sim R_*^{-3/4}$$

implying that the reattachment zone is inviscid asymptotically. The conclusion is that the Euler equations of motion are appropriate to the reattachment zone (outside the thin sublayer).

The results obtained here may be compared with a theory proposed by Lighthill for small perturbations on supersonic boundary layers in which the main part of the flow satisfies the inviscid equations of motion. In that case, if the unperturbed Mach number approaches zero at the wall, then no longitudinal pressure gradient is possible at the wall. Lighthill showed that this anomaly is due to neglect of the thin viscous sublayer $(\delta_v \sim L R_*^{-5/8})$. In our case the Mach number at the wall is not small, except near reattachment, allowing for a different scaling of the viscous sublayer. However, if Chapman's model is correct, the dividing streamline must reattach asymptotically far downstream, and Lighthill's theory would apply asymptotically for large x .

2. MATHEMATICAL FORMULATION

Let x_i denote the entry station to the reattachment zone (see Fig.1; $x_i \to -\infty$). At this station conditions for large y (distance from the wall) are those of the free shear layer, assumed known (Fig.2). Since the thin shear layer cannot support a transverse pressure variation, the pressure $p = p_i$ is uniform at x_i. Moreover, conservation of total pressure on streamlines requires that flow profiles for small y are just the reflection of flow profiles in that part of the shear layer below the dividing streamline. The entry flow angle is to be determined by the condition that the dividing streamline reattaches on the wall. Otherwise, all conditions are known at x_i.

For $x > x_i$ the usual condition $v = o$ is applied at $y = o$. At the outer edge of the shear layer, $y = y_e(x)$, matching to the external supersonic flow, assumed irrotational, requires a condition between pressure and flow angle. For convenience, the external stream is assumed to be uniform at x_i, so that the boundary condition at y_e is supplied by simple-wave theory (only left-running waves are allowed for $y > y_e$). Finally at the exit station downstream, the flow is required to be parallel to the wall corresponding to the uniform pressure field $p = p_f$. Hence the boundary condition at $x = x_f$ is $v = o$.

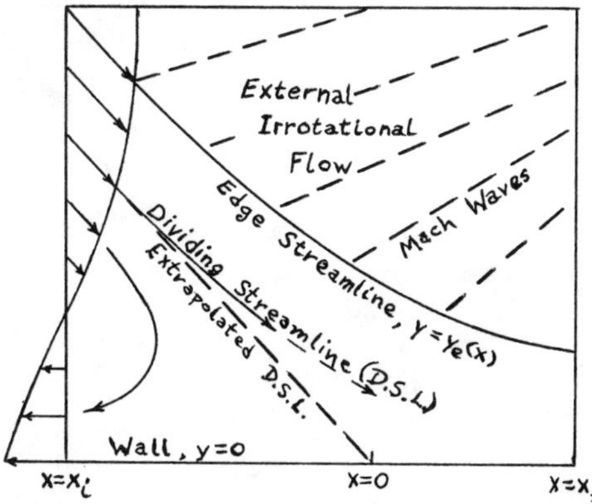

Fig. 1. Schematic of reattachment zone

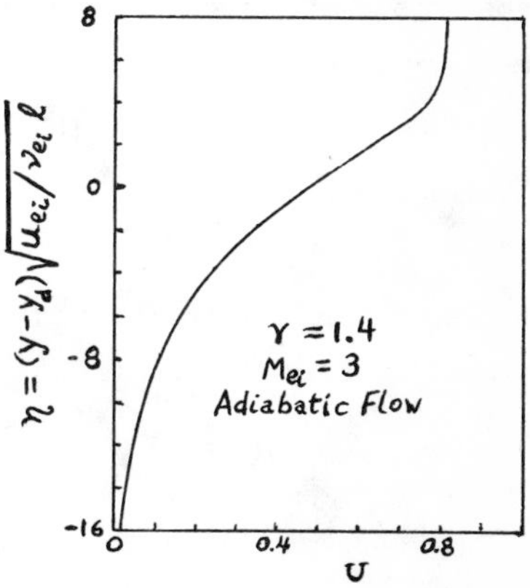

Fig. 2. Shear layer velocity profile

Clearly the boundary conditions are of mixed type. Moreover, the partial differential equations of motion are of mixed type, hyperbolic in the outer supersonic region and elliptic in the subsonic recirculating flow and near the reattachment point. Therefore severe instabilities would be expected to occur with many direct methods of solving the steady-flow equations of motion. To avoid these difficulties without resorting to a time-dependent method, an iterative method was formulated for solving the steady-flow equations of motion, which were expressed in the form of conservation integrals (total pressure, total temperature, entropy, vorticity/pressure). The first approximation was based on the observation that shear-layer flow angle predicted by Chapman's model, and confirmed by experimental base pressures, is quite small. In that case $v \ll u_e$ over the entire region and the transverse pressure gradient is small. Setting $\partial p/\partial y$ to zero as first approximation reduces the equations of motion to boundary-layer form (but excluding viscous terms). Thus the original mixed hyperbolic-elliptic problem has been replaced by a pseudo-parabolic problem in which the two Mach-line characteristics have been replaced by the single characteristic $dx = 0$, which applies even in the subsonic region. (The streamlines form a third family of characteristics for rotational flow, and these remain characteristics of the reduced problem.) Hence a stable solution can be obtained by forward integration through the reattachment zone. Once the flow field has been obtained, a revised pressure distribution is obtained by use of the y-momentum equation, and the flow-field calculation repeated with the new pressure. As will be seen, the results of the first iteration are a good approximation to the final converged solution. The explicit form of the equations used in the computations are summarized below.

The fluid is assumed to be a perfect gas with constant ratio of specific heats γ. For simplicity the total temperature was assumed to be uniform, although this is not essential to the procedure. Pressure and density were normalized with respect to their stagnation values in the external stream, and velocity components were normalized with respect to the limit velocity of the flow; i.e.

$$P = p/p_{t_e} \ , \qquad R = \rho/\rho_{t_e}$$

$$U = u/\sqrt{2c_pT_t} \ , \quad V = v/\sqrt{2c_pT_t}$$

The conservation integrals of the equations of motion were used; thus the total pressure distribution is

$$P_t = F(\psi) \tag{1}$$

where ψ is the stream function defined by

$$\frac{\partial\psi}{\partial y} = RU \ , \quad \psi(x,0) = 0 \tag{2}$$

The function $F(\psi)$ is given by the shear layer distribution at x_i. For isentropic flow of a perfect gas, the density given as

$$R = P/(1-U^2-V^2) \tag{3}$$

and the flow speed on the wall as *

$$U^2(x,o) = 1-(P/P_t)^{\frac{\gamma-1}{\gamma}} \tag{4}$$

For $y > o$, U is calculated from the vorticity

$$\frac{\partial U}{\partial y} = \frac{\partial V}{\partial x} - \omega \tag{5}$$

where the vorticity-conservation integral is

$$\omega = P\,G(\psi) \tag{6}$$

The function $G(\psi)$ is given by the shear layer distribution at x_i. In the first iteration V is neglected and P is set equal to its value P_e at the edge of the shear layer, $y = y_e(x)$, where the pressure is given by simple-wave theory

$$\frac{dP_e}{d\vartheta_e} = \gamma P_e M_e^2/\sqrt{M_e^2-1}\ , \quad \tan\vartheta_e = V_e/U_e \tag{7}$$

In second and later iterations V is computed from the stream function of the preceding iteration

$$RV = -\frac{\partial\psi}{\partial x} \tag{8}$$

and the pressure is computed from the y-momentum equation, using flow properties from the preceding iteration:

$$\frac{\partial P}{\partial y} = -\frac{2\gamma}{\gamma-1}\left[\frac{\partial}{\partial x}(RUV) + \frac{\partial}{\partial y}(RV^2)\right] \tag{9}$$

The independent variables were taken to be the normal coordinate y and the edge pressure P_e. The longitudinal coordinate x is computed from the resulting solution. For given entry flow angle ϑ_i and Mach number M_{e_i}, the pressures P_i and P_f at the entry and exit stations are known. The pressure interval$(P_f - P_i)$ is subdivided into equal subintervals (convergent solutions were obtained with both 20 and 40 subintervals). For fixed P_e, the system of equations (1) through (6) is solved on a gridwork of points uniformly spaced in the y-coordinate, using the Runge-Kutta integration method. This procedure is carried out for each pressure P_e in the range P_i to P_f, beginning at the entry station. The value of x corresponding to P_e is evaluated from the slope of the edge streamline, given from the pressure by simple wave theory:

$$\frac{dy_e}{dx} = \tan\vartheta_e$$

where y_e is given by the value of the stream function defining the edge streamline,

$$\psi(y_e) = \psi_e$$

Improved values of V and P are then calculated from (8) and (9), where the x-derivatives are evaluated using a three-point centered difference formula for the non-uniform x-mesh.

Because of the large gradients of flow properties transverse to the wall, the x-derivatives were evaluated by differentiating along streamlines; thus if f is

* The sign of $U(x,o)$ is taken as minus up to the reattachment point and plus beyond.

an arbitrary flow property,

$$\left(\frac{\partial f}{\partial x}\right)_y = \left(\frac{\partial f}{\partial x}\right)_\psi - \frac{V}{U}\left(\frac{\partial f}{\partial y}\right)_x \tag{10}$$

This scheme allows accurate calculation of the x-derivative using a fairly coarse x-mesh. However it is indeterminate on $U = 0$. Consequently for points below the line $U = 0$, $(\partial f/\partial x)$ was evaluated by differencing with y fixed, in the usual way. Eq.(10) is used for points above the dividing streamline, and for intermediate points, a linear mixture of the two is used to obtain smooth results. If the correct entry flow angle ϑ_i has been chosen, the iterative computation described above converges quite rapidly, as will be demonstrated. If ϑ_i is too small (in magnitude), the dividing streamline will not reattach. If ϑ_i is too large, the values of x become very large for some $P_e < P_f$ and the computation is terminated by exponent overflow. Hence the correct value of ϑ_i is determined easily and, for the range of conditions assumed here $(1.4 \leq M_{e_i} \leq 7)$, was found to be given quite accurately by Chapman's reattachment model.

3. DISCUSSION OF RESULTS

Computations of reattachment-flow fields have been made for several types of entry shear-layer profiles, with qualitatively similar results for all cases. All the results presented here correspond to entry shear-layer profiles defined by Chapman's (1956) self-similar solution; a sample is shown in Fig.2 for adiabatic flow with entry Mach number of the external stream $M_{e_i} = 3$. The reattachment-flow field corresponding to this condition is shown in Fig.3. In this figure, and others to follow, the x and y coordinates are normalized with respect to the total momentum-defect thickness of the upstream shear layer:

$$\theta = \int_{-\infty}^{\infty} \frac{\rho u}{\rho_e u_e}\left(1 - \frac{u}{u_e}\right) dy$$

The data plotted in Fig.3 resulted from 20 iterations, although convergence to one per cent accuracy occurred in only four or five iterations. Only a portion of the flow field is shown in the figure; the initial entry station was taken at $x_i \approx -100$ for convenience.* The computed value of x_f (the downstream exit station) is in excess of 200, although the precise value depends on the pressure-step size used, with larger values of x_f resulting for a finer grid. The results strongly suggest that $x_f \to \infty$ as the step size is reduced to zero.

The most significant conclusion to be drawn from Fig.3 is that the dividing streamline appears to approach the wall asymptotically as $x \to \infty$. This conclusion is borne out by the pressure distributions, which show that the wall pressure rises monotonically to its final value (Fig.5). This result is in agreement with Chapman's model of reattachment.

* The dividing streamline, extrapolated from x_i to the wall, defines $x = 0$ (Fig.1).

Also shown in Fig.3 are the results of the first iteration for the dividing streamline ($\psi^{(1)} = o$) and a sample isobar ($P^{(1)} = 0.040$). It is clear that the first iteration already leads to a good approximation to the converged flow field. The dashed portion of each of the converged isobars is just the Mach wave that would exist at the local flow conditions (the flow for $\psi > \psi_e$ was computed by simple-wave theory). It is evident from the slope of the isobar $P^{(1)}$ that the hyperbolic nature of the flow field has been restored after one cycle of iteration.

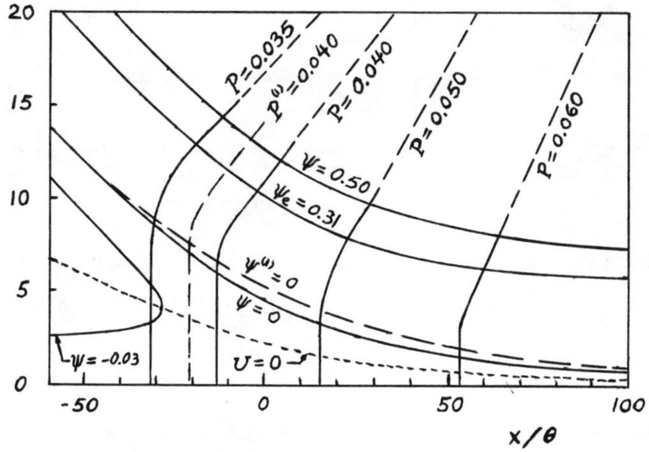

Fig. 3. Streamlines and isobars for $M_{e_i} = 3$, $\gamma = 1.4$

Fig.4 shows sample velocity profiles at various stations in the reattachment zone for the flow of Fig.3. The profiles for x near the entry station exhibit the double inflection associated with the shear layer and the backflow profile. The profile labelled $x/\theta = 200$ is the terminal reattachment profile and should be considered to hold for $x \to \infty$, as discussed above.

The wall pressure distributions are shown in Fig.5 for several Mach numbers. The length scale for reattachment is a strong function of Mach number, with a minimum at about $M_{e_i} = 3$. The controlling physical factors are entry flow angle and thickness of the low-speed tail of the shear layer. The length scales inversely as ϑ_i and proportional to the thickness of the shear-layer tail. The former is effective by a factor $(1+\frac{\gamma-1}{2}M^2)/\sqrt{M^2-1}$ according to linear theory. The latter produces another factor $(1+\frac{\gamma-1}{2}M^2)$. These factors are combined in the abcissa of Fig.5.

The transverse pressure variation is not shown. However, from Fig.3, it is seen to be significant only in the supersonic portion of the flow, above the dividing streamline. For $M_{e_i} = 3$, the maximum transverse pressure variation across the layer occurs near $x/\theta = 30$ and has a value of about ten percent of the overall reattachment pressure rise.

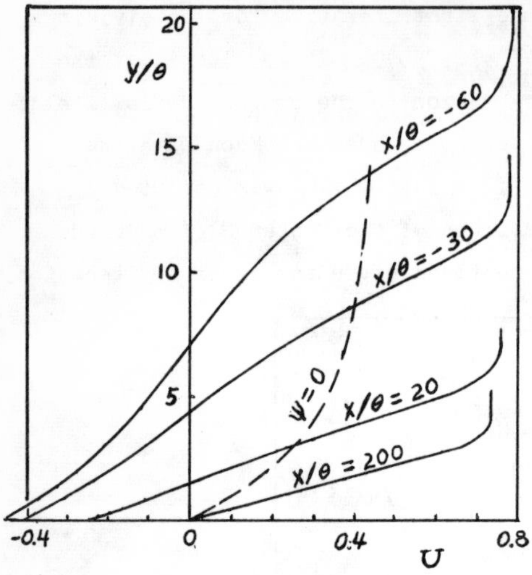

Fig. 4. Velocity profiles, $M_{e_i} = 3$

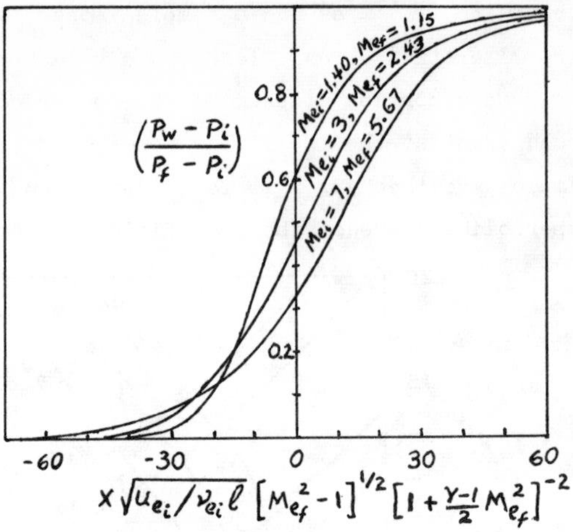

Fig. 5. Wall pressure distribution, $\gamma = 1.4$

4. CONCLUDING REMARKS

It was earlier remarked that the computational procedure was set up to allow stable forward integration with relatively large x-steps (actually P-steps), and this was borne out by the computations, which used as many as 240 y-steps together with as few as 20 pressure intervals. However in attempting to assess the effect of stepsize, it was observed that small oscillations in pressure developed when 40 P-intervals were used, and for 60 P-intervals, the oscillations would grow with each iteration, ultimately causing failure of the computation. These oscillations may be explained by making the analogy between a cycle of iteration and a time-step in an explicit time-dependent computation. In the latter case, instabilities are controlled by reducing the time step; a similar control might be accomplished in the iterative computation by use of under-relaxation. This has not been attempted, and from comparisons for 20, 30 and 40 P-intervals, it is felt that the results presented here are satisfactory. The conclusion that the dividing streamline reattaches asymptotically has analytical support in the fact that it is possible to set up an asymptotic expansion of the solution of the governing equations of motion in the form

$$U \sim U_f(y) - \frac{A}{x} U_f'(y) + \cdots$$

$$P \sim P_f - \frac{C}{x^2} + \cdots$$

where $U_f(y)$ is the terminal velocity profile ($x \to \infty$) and A and C are related constants. The structure of this expansion is not simple; an inner layer is required in terms of the variables x and $\eta = xy$. The details of the analysis will be given elsewhere.

5. ACKNOWLEDGEMENTS

The author is grateful for the financial support of the Air Force Aerospace Research Laboratories, Air Force Systems Command, United States Air Force, Contract No. F33615-68-C-1071, and of the Office of Naval Research, United States Navy, Contract No. N00014-67-A-0232-0014.

6. REFERENCES

Chapman, D.R.; 'A Theoretical Analysis of Heat Transfer in Regions of Separated Flow,' NACA TN 3792 (1956)

Chapman, D.R, Kuehn, D., and Larson, H.; 'Investigation of Separated Flows in Supersonic and Subsonic Streams with Emphasis on the Effect of Transition,' NACA Rpt. 1356 (1958)

Lees, L., and Reeves, B.; 'Supersonic Separated and Reattaching Laminar Flows: I. General Theory and Application to Adiabatic Boundary-Layer/Shock-wave Interactions,' AIAA J. 2, 1907 (1964)

Lighthill, M.J.; 'On Boundary Layers and Upstream Influence II. Supersonic Flows with Separation,' Proc. Roy. Soc. A, 217, 478 (1953)

TIME DEPENDENT CALCULATIONS FOR TRANSONIC FLOW*

by

Samuel Z. Burstein

Courant Institute of Mathematical Sciences

and

Arthur A. Mirin

Lawrence Livermore Laboratory

1. INTRODUCTION

Recently, several numerical methods have been developed for the calculation of steady transonic flow about an airfoil. Using small disturbance theory [1] and the exact formulation of a velocity potential in steady compressible flow [2], the flow field is obtained by a finite difference procedure which satisfactorily approximates the elliptic nature of the differential equations when the local sound speed exceeds the particle velocity and the hyperbolic nature when the fluid motion is locally supersonic. An extensive description including the programming of these methods can be found in [3].

Essentially we solve a system of partial differential equations in a bounded domain obtained by mapping the interior of the unit circle conformally onto the exterior of the airfoil. The boundary of the circle is mapped onto the boundary of the airfoil. Sells [4] has shown that a uniform polar coordinate grid in the transformed plane has the desired property that its image in the physical plane is most dense in the neighborhood of the nose and tail of the airfoil -- just where the flow gradients are largest.

We formulate the problem by using the physical conservation laws so that entropy changes can be accounted for. Even though the shock waves are weak, it is not clear if their position can be computed accurately by analysis which uses isentropy as a fundamental assumption. This is of special concern with wing shapes which are fairly flat so that Mach number variations over such spans do not vary markedly. Accurate determination of jump conditions will then become important. But most interesting is the question of the stability of the flow field about transonic airfoils; i.e., the characterization of the transient loading due to gusts superimposed on the free stream flow. It is our hope that the present formulation will have the capability of answering such suestions with sufficient accuracy.

*This research was supported by the U. S. Atomic Energy Commission, Contract No. AT(11-1)-3077. at New York University.

2. DIFFERENTIAL EQUATIONS

We assume, in this study, that the general equations of inviscid hydrodynamics, can be used to represent the flow about a plane airfoil. Since we wish to look at the general time dependent problem in which one or more shocks may be imbedded in the flow field, the governing differential equations, using the physically conserved variables, are written in divergence free form

$$w_t + f_x + g_y = 0 \tag{1}$$

where the vectors w, $f(w)$ and $g(w)$ are given by

$$w = \begin{pmatrix} \rho \\ \rho u \\ \rho v \\ E \end{pmatrix}, \quad f(w) = \begin{pmatrix} \rho u \\ p+\rho u^2 \\ \rho uv \\ (p+E)u \end{pmatrix}, \quad \text{and} \quad g(w) = \begin{pmatrix} \rho v \\ \rho vu \\ p+\rho v^2 \\ (p+E)v \end{pmatrix} \tag{2}$$

Here ρ is the density, ρu is the x-component of momentum, ρv is the y-component of momentum and the total energy $E = \rho(e + \frac{1}{2}(u^2+v^2))$ is seen to be the sum of the specific internal energy, e, and the kinetic energy. The pressure p is connected to the energy by an equation of state; we use the simple gas law

$$p = (\gamma-1)\rho e \tag{3}$$

System (1) is now transformed to the plane of integration; the coordinates of this plane are denoted by

$$\sigma = \xi + i\eta \tag{4}$$

and the map $F(\sigma)$ into the physical plane is of the form

$$z = x + iy = F(\sigma) = \frac{a_{-1}}{\sigma} + a_0 + a_1\sigma + \ldots + a_n\sigma^n + \ldots \tag{5}$$

A special case where $0 = a_0 = a_2 = a_3 = \ldots = a_n = \ldots$ is the Joukowski mapping.

Using the chain rule, and the Cauchy-Riemann equations the differential equations (1) can be written as

$$(B^2w)_t + (A_1f + A_2g)_\xi + (A_1g - A_2f)_\eta = 0 \tag{6}$$

where the coefficients $A_i = A_i(\xi,\eta)$ are given by the complex derivative

$$\frac{dz}{d\sigma} = A_1 + i\,A_2 \tag{7}$$

so that the modulus B of the transformation is computed by

$$B^2 = \left|\frac{dz}{d\sigma}\right|^2 = A_1^2 + A_2^2 \tag{8}$$

Now we write

$$\xi + i\eta = r\, e^{i\theta}$$

and again apply the chain rule; we obtain the transformed physical conservation law

$$(rB^2 w)_t + F_r + G_\theta = 0 \qquad (9)$$

from system (6). The transformed fluxes F and G are given by

$$\begin{pmatrix} F \\ G \end{pmatrix} = M \begin{pmatrix} f \\ g \end{pmatrix};$$

the matrix M being defined by

$$M = \begin{pmatrix} ra & rb \\ -b & a \end{pmatrix}$$

with entries $a = A_1 \cos\theta - A_2 \sin\theta$ and $b = A_1 \sin\theta + A_2 \cos\theta$.

Equation (9) is the differential equation which is solved over the domain $0 \leq r \leq 1$ and $0 \leq \theta \leq 2\pi$.

This equation has as unknowns the physical conservation variables Integration of this equation, with a suitable dissipative difference scheme, yields a weak solution of the differential equation (9); experience indicates that this solution is the one that approximates that which is found in nature. If, instead of the physical variables, the dependent variables chosen were those in which the momenta were parallel to the images of the uniform coordinates (r,θ) of the σ plane in the physical plane, then the equation would be expressed in general curvilinear orthogonal coordinates.

Let (x_1,x_2) represent an orthogonal coordinate system with metric tensor $ds^2 = h_1^2\, dx_1^2 + h_2^2\, dx_2^2$. The dependent variables are the same as in equation (1) but with ρu and ρv replaced by the corresponding momenta in the x_1 and x_2 directions respectively. Equation (1) in x_i space becomes

$$(h_1 h_2 w)_t + (h_2 f)_{x_1} + (h_1 g)_{x_2} + k = 0 \qquad (10)$$

where f and g are the same functions of w (even though w itself has been redefined) as given by equation (2). The inhomogeneous term has components

$$k = \begin{pmatrix} 0 \\ g_2\, \partial h_1/\partial x_2 - f_2\, \partial h_2/\partial x_1 \\ f_3\, \partial h_2/\partial x_1 - g_3\, \partial h_1/\partial x_2 \\ 0 \end{pmatrix}$$

and represent the Coriolis and centrifugal forces induced by the coordinate system. It is clear that we recapture our original system if we let $h_1 = h_2 = 1$. In the σ plane $x_1 = r$ and $x_2 = \theta$ while $h_1 = B$ and $h_2 = rB$.

Equation (10) is subject to the boundary condition, at $r = 1$, which requires the velocity normal to the surface of the airfoil, i.e., in the r direction, to vanish:

$$u = 0 \, . \tag{11}$$

The corresponding condition for equation (9)

$$v = u \tan \chi \tag{12}$$

where χ is the angle which the surface of the airfoil makes with respect to the x-axis.

At infinity, $r = 0$, the circulation Γ_∞, Mach number M_∞ and angle of attack, α are assumed to be known. In addition the pressure p_∞ and density ρ_∞ are assumed to be constants so that the sound speed $c_\infty^2 = \gamma\, p_\infty/\rho_\infty$ is defined.

In the next section we describe the procedure used to compute the mapping function $F(\sigma)$.

3. COMPUTATION OF THE MAPPING FUNCTION $F(\sigma)$

The results presented in this report deal with the flow about an ellipse where the mapping is known analytically. However, we describe the general case of the mapping of cusped airfoils since the numerical methods described will be extended to this important case.

We are given an airfoil with profile C, arclength s measured clockwise and curvature $\kappa(s) = d\chi/ds$ where χ is the angle that C makes with the x-axis. The cusp is assumed to be located at $s = 0$. The problem is to find an analytic map F sending the open unit disc in the σ plane conformally onto the exterior of the airfoil and the unit circle $|\sigma| = 1$ bijectively onto the profile C; F(1) locates the cusp, i.e., the point $r = 0$ and $\theta = 0$ is mapped onto the cusp. The existence of $F(\sigma)$ is postulated and assumed to be the form of equation (5) with the normalization $|a_{-1}| = 1$. Hence a_{-1} may be represented as

$$a_{-1} = e^{-i\alpha_1} \tag{13}$$

where α_1 is the angle of zero lift.

Now extend the definition of χ to all profiles obtained by considering values of $|\sigma| < 1$; such profiles are images of circles (r = constant) in the σ plane. Then the angle of the local tangent to all such profiles is given by

$$\chi = \arg\left[\frac{\partial}{\partial\theta} F(re^{i\theta})\right] \tag{14}$$

Performing the indicated differentiation we obtain for the bracketed expression of (14)

$$ire^{i\theta} F'(re^{i\theta}) = rBe^{i\chi} \tag{15}$$

Equating arguments, the angle of the tangent is

$$\chi = \frac{\pi}{2} + \theta + \arg F'(\sigma) \ . \tag{16}$$

Following reference [2] we substitute

$$U = \ln r^2 B \quad \text{and} \quad V = \chi + \theta \tag{17}$$

Note that

$$\exp\,(U + iV) = i\sigma^2 F'(\sigma) \tag{18}$$

is an analytic function which implies U and V are conjugate harmonic functions. The Cauchy Riemann equations can be used to obtain a boundary condition at r=1 for U:

$$\begin{aligned}\frac{\partial U}{\partial r} &= \frac{1}{r}\frac{\partial V}{\partial\theta} \\ &= 1 + \frac{\partial\chi}{\partial\theta} = 1 + \frac{\partial\chi}{\partial s}\frac{ds}{d\theta} = 1 + \kappa e^{U}\end{aligned} \tag{19}$$

Since the airfoil has a cusp at $\sigma = 1$ and since the mapping function is singular at $\sigma = 0$, equations (14) through (19) do not hold in general. In order to solve for U and V we must remove the singular -ities. Note that $F'(\sigma)$ has a zero of order 1 at $\sigma = 1$ which implies $F'(\sigma)/(1-\sigma)$ is analytic and nonzero near $\sigma = 1$. In addition $F'(\sigma)$ behaves like σ^{-2} as $\sigma \to 0$. We conclude that $\sigma^2 F'(\sigma)/(1-\sigma)$ is analytic and nonzero in a neighborhood of the closed unit disc.

Now introduce W and Y, the regularizations of U and V via

$$W + iY = \ln\left(\frac{\sigma^2 F'(\sigma)}{1-\sigma}\right) \tag{20}$$

Then $W = \ln|\sigma^2 F'(\sigma)/(1-\sigma)|$ which can then be expressed in terms of U as

$$W = U - \frac{1}{2}\ln\,(1 - 2r\cos\theta + r^2) \tag{21}$$

which is harmonic everywhere. The corresponding value of $Y = \arg[\sigma^2 F'(\sigma)/(1-\sigma)]$ yields

$$Y = 2\theta + \tan^{-1}[\,r\sin\theta/\,(1-r\cos\theta)] + \arg F'(\sigma) \tag{22}$$

with Y having a continuous branch over the closed unit disc.

Laplace's equation for $Q = \binom{W}{Y}$ is

$$Q_{\theta\theta} + r^2 Q_{rr} + rQ_r = 0 \tag{23}$$

subject to the boundary condition (19) for U expressed in terms of W

$$\frac{\partial W}{\partial r} = \frac{1}{2} + 2\kappa e^W \sin \theta/2 \tag{24}$$

and

$$Y = \chi + \theta/2 \tag{25}$$

at $|\sigma| = 1$.

The program of Garabedian and Korn [2] was used to solve for W. Additional programming was required to solve for Y and hence the arg $F'(\sigma)$. Successive over-relaxation was used to solve equation (23) subject to the boundary condition (25).

The leading coefficient of the mapping, equation (13) is computed using the mean value property:

$$\pi - \alpha_1 = \arg(-a_{-1}) = Y(0) = \frac{1}{2\pi}\int_0^{2\pi} Y(e^{i\theta})\, d\theta = \frac{1}{2\pi}\int_0^{2\pi} (\chi + \frac{\theta}{2})\, d\theta \tag{26}$$

For the case where there is no cusp the above procedure is modified by removing $(1-\sigma)$. In this case, equation (26) becomes

$$\alpha_1 = -\frac{1}{2\pi}\int_0^{2\pi} (\chi + \theta + \frac{\pi}{2})\, d\theta \ . \tag{27}$$

Figure 1 shows the image of the polar grid in the physical plane for an ellipse with thickness to chord ratio of 0.15 and for the NAE airfoil [8].

4. DIFFERENCE EQUATIONS

The differential equations (9) and (10) are solved on a grid defined by $\left\{(r_i,\theta_j) \mid r_i = (i-1)\Delta r,\ \theta_j = (j-1)\Delta\theta,\ 1 \le i \le I,\ 1 \le j \le J\right\}$ where $\Delta r = 1/(I-1)$ and $\Delta\theta = 2\pi/J$. We have used two sets of values for (I,J), namely (31,160) and (16,80). We represent both systems (9) and (10) in the general form

$$w_t + F_r + G_\theta + H = 0 \tag{28}$$

Using the notation $w^n_{i,j} = w(r_i,\theta_j,n\,\Delta t)$ the difference approximation to system (28) is

$$\begin{aligned} w^{n+1/2}_{i+1/2,j+1/2} &= \frac{1}{4}(w^n_{i,j} + w^n_{i+1,j} + w^n_{i,j+1} + w^n_{i+1,j+1}) \\ &- \frac{\lambda_r}{2}(F^n_{i+1,j+1/2} - F^n_{i,j+1/2}) - \frac{\lambda_\theta}{2}(G^n_{i+1/2,j+1} - G^n_{i+1/2,j}) \\ &- \frac{\Delta t}{8}(H^n_{i,j} + H^n_{i+1,j} + H^n_{i,j+1} + H^n_{i+1,j+1}) \end{aligned} \tag{29}$$

$$w_{i,j}^{n+1} = w_{i,j}^{n} - \lambda_r(F_{i+1/2,j}^{n+1/2} - F_{i-1/2,j}^{n+1/2}) - \lambda_\theta(G_{i,j+1/2}^{n+1/2} - G_{i,j-1/2}^{n+1/2}) - \Delta t\, H_{i,j}^{n+1/2} \quad . \tag{29}$$

Here $\lambda_r = \Delta t/\Delta r$, $\lambda_\theta = \Delta t/\Delta\theta$ and

$$F_{i+1,j+1/2}^{n} = F(\tfrac{1}{2}(w_{i+1,j}^{n} + w_{i+1,j+1}^{n}))$$

$$F_{i+1/2,j}^{n+1/2} = F(\tfrac{1}{2}(w_{i+1/2,j+1/2}^{n+1/2} + w_{i+1/2,j-1/2}^{n+1/2}))$$

$$H_{i,j}^{n+1/2} = H(\tfrac{1}{4}(w_{i-1/2,j-1/2}^{n+1/2} + w_{i+1/2,j-1/2}^{n+1/2} + w_{i-1/2,j+1/2}^{n+1/2} + w_{i+1/2,j+1/2}^{n+1/2})) \ ,$$

and similar definitions apply to G. It is clear that rather than applying space averaging to the arguments of F and G one could apply space averaging to the fluxes F and G directly. The above scheme is due to Richtmyer [5]. In [6] we have devised a similar scheme to (29) but with $\hat{w} = w((n+1)\Delta t)$ being computed at the first step. Here the step size is Δt rather than $\frac{1}{2}\Delta t$; the second step is replaced by

$$\begin{aligned} w_{i,j}^{n+1} = w_{i,j}^{n} &- \frac{\lambda_r}{4}(F_{i+1,j}^{n} - F_{i-1,j}^{n} + \hat{F}_{i+1/2,j+1/2}^{n+1} - \hat{F}_{i-1/2,j+1/2}^{n+1} \\ &\quad + \hat{F}_{i+1/2,j-1/2}^{n+1} - \hat{F}_{i-1/2,j-1/2}^{n+1}) \\ &- \frac{\lambda_\theta}{4}(G_{i,j+1}^{n} - G_{i,j-1}^{n} + \hat{G}_{i+1/2,j+1/2}^{n+1} - \hat{G}_{i+1/2,j-1/2}^{n+1} \\ &\quad + \hat{G}_{i-1/2,j+1/2}^{n+1} - \hat{G}_{i-1/2,j-1/2}^{n+1}) - \Delta t\, H_{i,j}^{n+1/2} \end{aligned} \tag{29'}$$

Here $\hat{F}_{i,j}^{n+1} = F(\hat{w}_{i,j}^{n+1})$ and the inhomogeneous term is given by

$$H_{i,j}^{n+1/2} = \frac{1}{2}(H_{i,j}^{n} + \frac{1}{4}(\hat{H}_{i+1/2,j+1/2}^{n+1} + \hat{H}_{i+1/2,j-1/2}^{n+1} + \hat{H}_{i-1/2,j+1/2}^{n+1} + \hat{H}_{i-1/2,j-1/2}^{n+1})) \ .$$

Stability of these schemes can be examined by carrying out the indicated differentiation of equation (28), after choosing the momentum representation, i.e. the dependent variable w. For equation (9), w is the same as equation (1) but <u>not</u> the same as for equation (10).

Introduce the matrices f_w and g_w obtained from the quasilinear form of equation (1). Then the quasilinear form of equation (28) shows that F_w and G_w are linear combinations of f_w and g_w. For equation (9)

$$\begin{pmatrix} F_w \\ G_w \end{pmatrix} = \frac{1}{rB^2} M \begin{pmatrix} f_w \\ g_w \end{pmatrix} \tag{30}$$

and equation (10)

$$\begin{pmatrix} F_w \\ G_w \end{pmatrix} = \begin{pmatrix} \frac{1}{h_1} f_w \\ \frac{1}{h_2} g_w \end{pmatrix} \tag{31}$$

It can be shown that f_w and g_w can be simultaneously symmetrized [7]. If c is the local sound speed then

$$f_w \sim \begin{pmatrix} u & c & 0 & 0 \\ c & u & 0 & 0 \\ 0 & 0 & u & 0 \\ 0 & 0 & 0 & u \end{pmatrix} \quad \text{and} \quad g_w \sim \begin{pmatrix} v & 0 & c & 0 \\ 0 & v & 0 & 0 \\ c & 0 & v & 0 \\ 0 & 0 & 0 & v \end{pmatrix}$$

Since these matrices are sparse, the computation of the eigenvalues of the amplification matrix of the associated difference operator is greatly simplified; here we note that the spectral radius μ corresponding to equation (30) is

$$\mu(F_w) = \frac{1}{B^2}\left\{|au + bv| + c\sqrt{a^2+b^2}\right\}$$

$$\mu(G_w) = \frac{1}{rB^2}\left\{|av - bu| + c\sqrt{a^2+b^2}\right\} \tag{32}$$

while for equation (31) it is simply

$$\mu(F_w) = |h_1^{-1}|\,(|u| + c)\,, \qquad \mu(G_w) = |h_2^{-1}|\,(|v| + c) \tag{33}$$

In either case the stability condition used is

$$\Delta t \leq k \cdot \inf\left\{\frac{\Delta r}{\mu(F_w)}, \frac{\Delta\theta}{\mu(G_w)}\right\} \tag{34}$$

with $k \sim 0.5$ although the theoretical maximal $k = 1$. It was found that the above difference schemes required an artificial viscosity for stability. The solution obtained from the difference schemes (29) or (29'), $w_{i,j}^{n+1}$, is modified so that the final solution after each cycle is just

$$w_{i,j}^{n+1} + \lambda_r[\omega_{i+1/2,j}^{n+1}(w_{i+1,j}^{n+1} - w_{i,j}^{n+1}) - \omega_{i-1/2,j}^{n+1}(w_{i,j}^{n+1} - w_{i-1,j}^{n+1})] + \tag{35}$$

$$+ \lambda_\theta[\omega^{n+1}_{i,j+1/2}(w^{n+1}_{i,j+1} - w^{n+1}_{i,j}) - \omega^{n+1}_{i,j-1/2}(w^{n+1}_{i,j} - w^{n+1}_{i,j-1})]$$. See [9,10].

To keep second order accuracy ω is chosen to be an undivided space difference of the velocity in each direction, i.e. $\omega^{n+1}_{i+1/2,j} \sim |u^{n+1}_{i+1,j} - u^{n+1}_{i,j}|$ and $\omega^{n+1}_{i,j+1/2} \sim |v^{n+1}_{i,j+1} - v^{n+1}_{i,j}|$, etc.

We handle the boundary condition at $|\sigma| = 1$ by using the normal, u, and tangential, v, velocities in the following rules

$$\begin{aligned} u_{I+1,j} &= -u_{I-1,j} \\ v_{I+1,j} &= v_{I-1,j} \end{aligned} \tag{36}$$

The simple reflection rules for density and pressure

$$\begin{aligned} \rho_{I+1,j} &= \rho_{I-1,j} \\ p_{I+1,j} &= p_{I-1,j} \end{aligned} \tag{37}$$

are also used. In equations (36) and (37) $j = 1,2,\ldots,J$.

It was found that, where the curvature of the coordinate system is greatest, i.e., the nose and tail of the airfoil, the greater the difficulty in satisfying (12) using (9). However, since the momenta in (10) are alligned with the polar coordinates in the σ-plane it may be possible to improve the accuracy at the body locally at the nose and tail by using (10). If shocks are not present in such regions then (9) will still yield the correct jump conditions for the shocked flow field.

5. RESULTS AND DISCUSSION

The numerical methods described wre programmed for the CDC 7600 at Lawrence Radiation Livermore Laboratory and the CDC 6600 at New York University. The computational speed ratio was found, for the same program, to be between four and five to one in favor of the 7600. Computation time for the time dependent elliptical flow cases with the larger of the two mesh sizes, $\Delta r = 1/15$ and $\Delta\theta = 2\pi/80$ was about a half hour on the 7600. This corresponds to the time a wave with velocity $|\vec{u}| + c$ will take to travel approximately ten s_0 where s_0 is the half arc length of the ellipse. Most of the transient stage is completed in about half that time, i.e. 5 s_0. There is always a non-steady wave propagation observed in the neighborhood of infinity as should be the case for a true time dependent formulation.

Each compuation was carried out with a maximum cycle count of 2500; the artificial viscosity coefficient $\lambda_r = \lambda_\theta = 2$ and $\kappa = 0.5$ in equation (34) were used.

We computed the distance a wave with speed $|\vec{u}|$ + c moves in time N Δt over the surface of the airfoil with starting position $\sigma = -1$ (airfoil nose). At 1400 cycles approximately 5 wave traversals were accumulated; a maximum of 9.5 at 2500 cycles was recorded.

The potential solution predicts $(c_L, c_D) = (3.624, -.098)$ while system (29) yields (3.36, -.028) and (29') yields (3.35, -.029). Here, it was assumed that the velocity at the point of maximum curvature vanished. This is the Kutta-Joukowski condition when B, the modulus, does not vanish. When B does vanish, as is the case for the N.A.E. airfoil with a cusped tail, the ratio of the gradient of the potential to B at $\sigma = 1$ must be bounded.

For the case where there is circulation about the ellipse at two degree angle of attack, a shock is positioned at 62.5% chord on the upper surface. The computed pressure ratio of 1.437 corresponds to a Mach number which is approximately 1.17; the exact pressure jump for a normal shock at this Mach number is 1.430. The detailed results, which were presented in part through a seven minute motion picture in Paris on July 7, 1972 will be given in future publications.

We conclude our paper by noting that additional work must be done in formulating the boundary conditions (36) and (37) prescribed at the airfoil surface, e.g. one sided differencing using the conservation variables or characteristic variables. This is due to the effect of the singularity in the mapping function for the elliptical airfoil on the values of the reflected conservation variables required when $r>1$.

REFERENCES

1. Murman, E. M., and Cole, J. D., Calculation of Plane Steady Transonic Flows, AIAA Jour. 9, 114-121 (1971).
2. Garabedian, P. R., and Korn, D. G., Analysis of Transonic Air-Foils, Comm. Pure Appl. Math., 24, 841-851 (1971).
3. Bauer, F., Garabedian, P., and Korn, D., Supercritical Wing Sections, Lecture Notes in Economics and Mathematical Systems, 66, Springer-Verlag (1972).
4. Sells, C. C., Plane Subcritical Flow Past a Lifting Airfoil, Proc. Roy. Soc. A. 308, 377-401 (1968).
5. Richtmyer, R. D., A Survey of Difference Methods for Nonsteady Gas Dynamics, NCAR Tech. Note 63-2.
6. Burstein, S. Z., F nite Difference Calculations for Hydrodynamic Flows Containing Discontinuities, J. Comp. Physics, 1, (1966), pp. 198-222.
7. Turkel, E., Symmetrization of the Fluid Dynamic Matrices with Applications, (to appear).
8. Kacprzynski, J. J., Ohman, L. H., Garabedian, P. R., and Korn, D. G., Analysis of the Flow Past a Shockless Lifting Airfoil in Design and Off-Design Conditions, N.R.C. of Canada Aeronautical Report LR-554, Ottawa, (1971).
9. Harten, A., and Zwas, G., Switched Numerical Shuman Filters for Shock Calculations, Jour. of Eng. Math., 6, (2), (April 1972).
10. Lapidus, A., A Detached Shock Calculation by Second-Order Finite Differences, J. of Compuational Physics 2, 154 (1967).

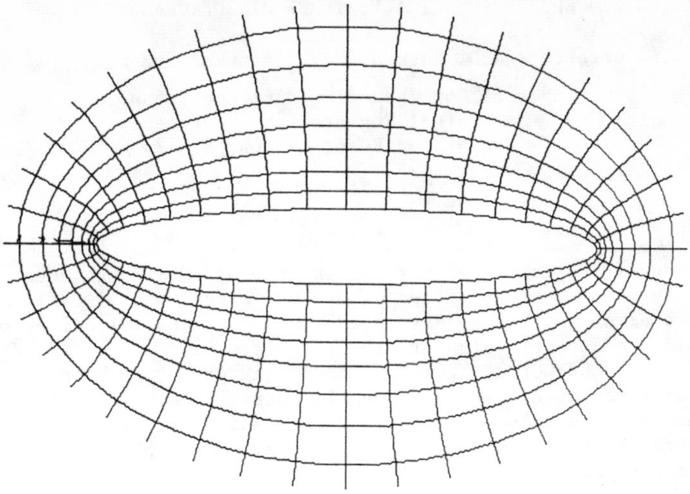

Fig. 1a - Ellipse T/C = 0.15

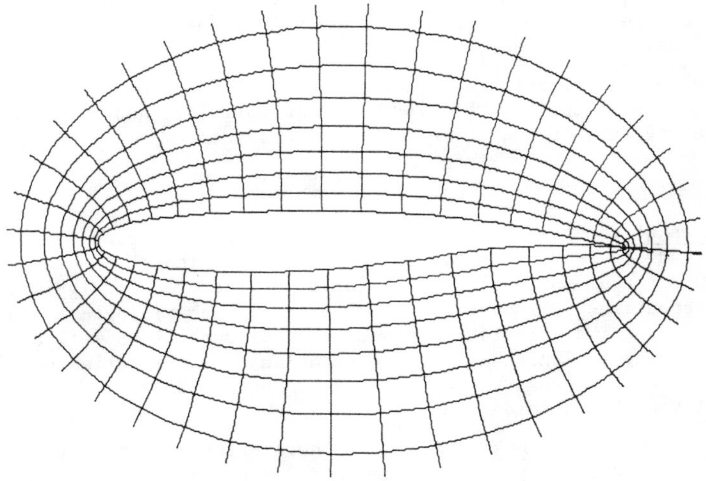

Fig. 1b - N.A.E. Airfoil

TWO CALCULATION PROCEDURES FOR STEADY, THREE-DIMENSIONAL FLOWS WITH RECIRCULATION

L.S. Caretto,* A.D. Gosman, S.V. Patankar and D.B. Spalding

Imperial College of Science and Technology
Mechanical Engineering Department
Exhibition Road, London, S.W.7.

ABSTRACT

Two procedures are described for solving the Navier-Stokes equations for steady, fully three-dimensional flows: both are extensions of earlier methods devised for three-dimensional boundary layers, and have the following common features: (i) the main dependent variables are the velocities and pressure; (ii) the latter are computed on a number of staggered, interlacing grids, each of which is associated with a particular variable; (iii) a hybrid central-upwind difference scheme is employed; and (iv) the solution algorithms are sufficiently implicit to obviate the need to approach the steady state via the time evolution of the flow, as is required by wholly explicit methods.

The procedures differ in their manner of solving the difference equations. The SIVA (for SImultaneous Variable Adjustment) procedure, which is fully-implicit, uses a combination of algebraic elimination and point-successive substitution, wherein *simultaneous* adjustments are made to a point pressure, and the six surrounding velocities, such that the equations for mass and (linearised) momentum are locally satisfied.

The SIMPLE (for Semi-Implicit Method for Pressure-Linked Equations) method proceeds in a *successive* guess-and-correct fashion. Each cycle of iteration entails firstly the calculation of an intermediate velocity field which satisfies the linearised momentum equations for a guessed pressure distribution: then the mass conservation principle is invoked to adjust the velocities and pressures, such that all of the equations are in balance.

By way of an illustration of the capabilities of the methods, results are given of the calculation of the flow of wind around a building, and the simultaneous dispersal of the effluent from a chimney located upstream.

1. INTRODUCTION

1.1 Objectives of the present research. We are here concerned with prediction methods for that class of convective-flow phenomena which are steady, recirculating, low-speed and three-dimensional: the majority of the practically-important flow situations encountered in industrial, environmental, physiological and other fields are of this kind. Two calculation procedures for such flows will be described: both proceed by way of finite-difference solution of the Eulerian partial-differential equations for the conservation of mass, momentum, energy and other properties; and both employ the velocities and pressure as the main hydrodynamic variables.

1.2 Relation to previous work. Although there exist a number of finite-difference procedures which could, in principle, be used for the present class of problems, none appear to be well-suited for this purpose. Thus, for example, nearly all of the available methods attempt to follow the time evolution of the flow in arriving at the steady-state solution. When however the latter is the *only* feature of interest, this is usually needlessly expensive, especially when an *explicit* formulation is employed.

The procedures to be described here contain a number of innovations, allowing particularly economical routes to the steady state: they also however incorporate many known features including:- the displaced grids for velocity and pressure employed by Harlow and Welch (1965); the concept of a guess-and-correct procedure for

---

* L.S. Caretto is currently at California State University, Northridge, California.

the velocity field, used by Amsden and Harlow (1970) and Chorin (1968); and the implicit calculation of velocities, along the lines of the Pracht (1970) version of the Harlow-Welch (1965) procedure. Additional guidance in the formulation of the new procedures has been derived from earlier work by the authors and their colleagues on methods for two-dimensional flows (Patankar and Spalding, 1970; Gosman et al., 1969), and three-dimensional boundary layers (Patankar and Spalding, 1972a; Caretto et al., 1972).

1.3 Contents of the paper. Section 2 of the paper is devoted to the description of the two procedures, code-named SIMPLE and SIVA. Because the point of departure between the two methods is in the manner of solving the finite-difference equations, the latter are described first; then details are given of the individual solution paths.

In Section 3, we provide a summary of the experience gained from application of the procedures to a variety of test cases. Then, by way of a demonstration, we present the results of a computer simulation of the flow of wind past a building, and the simultaneous dispersal of the effluent from a chimney located upwind of the building. Finally, in Section 4 we present our conclusions about the relative merits of the two procedures, and the prospects for further development.

2. ANALYSIS

2.1 The equations to be solved. The mathematical problem may be compactly expressed, with the aid of Cartesian tensor notation, in terms of the following set of differential equations:

$$\partial(\rho u_i)/\partial x_i = 0 \quad ; \quad (1)$$

$$\partial(\rho u_i u_j)/\partial x_i - \partial(\mu_{eff} \partial u_j/\partial x_i)/\partial x_i + \partial P/\partial x_j - s_j = 0 \quad ; \quad (2)$$

$$\partial(\rho u_i \phi)/\partial x_i - \partial(\Gamma_{\phi,eff} \partial\phi/\partial x_i)/\partial x_i - s_\phi = 0 \quad ; \quad (3)$$

which express the laws of conservation of mass, momentum and a scalar property ϕ respectively. Here the dependent variables are the (time-average) values of: the velocities u_j; the pressure P; and ϕ, which stands for such scalar quantities as enthalpy, concentration, kinetic energy or dissipation rate of turbulence (Launder and Spalding, 1971) and radiation flux (Spalding, 1972a) etc. The symbols s_j and s_ϕ stand for additional sources (or sinks) associated with such phenomena as natural convection, chemical reaction and non-uniformity of transport coefficients, while ρ, μ_{eff} and $\Gamma_{\phi,eff}$ are respectively the density, viscosity and exchange coefficient for ϕ. The subscript 'eff' appended to the latter two indicates that, for turbulent flows, they are sometimes ascribed 'effective' values, deduced from turbulence quantities.

2.2 Finite-difference equations

(a) Grid and notation. The staggered-grid system employed for both methods is depicted in Fig. 1: this shows only the xy plane, but the treatment in the other planes follows identical lines.

The intersections of the solid lines mark the grid nodes, where all variables except the velocity components are stored. The latter are stored at points which are denoted by the arrows and located mid-way between the grid intersections. A considered node and its immediate neighbours are denoted by the subscripts P, x+, x-, y+, y-, z+ and z-: the significance of these can be perceived from Fig. 1. The velocities are similarly referenced, with the convention that P (and each of the other subscripts) now refers to a *cluster* of variables, as indicated in the diagram.

(b) Differencing practices. Attention will first be focussed on the differential conservation equation (3) for a scalar property ϕ. A difference equation relating ϕ_P to the surrounding ϕ's is obtained by integration of (3) over the control volume enclosing P, with the aid of flux expressions derived from one-dimensional flow theory. Some details will now be given.

We represent the net x-direction convection and diffusion of ϕ through the control volume (Fig. 2) by:

$$C^{\phi}_{x+}(\phi_{x+} - \phi_P) + C^{\phi}_{x-}(\phi_{x-} - \phi_P) \tag{4}$$

where, e.g.:

$$C^{\phi}_{x+} = \begin{cases} 0, \text{ when } F_{x+} > D_{x+} & ; \\ -2F_{x+}, \text{ when } F_{x+} < -D_{x+} & ; \\ D_{x+} - F_{x+}, \text{ in all other circumstances.} \end{cases}$$

$$F_{x+} \equiv \dot{m}''_{+} A_x/2 \quad ; \qquad D_{x+} \equiv \overline{\Gamma}_{\phi+} A_x/\delta x_{+} \quad ;$$

and $\dot{m}''_{+}$, A_x and $\overline{\Gamma}_{\phi+}$ respectively stand for the mass flux, cross-sectional area and average exchange coefficient at the boundary in question. The other quantities in (4) are similarly defined.

The above expression may be regarded as a hybrid of central- and upwind-difference schemes, in that it reduces to the former when the ratio $|F/D|$ (a local Peclet number) is less than unity; and it yields the large- (F/D) asymptote of the latter for $|F/D|$ greater than unity. The hybrid scheme has the advantages of being more accurate over a wide range of F/D (Spalding, 1972b; Runchal, 1970), than either of its components, and of yielding a diagonally-dominant matrix of coefficients for all F/D.

(c) <u>The difference equations</u>. When the fluxes in the y and z directions are expressed in a similar manner, the resultant finite-difference equation is:

$$C^{\phi}_P \phi_P = C^{\phi}_{x+}\phi_{x+} + C^{\phi}_{x-}\phi_{x-} + C^{\phi}_{y+}\phi_{y+} + C^{\phi}_{y-}\phi_{y-} + C^{\phi}_{z+}\phi_{z+} + C^{\phi}_{z-}\phi_{z-} + S^{\phi} \quad ; \tag{5}$$

where S^{ϕ} represents the integral of the source s_{ϕ} over the control volume and:

$$C^{\phi}_P \equiv C^{\phi}_{x+} + C^{\phi}_{x-} + C^{\phi}_{y+} + C^{\phi}_{y-} + C^{\phi}_{z+} + C^{\phi}_{z-} \quad .$$

The treatment of the momentum equations is essentially the same as that above. The control volumes for the velocities are of course displaced from those for ϕ. Interpolation is sometimes necessary to obtain convection velocities, densities, viscosities etc. at the required locations. In all cases our choice of interpolation practices is guided by the requirement that the resulting difference equation be conservative. If we denote the velocities in the x,y and z co-ordinate directions by u, v and w respectively, then the difference equations for momentum may be written:

$$C^u_P u_P = \sum_n C^u_n u_n + A_x (P_{x-} - P_P) + S^u \quad ; \tag{6}$$

$$C^v_P v_P = \sum_n C^v_n v_n + A_y (P_{y-} - P_P) + S^v \quad ; \tag{7}$$

$$C^w_P w_P = \sum_n C^w_n w_n + A_z (P_{z-} - P_P) + S^w \quad . \tag{8}$$

Here, the summations are over the six neighbouring velocities; and the coefficients in the equations are defined in an analogous fashion to those in (5). Finally, we complete the transformation to difference form by expressing the continuity relation (1) as:

$$[(\rho u)_{x+} - (\rho u)_P] A_x + [(\rho v)_{y+} - (\rho v)_P] A_y + [(\rho w)_{z+} - (\rho w)_P] A_z = 0. \tag{9}$$

<u>2.3 The SIMPLE procedure</u>. This 'Semi-Implicit Method for Pressure-Linked Equations' solves the set (6) to (9) by a cyclic series of guess-and-correct operations, wherein the velocities are first calculated by way of the momentum equations for a

guessed pressure field, and then the latter, and later the velocities, are adjusted so as to satisfy continuity.

The first step in the cycle is straightforward: thus the guessed pressures (which may be initial guesses, or values from a previous cycle), denoted by P*, are substituted into linearised† versions of (6) - (8). These are then solved to yield a field of intermediate velocities u*, v* and w* which will not, unless the solution has been reached, satisfy continuity.

It is here that the main novelties of the procedure enter, in the manner of satisfying the continuity requirement. The approach is to substitute for the velocities in eqn. (9) relations of the form:

$$u_P = u^*_P + A^u_P (P'_{x-} - P'_P) \quad ; \qquad (10)$$

$$v_P = v^*_P + A^v_P (P'_{y-} - P'_P) \quad ; \qquad (11)$$

$$w_P = w^*_P + A^w_P (P'_{z-} - P'_P) \quad ; \qquad (12)$$

where P' is a pressure correction, and the A's bear the following relation to coefficients in the momentum equations:

$$A^u_P \equiv A_x/C^u_P \ ; \qquad A^v_P \equiv A_y/C^v_P \ ; \qquad \text{and} \qquad A^w_P \equiv A_z/C^w_P \ .$$

The result is the finite-difference equivalent of a Poisson equation for P', viz:

$$C^P_P P'_P = \sum_n C^P_n P'_n + S^P \quad . \qquad (13)$$

Here the summation sign has the usual meaning, and the coefficients are given by:

$$S^P \equiv [(\rho u^*)_{x+} - (\rho u^*)_P]\, A_x + [(\rho v^*)_{y+} - (\rho v^*)_P]\, A_y + [(\rho w^*)_{z+} - (\rho w^*)_P]\, A_z \ ;$$

$$C^P_P \equiv \sum_n C^P_n \ ; \qquad C^P_{x-} \equiv \rho_{x-} A_x A^u_P \ ; \qquad (14)$$

with similar definitions for the other terms. S^P, it should be noted, is nothing more than the local mass imbalance of the intermediate velocity field: so, when continuity is everywhere satisfied, the pressure correction goes to zero, as would be expected.

Once the P' field has been obtained from (13), it is a straightforward matter to update the pressures and velocities (from eqns. 10-12): then, if necessary, they may be used as guesses for a new cycle. If there are ϕ's to be calculated, they may be fitted in at a convenient stage in the cycle: often the choice is arbitrary.

Because the SIMPLE procedure computes the variable fields successively, rather than simultaneously, it is highly flexible in respect of the methods of solution which it will admit for the difference equations. For the present calculations, we have employed a line-iteration method, wherein the unknown variables along each grid line are calculated by application of the tridiagonal matrix algorithm, on the assumption that values on neighbouring lines are known. This operation is performed in turn on the sets of lines lying in the x, y and z directions: it usually suffices to perform one such 'triple sweep' on the velocities and ϕ's, and three sweeps on P', per cycle of calculation. This method is substantially faster than point iteration; however it must be stressed that when even more economical methods become available, they may readily be incorporated into the procedure.

2.4 The SIVA procedure. This procedure derives its name from the novel way in which it combines point iteration with SImultaneous Variable Adjustment. With this combination, it is possible to satisfy simultaneously, on a local basis, the equations

---

† The coefficients and source terms are evaluated from the previous cycle, and held constant.

for momentum and continuity: although this balance is later destroyed when neighbouring nodes are visited, the net effect is to reduce the residual sources, and so procure convergence.

The procedure involves the adjustment, as each node is visited, of 7 variables, namely the pressure P, and the 6 surrounding velocity components, u_P, u_{z+}, v_P, v_{y+}, w_P and w_{z+}. The formulae for the variable adjustments are obtained by *algebraic* solution of: the continuity equation (9); and linearised versions of the momentum equations for the six velocities, expressed in the following form:

$$u_P = \alpha_P^u \, u_{x+} + \beta_P^u \, P_P + \gamma_P^u \quad ; \tag{15}$$

$$v_P = \alpha_P^v \, v_{y+} + \beta_P^v \, P_P + \gamma_P^v \quad ; \tag{16}$$

$$w_P = \alpha_P^w \, w_{z+} + \beta_P^w \, P_P + \gamma_P^w \quad ; \tag{17}$$

with similar expressions for u_{x+}, v_{y+} and w_{z+}. The quantities α, β and γ in these equations are readily deducible from the parent equations (6)-(9), whose terms involving variables outside of the 'SIVA cluster' have been swept into the γ's, and regarded (temporarily) as knowns. It is a straightforward matter to manipulate this set into equations which contain only the known coefficients on the right-hand sides: details will not be given here.

SIVA proceeds in all other respects in the manner of a normal point-iteration procedure: thus the grid is repeatedly swept, until the residual sources of the difference equations are reduced to acceptably small values. As with the SIMPLE method, the calculation of ϕ's is fitted in where appropriate.

3. APPLICATIONS

3.1 Test calculations. The SIMPLE and SIVA procedures were initially tested by application to a class of problems involving the laminar motion of a fluid in a cubic enclosure of side H, which has one wall moving at a steady velocity V in its own plane.

For a coarse mesh of 10 equally-spaced intervals, in each direction, convergent solutions were obtained for artifically high Reynolds numbers (based on V and H) in excess of 10^6. The SIMPLE procedure did exhibit some signs of instability in the initial stages of the calculations at the higher Reynolds numbers: this however could easily be cured by straightforward under-relaxation of P' (with a factor of about 0.2), often in the initial stages only. Although no other solutions were available for comparison, the predictions were entirely plausible, and two-dimensional versions of both methods agreed to within a few percent with Burggraf's (1966) fine-mesh computations. The initial studies confirmed that the two methods gave equal accuracy and numerical stability, but the SIMPLE method proved to be appreciably more economical of computing time than SIVA. It is therefore the former which we currently favour in our work.

In subsequent studies, SIMPLE has been successfully applied to several problems of practical interest, including the prediction of flow, heat transfer and chemical reaction in a three-dimensional furnace (Patankar and Spalding, 1972b) and the calculation of the steady-state and transient behaviour of a shell-and-tube heat exchanger (Patankar and Spalding, 1972c). Flows with strong effect of compressibility, and with distributed internal resistances, have also been predicted by the SIMPLE method.

3.2 The building problem. As a further example of the type of problem for which the SIMPLE method is well-suited, we here present calculations of the simulated (laminar) flow of wind past the slab-sided 'building', depicted in Fig. 3. The oncoming wind varies in strength in a parabolic fashion with distance from the ground, and is directed normal to the face of the building. An additional feature is a chimney located upwind of the building: the path of the effluent from this is also followed numerically.

The grid employed for the calculations had 10 nodes in each direction: non-uniform spacing was employed so as to cause the nodes to be concentrated near the building, and more widely-spaced elsewhere. The domain of solution, measured in building heights H, extended approximately ±5H in the mainstream (z) direction, and 8H in both the vertical (y) and lateral (x) directions. The plane x=0 was prescribed as a plane of symmetry, while at all other free boundaries the flow was presumed to be undisturbed by the presence of the building. The Reynolds number, based on H and the undisturbed velocity w_B at y=H, was approximately 100, in this purely illustrative example.

The results are displayed in Fig. 4, in the form of plots, at a number of constant-z planes, of: contours of constant mainstream velocity; vectors representing the direction and magnitude of the resultant velocities in the xy planes; isobars; and contours of the effluent concentration.

Taken together, the velocity and pressure plots reveal a consistent and plausible pattern of behaviour: thus the build-up of pressure in front of the building provokes reverse flow (indicated by the negative-w contour) in the low-velocity region near the ground, and deflects the wind away from the building. Downstream, the low-pressure zone behind the building also gives rise to reverse and lateral flows: now however the fluid is drawn inwards.

The concentration contours show that the effluent plume initially spreads downwards, thereby causing relatively high concentrations at the upwind face of the building. The flow around the latter then deflects the plume upwards, so that the concentration on the downwind face is lower, although still appreciable.

Although it cannot be claimed that a laminar-flow calculation on a relatively sparse grid is quantitatively representative of the real situation, the above results are probably at least qualitatively correct: moreover, they were obtained at a quite modest cost (approximately 100 seconds on a CDC 6600 machine).

4. DISCUSSION AND CONCLUSIONS

4.1 Assessment of the procedures. Experience with the SIVA and SIMPLE algorithms, which have now been applied to a large number of flow situations of varied type, has demonstrated the great flexibility and stability that results from using implicit finite-difference formulations, with the hybrid difference scheme. It has also shown that the line-by-line nature of the SIMPLE adjustment procedure makes for greater economy of computer time than the point-by-point SIVA adjustment. The slightly-reduced stability of SIMPLE can be rectified by an inexpensive under-relaxation. The authors therefore intend to concentrate on SIMPLE in their future work.

4.2 Prospects for future development. The example of Fig. 3 shows that the calculation procedure can be employed for predicting practically-important phenomena which, at present, can be predicted only by way of rather expensive and time-consuming experiments. However, a consideration of the shortcomings of that example shows also much development still to be done. First of all, the calculation was performed for a low-Reynolds-number laminar flow; but flows over real buildings are of high Reynolds number, and turbulent. It is therefore necessary to incorporate into the calculation procedure "turbulence models", of the kind recently surveyed by Launder and Spalding (1971).

Secondly, it will have been observed that the calculation task was made especially easy by the fact that four of the boundaries of the domain of integration were treated as impervious to matter, while the inlet boundary was as one at which the velocity distribution was known. In reality, the elliptic nature of the flow ensures that the presence of the building modifies the velocity distribution at these boundaries: some economical means of calculating this modification needs to be built into the calculation procedure.

Finally, buildings are not simply rectangular blocks; sometimes the departures from simplicity of form may have significant aerodynamic effects. It is therefore necessary to arrange that significant minor details of the surface, for example its distribution of roughness, can be allowed with the calculation scheme, without necessitating excessive refinement of the grid.

If these problems can be speedily surmounted, there is every reason to expect that numerical computations will replace model experiments for civil-engineering aerodynamics, furnace design, and many areas of hydraulic and aeronautical engineering. No difficulties of principle appear to stand in the way of these developments, and none of the difficulties of detail is of a kind which has not been surmounted elsewhere.

5. REFERENCES

Amsden A.A. and Harlow F.H. (1970). "The SMAC Method". Los Alamos Scientific Laboratory Report No. LA-4370.

Caretto L.S., Curr R.M. and Spalding D.B. (1972). "Two numerical methods for three-dimensional boundary layers". Comp. Methods in Appl. Mech. and Eng., 1, pp.39-57.

Chorin A.J. (1968) "Numerical solution of the Navier-Stokes equations". Maths. of Computation, 22, No.104, pp. 745-762.

Gosman A.D., Pun W.M., Runchal A.K., Spalding D.B. and Wolfshtein M. (1969). Heat and Mass Transfer in Recirculating Flows. Academic Press, London.

Harlow F.H. and Welch J.E. (1965). "Numerical calculation of time-dependent viscous incompressible flow of fluid with free surface". Physics of Fluids, 8, No. 12, pp. 2182-2189.

Launder B.E. and Spalding D.B. (1971). "Turbulence models and their application to the prediction of internal flows". Proc. I. Mech. Eng. Symposium on Internal Flows, Salford.

Patankar S.V. and Spalding D.B. (1970). "Heat and Mass Transfer in Boundary Layers". Intertext Books, London, Second Ed.

Patankar S.V. and Spalding D.B. (1972a). "A calculation procedure for heat, mass and momentum transfer in three-dimensional parabolic flows". To be published in Int. J. Heat and Mass Transfer.

Patankar S.V. and Spalding D.B. (1972b). "A computer model for three-dimensional flow in furnaces". To be presented at 14th Symposium on Combustion.

Patankar S.V. and Spalding D.B. (1972c). "A calculation procedure for steady and transient performance of a shell-and-tube heat exchanger". To be presented at the Int. Summer School at Trogir, Yugoslavia.

Pracht W.E. (1970). "A numerical method for calculating transient creep flows". Los Alamos Scientific Lab. Rept. LA-DC-11312.

Runchal A.K. (1970). "Convergence and accuracy of three finite-difference schemes for a two-dimensional conduction and convection problem". Imperial College Mech. Eng. Dept. Rept. EF/TN/A/24.

Spalding D.B. (1972a). "Mathematical models of continuous combustion". Proc. Gen. Motors Symposium on Emissions from Continuous Combustion Systems, Detroit.

Spalding D.B. (1972b). "A novel finite-difference formulation for differential expressions involving both first and second derivatives". To be published in Int. J. Num. Methods in Eng., 4.

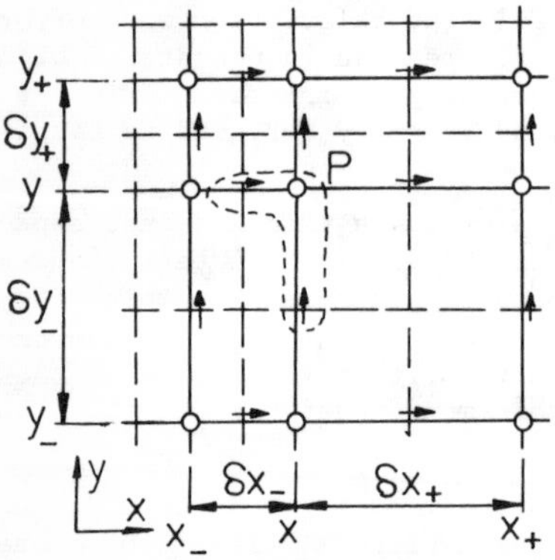

Fig. 1. The Staggered Grid

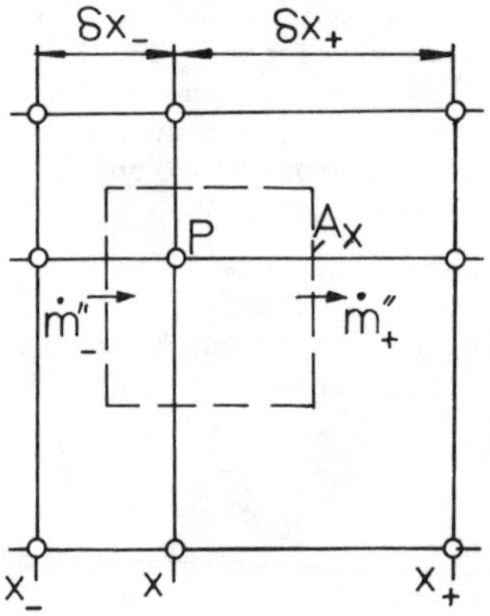

Fig. 2. Notation for x-direction fluxes

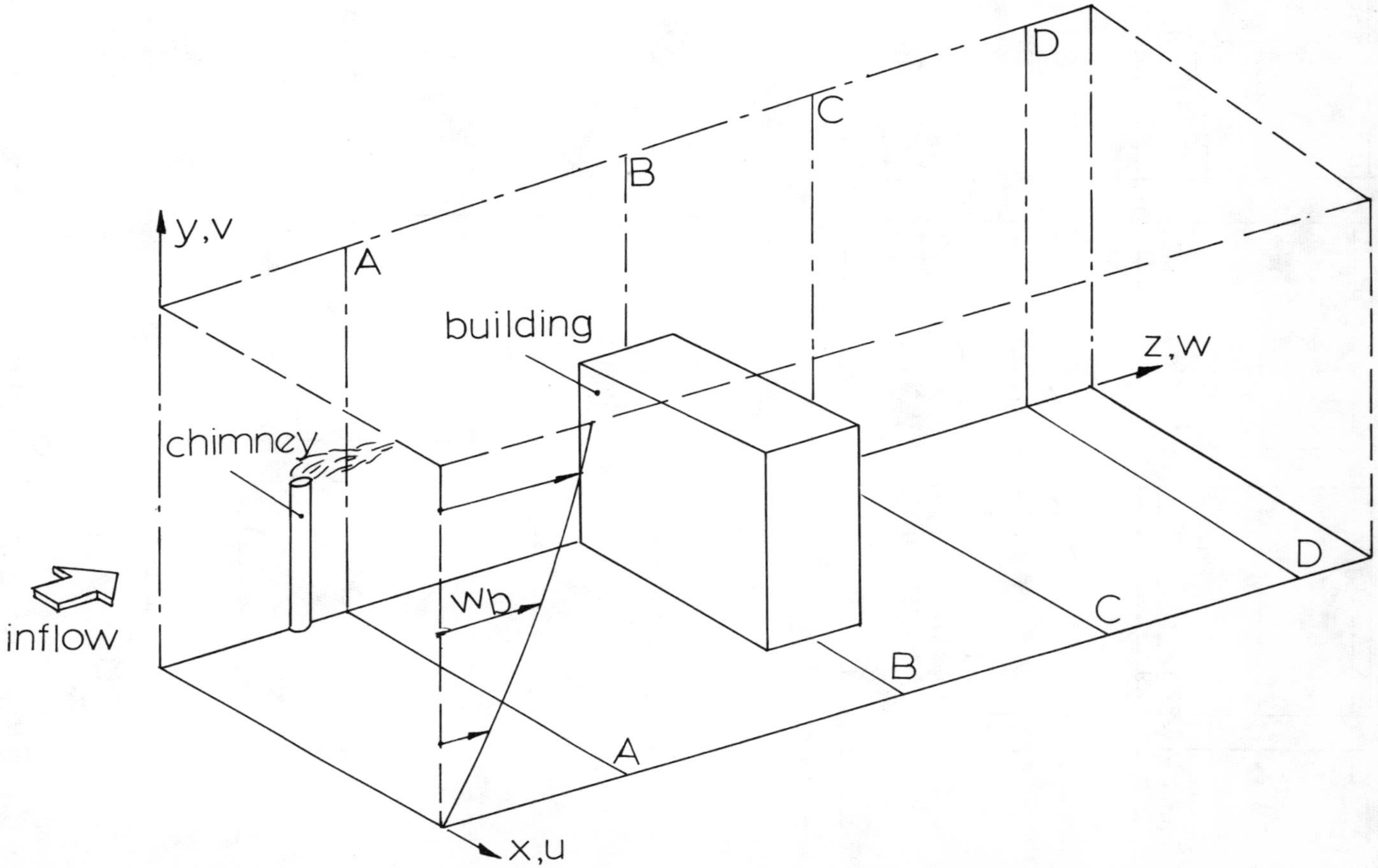

Fig. 3. Illustration of the flow-past-building problem

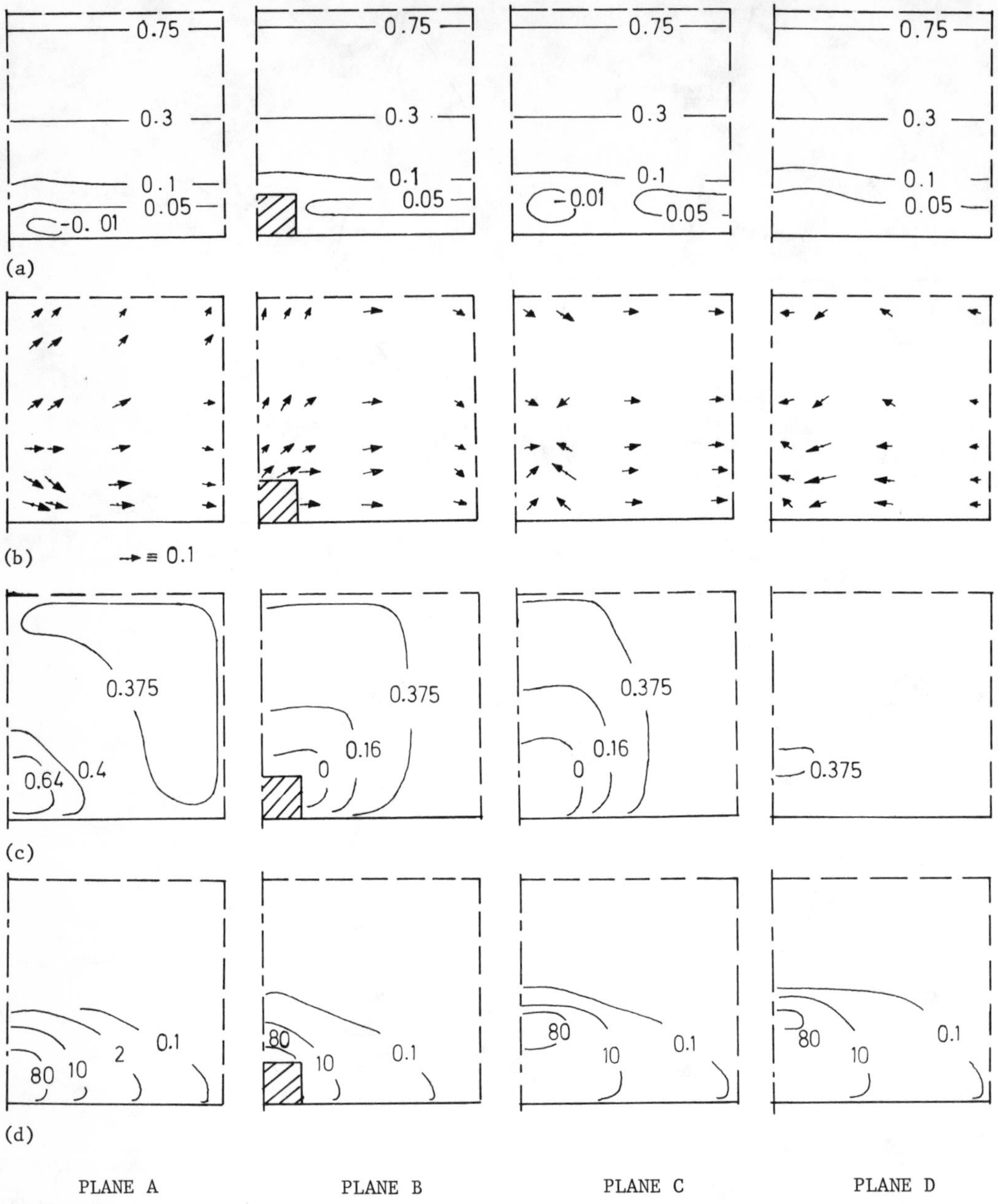

Fig. 4. Results of the flow-past-building problem. (a) main-flow velocity contours (w/w_{max}); (b) velocity vectors in cross-stream planes; (c) static-pressure contours [$2(P-P_{ref})/\rho w_B^2$]; (d) effluent concentration contours (source concentration = 100)

NUMERICAL SOLUTIONS OF THE SUPERSONIC, LAMINAR FLOW OVER A TWO-DIMENSIONAL COMPRESSION CORNER

James E. Carter

NASA Langley Research Center
Hampton, Virginia

INTRODUCTION

In the present paper numerical solutions of the Navier-Stokes equations are presented for the supersonic laminar flow over a two-dimensional compression corner. Figure 1 shows a schematic diagram in which the characteristic features of the compression corner flow field are denoted. The well-known time dependent method was used wherein the asymptotic steady solutions of the unsteady Navier-Stokes equations were obtained with the Brailovskaya (1965) finite-difference scheme. Similar investigations have been made by Allen and Cheng (1970) who used a modified Brailovskaya scheme for the flow over a rearward facing step and by MacCormack (1971) who developed an explicit, second order accurate scheme to solve the flow field for a shock impinging on a laminar flat-plate boundary layer.

PROBLEM FORMULATION

The Navier-Stokes equations may be written with respect to Cartesian coordinates x and y in the following vector form;

$$\frac{\partial w}{\partial t} + \frac{\partial F}{\partial x} + \frac{\partial G}{\partial y} = S \tag{1}$$

where

$$w = \begin{Bmatrix} \rho \\ \rho u \\ \rho v \\ E \end{Bmatrix} \quad F = \begin{Bmatrix} \rho u \\ p+\rho u^2 \\ \rho uv \\ u(E+p) \end{Bmatrix} \quad G = \begin{Bmatrix} \rho u \\ \rho uv \\ p+\rho v^2 \\ v(E+p) \end{Bmatrix}$$

$$S = \begin{Bmatrix} 0 \\ \dfrac{\partial \tau_{xx}}{\partial x} + \dfrac{\partial \tau_{xy}}{\partial y} \\ \dfrac{\partial \tau_{xy}}{\partial x} + \dfrac{\partial \tau_{yy}}{\partial y} \\ \dfrac{\partial}{\partial x}\left(k\dfrac{\partial T}{\partial x}\right) + \dfrac{\partial}{\partial y}\left(k\dfrac{\partial T}{\partial y}\right) + \dfrac{\partial}{\partial x}\left(u\tau_{xx} + v\tau_{xy}\right) + \dfrac{\partial}{\partial y}\left(u\tau_{xy} + v\tau_{yy}\right) \end{Bmatrix} \tag{2}$$

and

$$\begin{aligned} \tau_{xx} &= 2\mu\frac{\partial u}{\partial x} - \frac{2\mu}{3}\left(\frac{\partial u}{\partial x} + \frac{\partial v}{\partial y}\right) \\ \tau_{xy} &= \mu\left(\frac{\partial u}{\partial y} + \frac{\partial v}{\partial x}\right) \\ \tau_{yy} &= 2\mu\frac{\partial v}{\partial y} - \frac{2\mu}{3}\left(\frac{\partial u}{\partial x} + \frac{2v}{\partial y}\right) \end{aligned} \tag{3}$$

In equations (2) and (3) ρ represents the density, u and v, the velocity components in the x and y direction, E, the total energy, and μ, the viscosity which is related to the temperature T by the Sutherland relation

$$\frac{\mu}{\mu_r} = \left(\frac{T}{T_r}\right)^{3/2} \frac{T_r + S}{T + S} \tag{4}$$

where the subscript r refers to the reference conditions and S is a constant which is 110° K for air. The conductivity k is expressed in terms of $\frac{C_p \mu}{Pr}$ where C_p is the specific heat at constant pressure and Pr is the Prandtl number. Both C_p and Pr were assumed constant. The pressure p is found from the state equation $p = \rho RT$.

Finite-Difference Technique

Application of the Brailovskaya finite-difference scheme to equation (1) with $t = n\Delta t$, $x = j\Delta x$, and $y = k\Delta y$ results in the following difference equations where the grid spacing has been assumed constant in the x and y directions:

1st step:

$$\frac{\overline{w}^{n+1}_{j,k} - w^n_{j,k}}{\Delta t} = - \frac{F^n_{j+1,k} - F^n_{j-1,k}}{2\Delta x} - \frac{G^n_{j,k+1} - G^n_{j,k-1}}{2\Delta y} + S^n_{j,k} + O(\Delta t + \Delta x^2 + \Delta y^2) \tag{5}$$

2nd step:

$$\frac{w^{n+1}_{j,k} - w^n_{j,k}}{\Delta t} = - \frac{\overline{F}^{n+1}_{j+1,k} - \overline{F}^{n+1}_{j-1,k}}{2\Delta x} - \frac{\overline{G}^{n+1}_{j,k+1} - \overline{G}^{n+1}_{j,k-1}}{2\Delta y} + S^n_{j,k} + O(\Delta t + \Delta x^2 + \Delta y^2) \tag{6}$$

In the first step, a temporary value of $w \left(\text{denoted by } \overline{w}^{n+1}_{j,k} \right)$ is calculated at the new time step; this value is improved in the second step by reevaluating the convection terms with the temporary values of w. The stress term S is approximated by central differences in the first step and is simply repeated in the second step. This repetition results in savings in computation which is an advantage of the Brailovskaya scheme over other two-step schemes such as that used by Thommen (1965) or MacCormack (1971) where the stress term is reevaluated in each step.

There was initial concern that the Brailovskaya scheme would not give steady-state results as accurate as a Lax-Wendroff* scheme since the Brailovskaya scheme is not fully second order accurate. Carter (1971) has made comparisons between results obtained with the Brailovskaya scheme and the Lax-Wendroff scheme used by Thommen for both the steady-state solutions to Burgers' equation and for the supersonic flow field near the leading edge of a flat plate. In both problems only small differences in the asymptotic solutions were found. Additional comparisons were made by Carter (1971) for solutions to Burgers' equation for the so-called "windward" scheme which uses one-sided differences on the convection terms. This scheme has only first-order accuracy and comparisons with the exact solution and with the second-order schemes reflect this increase in truncation error.

An approximate stability criterion was derived by applying the von Neumann stability analysis to the linearized difference equations. The inviscid and viscous parts of the difference equations were analyzed separately and the time step Δt was chosen as the minimum of the following:

$$\Delta t \le \frac{1}{\frac{|u|}{\Delta x} + \frac{|v|}{\Delta y} + c\sqrt{\frac{1}{\Delta x^2} + \frac{1}{\Delta y^2}}} \tag{7}$$

*A Lax-Wendroff scheme is referred to here as an explicit scheme which has a truncation error of second order.

$$\Delta t \leq \frac{\Pr R_{\infty,L}}{2\gamma \frac{\mu}{\rho}\left(\frac{1}{\Delta x^2} + \frac{1}{\Delta y^2}\right)} \tag{8}$$

Equation (7) is recognized as the CFL (Courant-Friedrichs-Lewy) condition and equation (8) is the stability criterion for the two-dimensional heat diffusion equation. Numerical experiments which resulted in stable calculations when conditions (7) and (8) were used tend to verify splitting the stability analysis into two parts for the Brailovskaya scheme.

In the compression corner calculations the equations were transformed to a skewed coordinate system which is shown in Figure 2 so that the grid points could be placed on the ramp. Second-order accuracy was maintained by using special difference quotients for the derivatives with respect to x along the interface between the Cartesian and skewed coordinate systems. For more details see Carter (1972).

Boundary Conditions

Solutions of the Navier-Stokes equations require the flow field to be enclosed by a computational box along which boundary conditions are specified. It was necessary to break the flat plate flow field into several successive calculations because of the small grid required in the leading-edge region. Numerical instability resulted when the Reynolds number based on the grid spacing and free-stream conditions exceeded 250. Figure 2 shows the arrangement of the computational boxes used to calculate the flow field over the flat plate and compression corner. It should be noted that MacCormack performed similar calculations but was able to use a much larger grid in the leading-edge region than that in the present case and, therefore, was able to calculate from the leading edge to downstream of the incident shock in one calculation. This difference may be due to a splitting technique of the difference equations that MacCormack used near the surface.

For each of the calculations it is necessary to prescribe the boundary conditions along the boundaries of the box. Free-stream conditions were specified along the outer boundary, provided it was located outside of the leading-edge shock wave. For the computational box enclosing the compression corner the outer boundary was placed about three boundary-layer thicknesses from the wall in the disturbed portion of the flow field, beneath the point of formation of the compression corner shock. The flow along this boundary was assumed to be of simple-wave type so that the flow variables could be computed from the first row of grid points inside the boundary by simply extending the left running characteristics. Naturally the success of the simple-wave extrapolation depends on the formation point of the compression-corner shock as well as the proximity of the leading-edge shock to the outer boundary. Numerical tests were made to deduce the level of error introduced by this approximation.

Along the wall boundary $u = v = o$ and $T = T_w$ where the isothermal wall temperature was generally chosen as approximately the recovery temperature. The pressure at the wall was determined by quadratic extrapolation normal to the wall. Several other methods were tried but the extrapolation worked best for the present calculations. For a cold wall, where the density variation is large near the wall, some other technique would probably have to be used.

Quadratic extrapolation in the x direction was used to continuously update the flow variables at the downstream boundary. Here also numerical tests were used to deduce that the errors introduced by this extrapolation were small; therefore, it was unnecessary to overlap the computational boxes in making the flat-plate calculations.

RESULTS AND DISCUSSION

The computational rate of the present program as applied to the compression-corner computational box was found to be 2.25×10^6 grid points/hour on the CDC 6600 computer. In these calculations the number of grid points was typically of the order of 3000 and the number of cycles required for convergence varied from 1500 to 3000 depending on the initial conditions. These calculations took 2 to 4 hours. Convergence was assumed when all the variables ceased to change in their fifth significant digit.

Flat Plate

Calculations were made for the flow over a flat plate with $M_\infty = 6.06$, $T_\infty = 88^\circ$ R, and a constant wall temperature $T_w = \overline{T}_{aw}$ where

$$\overline{T}_{aw} = T_\infty \left(1 + \frac{\gamma - 1}{2} \sqrt{Pr}\, M_\infty^{\;2}\right) \tag{9}$$

These calculations were computed using five tandem computational boxes which extend from the leading edge to the point where $R_{\infty,x} = 10^5$. Variable grid in both the x and y directions was used to increase the grid spacing while keeping the number of grid points constant (approximately 1300) in each box. Figure 3 gives a flow field map deduced from these calculations which shows the characteristic features of the supersonic flat-plate flow field: the boundary-layer-induced shock wave, displacement thickness, streamline pattern, and contours of constant pressure. The shock position deduced from the pressure distributions is in excellent agreement with that found by Lewis (1967) from pitot measurements. Also shown is the displacement thickness predicted by the Kubota and Ko (1967) weak interaction analysis. The difference between their approximate result and the present numerical solution is expected since the assumption by Kubota and Ko that $Pr = 1$ results in a larger recovery temperature and a thicker boundary layer.

Compression Corner

A calculation was made for the $M_\infty = 3.0$ flow over a 10° compression corner for which $T_w = T_{o,\infty}$ and the Reynolds number based on free-stream conditions and the distance from the leading edge to the corner x_c is $R_{\infty,x_c} = 1.68 \times 10^4$. The computational box in the corner region extended from $x/x_c = 0.357$ to $x/x_c = 1.99$ and from the wall to $y/x_c = 0.0868$, which is approximately one-half of the shock-layer thickness and twice the boundary-layer thickness at the upstream boundary. The total number of grid points used was 2156; 77 in the x direction and 28 in the y direction, thereby resulting in a constant grid spacing in the x and y directions of 0.0214 and 0.00321, respectively.

Figure 4(a) gives the computed flow field in the immediate region of the corner showing the streamline pattern both inside and exterior to the separated region. In the separation bubble, the locus of $u = o$ is shown to indicate the regions of forward and reverse flow.

Effect of Wall Suction

Figure 4(b) shows the computed flow field which results when a wall suction velocity of $v/u_\infty = -0.01$ is applied in the region of $0.786 \leq x/x_c \leq 1.214$. It was found that with this wall suction, the amount of fluid removed in the corner region was 14 percent of that flowing in the boundary layer just upstream of the start of the corner interaction. Comparison of the streamlines and displacement thickness in figures 4(a) and 4(b) demonstrates the large effects of suction on the compression corner flow field.

Numerical Test of Simple-Wave Extrapolation

Several calculations were made to determine the sensitivity of the computed results to the position of the outer boundary along which simple-wave extrapolation was used. The four computational boxes used are shown in figure 5. For boxes I, II, and III, simple-wave extrapolation was used all along the outer boundary, whereas box IV served as the reference calculation since free-stream conditions were imposed along the outer boundary to the approximate position of the intersection of this boundary and the leading-edge shock. Downstream of that point simple-wave extrapolation was used, but a check on the Mach line inclinations shows that the wall pressure is unaffected by this extrapolation. Comparison of the wall pressure distributions corresponding to the four computations is given in figure 6. The result from box III is clearly in error since the overall pressure rise is less than the inviscid value. This result is not surprising since simple-wave extrapolation was used in this calculation in the vicinity of the leading-edge shock wave. Test calculations were also made on the flat-plate flow field and it was again found that simple-wave extrapolation near the leading-edge shock causes erroneous expansion waves to be reflected back into the flow field. Without the fourth calculation to serve as a reference, it is not as simple to deduce which of calculations I or II is the more correct; hence, this difference is indicative of the uncertainty introduced into the calculations by the simple-wave extrapolation at the outer boundary.

Comparison with Experiment

Calculations were made for the $M_\infty = 6.06$ and $R_{\infty,x_c} = 1.5 \times 10^5$ flow over a 10.25° adiabatic compression corner for which Lewis (1967) obtained experimental measurements. The computational box enclosing the corner extends from $x/x_c = 0.5$ to $x/x_c = 2.1$ and from the wall to $y/x_c = 0.105$, which is approximately two-thirds of the distance between the boundary-layer edge and the leading-edge shock wave. The x and y grid spacings were $\Delta x/x_c = 0.0125$ and $\Delta y/x_c = 0.003$, respectively, resulting in a total number of grid points of 4644.

Figure 7 shows a comparison of the wall pressure distribution determined from the present calculations with that found experimentally. Upstream of the corner, the present solution and the experiment predict approximately the same degree of compression when the upstream pressure difference is taken into account. Downstream of the corner, the agreement becomes better as the effects of the upstream pressure difference become smaller. Also shown for comparison is the theoretical result obtained by Klineberg (1968) using an integral method to solve the boundary-layer equations. His result agrees well with the experiment up to the approximate point of separation, but downstream it falls below the experimental distribution, indicating a longer interaction region.

In figure 8 the skin-friction distribution obtained in the present investigation is compared to that found by Klineberg using an integral method. The points of separation and reattachment predicted by these two approaches differ by about 5 percent of the distance from the leading edge. (Their locations were not determined experimentally.) Elsewhere, the agreement is reasonably good up to the reattachment region; however, downstream of that region, the two distributions differ considerably. The difference is consistent with that in figure 7 in that the integral approach shows the interaction to extend farther downstream than that found in the present investigation.

Free Interaction

Stewartson and Williams (1969) used the method of matched asymptotic expansions to show that solving the incompressible boundary-layer equations subject to novel boundary conditions results in a universal solution in the free-interaction region. Previously Chapman, Kuehn, and Larson (1958) had observed experimentally that in a supersonic flow field the initial pressure rise through separation to the plateau value was independent of the details of the disturbing mechanism (and hence is referred to as a "free-interaction"), whether it be, for example, an incident shock, a forward-facing step, or a ramp. In the Stewartson and Williams analysis the pressure is given by

$$\tilde{p}_2 = \frac{(M_o^2 - 1)^{1/4}\left(\frac{p}{p_o} - 1\right)}{\left(\frac{C_{f_o}}{2}\right)^{1/2} \gamma M_o^{\,2}} \tag{10}$$

$$\tilde{X} = \frac{\left(M_o^2 - 1\right)^{3/8} R_o \left(\frac{C_{f_o}}{2}\right)^{5/4} \left(\frac{x - x_o}{x_o}\right)}{\left(\frac{T_w}{T_e}\right)_o^{3/2} C} \tag{11}$$

where the subscript o refers to the conditions at the start of the interaction.

Figure 9 shows a comparison of the present calculations plotted in Stewartson and Williams coordinates to the universal pressure distribution which they obtained. The origin of the $\tilde{X}$ scale has been shifted to the point of separation (denoted by subscript s). The present calculations, with the exception of that at $M_\infty = 6.06$, are well correlated by the Stewartson and Williams coordinates, but this correlation does not agree with the numerical solution obtained by Stewartson and Williams. At the present time the cause of this disagreement is not known.

ACKNOWLEDGEMENTS

The author wishes to express his gratitude to Ruby Davis of Langley Research Center, NASA, for the excellent job which she did in writing the computer program for these calculations.

REFERENCES

Allen, J. S., and Cheng, S. I.: The Physics of Fluids, Vol. 13, No. 1, 1970 pp. 37-52.

Brailovskaya, I. Yu.: Sov. Phys. - Doklady, Vol. 10, No. 2, August, 1965, pp. 107-110.

Carter, James E.: Ph. D. Dissertation. Virginia Polytechnic Institute and State University, Aug. 1971.

Carter, James E.: NASA TR R-385, 1972.

Chapman, Dean R., Kuehn; Donald M.; and Larson, Howard K.: NACA Rept. 1358, 1958.

Klineberg, J. M.: Ph. D. Dissertation, California Institute of Technology, 1968.

Kubota, T. and Ko, Denny, R. S.: AIAA Journal, Vol. 5, No. 10, 1967, pp. 1915-1917.

Lewis, J. E.: Ph. D. Dissertation, California Institute of Technology, 1967.

MacCormack, Robert W.: Proceedings of the Second International Conference on Numerical Methods in Fluid Dynamics, ed. by Maurice Holt, 1971, pp. 151-163.

Stewartson, K. and Williams, P. G.: Pro. Roy. Soc. (London), Vol. A-312, Sept. 1969, pp. 181-206.

Thommen, H. V.: GDC-ERR-AN733, General Dynamica/Convair, 1965.

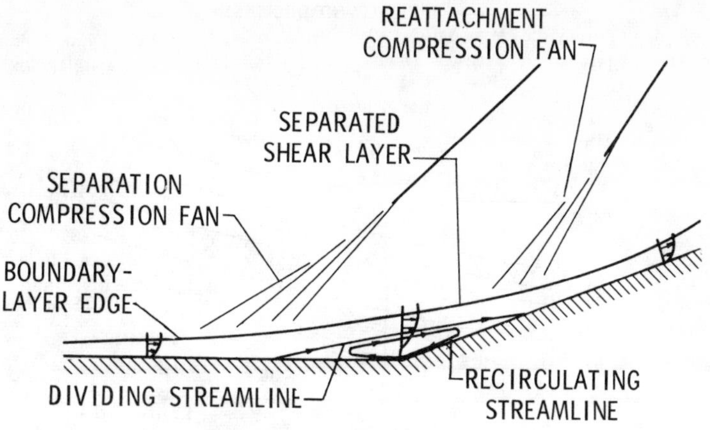

Fig. 1. Schematic diagram of a supersonic flow field in a compression corner

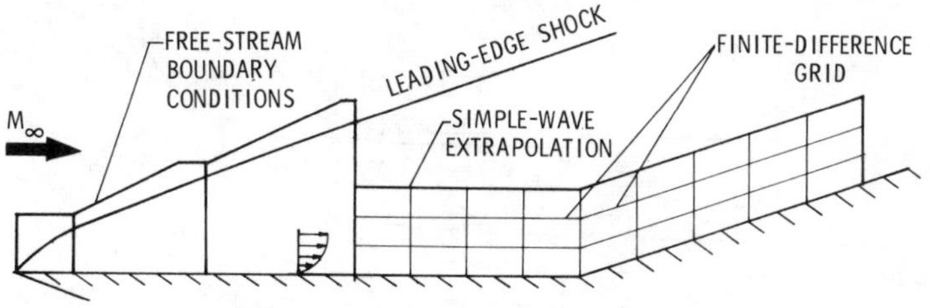

Fig. 2. Schematic diagram of computational boxes used for flat-plate and compression-corner calculations

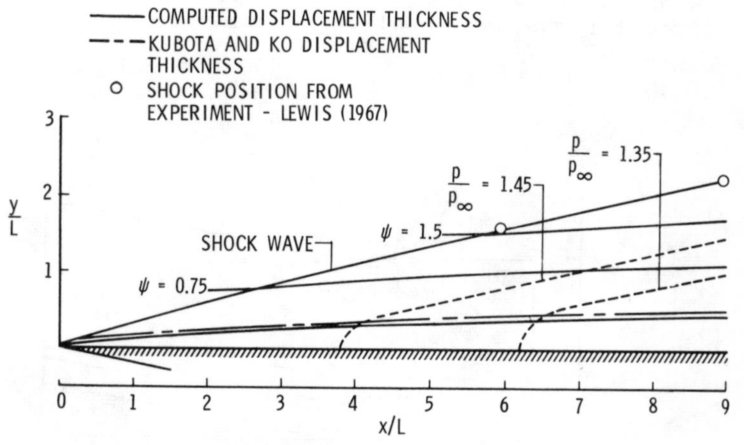

Fig. 3. Computed flow filed over a flat plate with $M_\infty = 6.06$, $T_w = \overline{T}_{aw}$, and $R_{\infty,L} = 10^4$

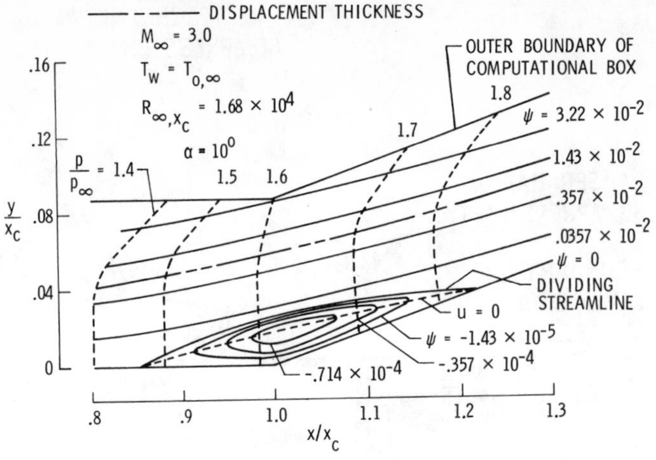

Fig. 4 a. Computed flow field over a 10° compression corner for $M_{\infty} = 3.0$

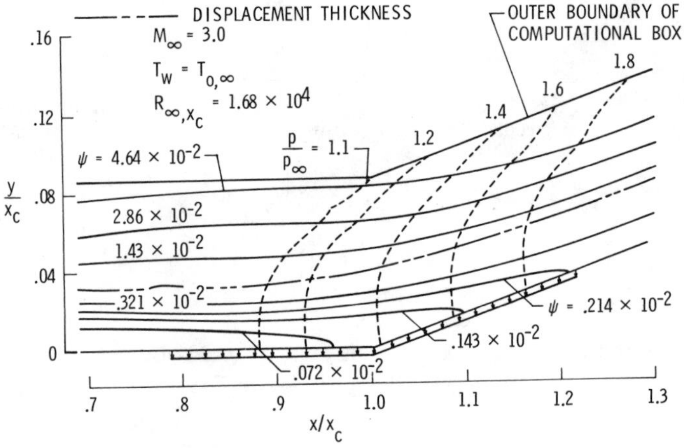

Fig. 4 b. Computed flow field over a 10° compression corner with a discontinuous wall-suction velocity of $v_w/U_{\infty} = -0.01$

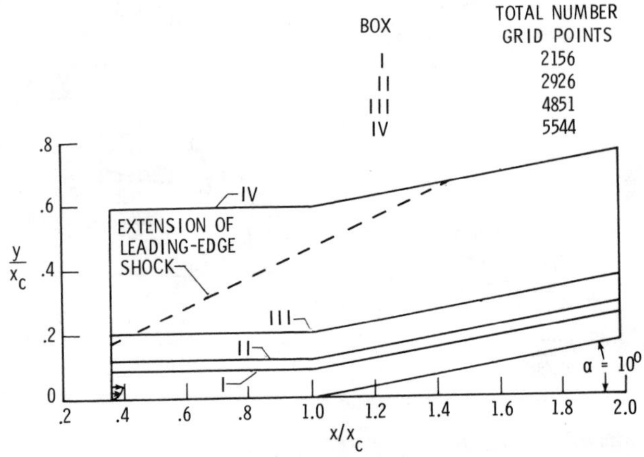

Fig. 5. Computational boxes used to test the simple-wave extrapolation for $M_{\infty} = 3.0$ and $\alpha = 10^{\circ}$

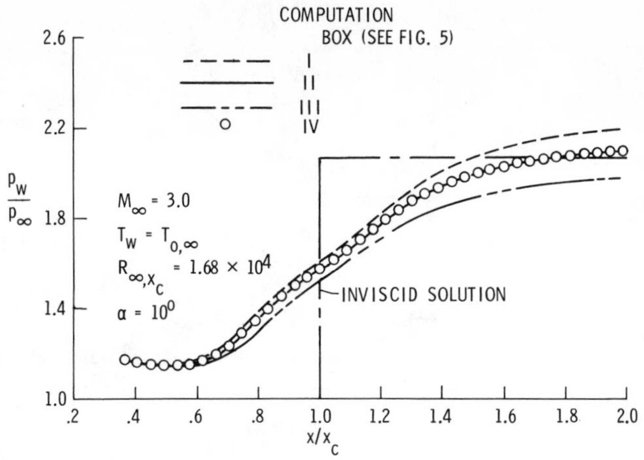

Fig. 6. Comparison of wall-pressure distributions for four position of the outer boundary of the computational box

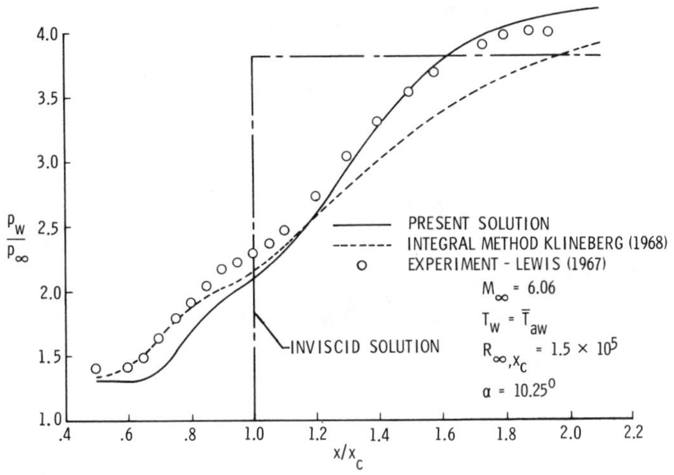

Fig. 7. Comparison of theoretical and experimental wall-pressure distributions

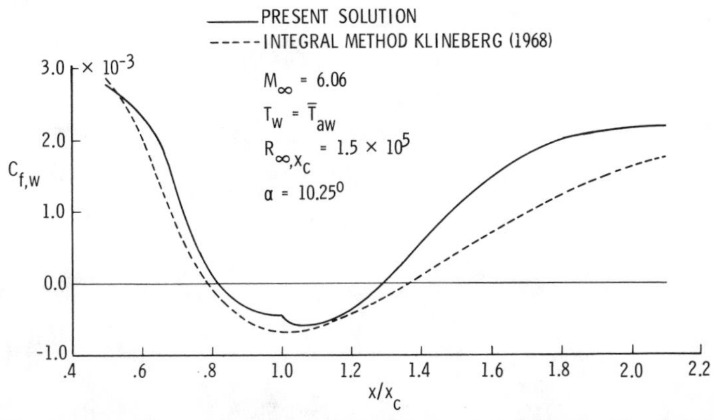

Fig. 8. Comparison of wall skin-friction coefficient obtained from th Navier-Stokes equations with that from the boundary-layer equations

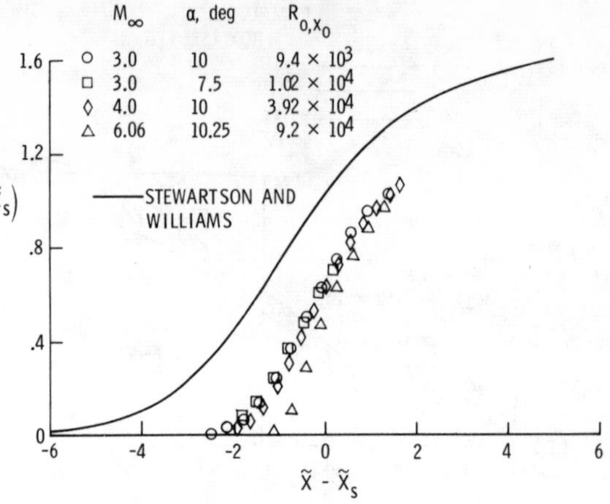

Fig. 9. Comparison of the pressure from the present numerical results with the free-interaction theoretical solution of Stewartson and Williams

LAMINAR BOUNDARY LAYERS WITH ASSIGNED WALL SHEAR*

Tuncer Cebeci
California State University, Long Beach, California
Herbert B. Keller
California Institute of Technology, Pasadena, California

1. INTRODUCTION

The friction drag of a body may be reduced by fluid injection at the wall or by manipulation of the pressure gradient. The former approach can be used when the pressure distribution around the body is predetermined by other requirements. However, beyond some station along the wall it may be both feasible and desirable to control skin friction by the manipulation of the pressure gradient alone. Similarly in some problems, when the flow separates at some station along the body, it is useful to know how much one can manipulate the pressure distribution to prevent the flow from separating. Mathematically this leads to a form of inverse problem in which a coefficient (scalar or function) in an (ordinary or partial) differential equation is to be determined so that the solution satisfies an overdetermined set of boundary conditions. We shall consider the numerical solution of such problems for both similar and non-similar two dimensional laminar flows. In the former case our technique can also be used to obtain the reverse-flow solutions of the Falkner-Skan equation. There is no difficulty in adopting our methods to turbulent flows (using an eddy viscosity formulation) but we do not include such calculations here.

We show two different approaches to the problem. In the first approach the unknown pressure distribution is treated as an eigenvalue at each streamwise station and it is computed by means of a Newton iteration scheme based on satisfying the excess boundary condition. It turns out that for each iteration a standard boundary-layer flow problem must be solved. Thus a key element in this procedure is a very accurate and efficient difference scheme for computing similar and non-similar boundary-layer flows (in which the pressure distribution is given). This nonlinear eigenvalue approach has previously been used to get reverse-flow solutions of the Falkner-Skan equation by means of shooting techniques. Even for that problem, however, the present finite difference method seems superior.

In the second approach (the mechul-function scheme) the unknown pressure distribution is treated as an unknown function which, for the correct solution, turns out to be independent of the boundary layer variable. This version of the problem is solved by a procedure very closely related to the above cited accurate scheme for standard boundary-layer flows.

2. THE INVERSE PROBLEMS

The boundary-layer equations for incompressible laminar flows over a plane surface can be written as [4]

$$\frac{\partial^3 f}{\partial\eta^3} + f\frac{\partial^2 f}{\partial\eta^2} + \beta(\xi)\left[1 - \left(\frac{\partial f}{\partial\eta}\right)^2\right] = 2\xi\left[\frac{\partial f}{\partial\eta}\frac{\partial^2 f}{\partial\xi\partial\eta} - \frac{\partial^2 f}{\partial\eta^2}\frac{\partial f}{\partial\xi}\right] \tag{2.1}$$

Here $f(\xi,\eta)$ is a dimensionless stream function, $\xi \geqslant 0$ is the streamwise variable η is the similarity variable and $\beta(\xi)$ is the pressure gradient. That is

*This work was partially supported by the National Science Foundation Grant No. GK-30981 and by the U.S. Army Research Office, Durham under Contract DAHC 04-68-0006.

$\beta(\xi) \equiv [2\xi/u_e(\xi)]\, du_e(\xi)/d\xi$ where $u_e(\xi)$ is the external velocity field which is specified in most problems. General boundary conditions are of the form:

$$\left.\begin{array}{lll} \text{a)} & f(\xi,0) = f_w(\xi) \quad , & \dfrac{\partial f}{\partial \eta}(\xi,0) = u_w(\xi) \\ \text{b)} & \dfrac{\partial f}{\partial \eta}(\xi,\eta_\infty) = 1 & \end{array}\right\} \ \xi \geq 0 \tag{2.2}$$

Here $f_w(\xi)$ and $u_w(\xi)$ allow us to include mass transfer and moving walls, respectively, while $\eta_\infty = \eta_\infty(\xi)$ is the outer edge of the boundary layer. Eqs. (2.1) and (2.2) with $\beta(\xi)$ given will be called the STANDARD PROBLEM. It is easy to include the determination of $\eta_\infty(\xi)$ as part of the problem but we do not discuss this here and take $\eta_\infty(\xi) \equiv$ const.

It is frequently desired to specify the wall shear, say as

$$\frac{\partial^2 f}{\partial \eta^2}(\xi,0) = S(\xi) \ , \quad \xi \geq 0 \tag{2.3}$$

Then (2.1)-(2.3) is overdetermined. Thus in this case we do not specify $\beta(\xi)$ but seek it along with $f(\xi,\eta)$ -this is a form of inverse problem.

Another way to formulate this inverse problem is to let $\beta(\xi) = \beta(\xi,\eta)$ and to require that

$$\frac{\partial \beta}{\partial \eta} = 0 \ , \quad \xi \geq 0 \tag{2.4}$$

This is known as the mechul function method in which we must solve the two partial differential equations (2.1) and (2.4) subject to the boundary conditions (2.2) and (2.3). For similar flows in which $\partial f/\partial \xi \equiv 0$ and the boundary conditions $f_w(\xi)$, $u_w(\xi)$ and $S(\xi)$ are independent of ξ we can also take β independent of ξ. In this case the mechul function method reduces to the system of ordinary differential equations (with $' \equiv d/d\eta$):

$$\text{a)} \quad f''' + ff'' + \beta[1-(f')^2] = 0 \ , \qquad \text{b)} \quad \beta' = 0 \tag{2.5}$$

subject to the boundary conditions:

$$\text{a)} \quad f(0) = f_w, \ f'(0) = u_w, \ f''(0) = S, \quad \text{b)} \quad f'(\eta_\infty) = 1 \tag{2.6}$$

3. NONLINEAR EIGENVALUE SCHEME

Using a very accurate and efficient numerical scheme [4], [5] for solving the STANDARD PROBLEM, (2.1)-(2.2) with $\beta(\xi)$ assumed known, we solve the inverse problem by a "nonlinear-eigenvalue" approach. That is, in summary, the equation (2.1) is first written as a first order system of partial differential equations by introducing $u \equiv \partial f/\partial\eta$ and $v = \partial u/\partial\eta$. Then on the rectangular mesh $\{\xi_i, \eta_j\}$ with arbitrary nonuniform spacing $k_i = \xi_i - \xi_{i-1}$ and $h_j = \eta_j - \eta_{j-1}$ these differential equations are replaced by difference equations using two point centered differences. The system with $\xi_0 = 0$ is first solved as a special case of the general scheme. Then assuming the solution known at any ξ_{i-1} the difference equations centered at $\xi_{i-1/2}$ yield a nonlinear system of 3J algebraic equations for the 3J+3 unknowns $\{f_{ij}, u_{ij}, v_{ij}\}$, $0 \leqslant j \leqslant J$ and i fixed. The boundary conditions (2.2) and (2.3) yield four additional conditions. Thus we consider β_i which enters the 3J algebraic equations to be an "eigenvalue" whose determination enables us to satisfy the overdetermined system.

The iteration scheme for determining β_i employs Newton's method on the excess boundary condition: $v_{i0}(\beta_i) = S(\xi_i)$, to give

$$\beta_i^{(\nu+1)} = \beta_i^{(\nu)} - \frac{[v_{i0}(\beta_i^{(\nu)}) - S(\xi_i)]}{\partial v_{i0}(\beta_i^{(\nu)})/\partial\beta} \tag{3.1}$$

The derivative $\partial v/\partial\beta$ is computed by solving a difference form of the variational equations. An application of (3.1) is called an <u>outer</u> iteration. Before this can be done one or more <u>inner</u> iterations must first be performed to compute $v_{i0}(\beta_i^{(\nu)})$. These are just the Newton iterates for solving the STANDARD PROBLEM with specified $\beta(\xi_i) = \beta_i^{(\nu)}$. The results of these inner iterations are also used to determine the coefficients in the variational equations for $\{\partial f/\partial\beta, \partial u/\partial\beta, \partial v/\partial\beta\}$. This scheme is explained in detail in [3].

4. MECHUL FUNCTION SCHEME

We write the equations (2.1) and (2.4) as the first order system:

$$\begin{aligned}
&\text{a)} \quad \partial f/\partial\eta = u \\
&\text{b)} \quad \partial u/\partial\eta = v \\
&\text{c)} \quad \partial\beta/\partial\eta = 0 \\
&\text{d)} \quad \partial v/\partial\eta = 2\xi(u\,\partial u/\partial\xi - v\,\partial f/\partial\xi) - fv - \beta(1-u^2)
\end{aligned} \tag{4.1}$$

The boundary conditions (2.2) and (2.3) are:

$$\begin{aligned}
&\text{a)} \quad f(\xi,0) = f_w(\xi),\ u(\xi,0) = u_w(\xi),\ v(\xi,0) = S(\xi),\ \xi \geqslant 0 \\
&\text{b)} \quad u(\xi,\eta_\infty) = 1
\end{aligned} \tag{4.2}$$

"Initial" conditions, at $\xi = 0$, are obtained by setting $\xi = 0$ in (4.1) and (4.2) to get precisely the similar flow problem formulated in (2.5)-(2.6). This problem is solved by a special case of the numerical scheme used to solve (4.1)-(4.2).

On a rectangular net of points $\{\xi_i, \eta_j\}$ with arbitrary spacing (k_i, h_j) we replace (4.1) by the "obvious" centered difference approximations on each net rectangle. The resulting nonlinear algebraic system of difference equations is solved by Newton's method. The $4J+4$ variables $\{f_{ij}, u_{ij}, v_{ij}, \beta_{ij}\}$, $0 \leq j \leq J$ are computed successively for $i = 0, 1, 2, \cdots$. When $i = 0$ we are in the similar flow case (i.e. $\xi = 0$). For $i \geq 1$ the nonlinear equations differ from those for $i = 0$ only in the inhomogeneous terms. The Newton iterates are obtained, for all i, as the solution of a linear algebraic system with coefficient matrix in block-tridiagonal form. The blocks are 4×4 and there are $(J+1)$ of them. These linear systems are solved in a very efficient manner using block - LU - decomposition (see [2]).

Newton's method was observed to converge quadratically in all the mechul function applications. However in the nonsimilar flow problems a small disturbance appears at the outer edge of the boundary layer, $\eta = \eta_\infty$, and slowly propagates towards the wall as the computations proceed downstream. This is clearly a weak numerical instability whose cause and elimination have not yet been determined. It does not destroy the accuracy of the results on the wall up to 75% of the distance to separation. Furthermore for similar flow calculations the mechul function scheme is extremely accurate, stable and efficient.

5. COMPUTATIONS FOR SIMILAR FLOWS

For similar flows computations were done for positive and negative wall shear using both the nonlinear eigenvalue scheme and the mechul function scheme. For positive wall-shear we use $\eta_\infty = 6$ and $\Delta\eta = 0.1$; for negative wall shear we use $\eta_\infty = 10$ and $\Delta\eta = 0.1$. All the calculations used the convergence test: $|\beta^{(\nu+1)}(\xi_i) - \beta^{(\nu)}(\xi_i)| \leq 10^{-4}$.

Table I

Comparison of positive wall-shear solutions for similar flows
(ν denotes iteration number)

f''_w	Nonlinear eigen-value scheme		Mechul function scheme		Smith
	$-\beta$	ν	$-\beta$	ν	$-\beta$
.40032	.05031	3	.05025	3	.05
.31927	.10017	3	.10021	3	.10
.23974	.14024	3	.14019	3	.14
.19078	.16016	3	.16019	3	.16
.12864	.18025	3	.18020	3	.18
.05517	.19528	3	.19524	3	.195
0	.20259	2	.19917	3	.198834

In Table I we present the computed values of β for seven given values of $f''_w \geq 0$ and compare the results with those of Smith [6]. The agreement is quite good. Table I also shows the number of iterations, ν. Both schemes require the same number of iterations (except for the last one) and converged quadratically in all cases.

Table II

Comparison of reverse-flow solutions for similar flows

f''_w	Nonlinear eigen-value scheme		Mechul function scheme		Stewartson	Cebeci-Keller
	-β	ν	-β	ν	-β	-β
-.097	.18143	2	.18074	4	.18	.18055
-.132	.15416	6	.15234	4	.15	.15212
-.141	.13545	6	.13412	4	--	–

In Table II we present the computed values of β for three given values of $f''_w < 0$ and compare them with those computed by Stewartson [7] and by Cebeci and Keller [1]. The comparison shows that the results obtained by the mechul function scheme are in better agreement with the earlier results of Cebeci and Keller [1] than those obtained by the nonlinear eigenvalue scheme. Figure 1 shows the separation profile ($f''_w = 0$) together with the three reverse-flow profiles.

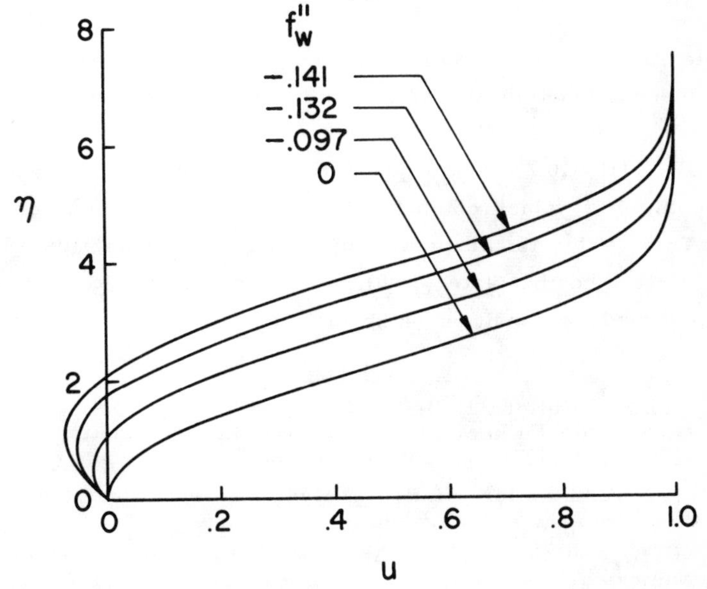

Figure 1. Reverse-flow profiles for similar flows computed by the mechul function scheme

6. COMPUTATIONS FOR NONSIMILAR FLOWS

For nonsimilar flows the wall shear distribution was taken to be

$$S(\xi) = 0.4696(1-\xi) \tag{6.1}$$

At $\xi = 0$ this corresponds to a flow with zero pressure gradient. At $\xi = 1$ the wall shear changes sign corresponding to a point where the flow separates from the surface.

Table III

Computed pressure-gradient parameter β as a function of ξ for the wall shear distribution given by (6.1)

ξ	Nonlinear eigenvalue scheme		Mechul function scheme	
	$-\beta$	ν	$-\beta$	ν
0	.00179	3	.00179	3
.10	.04532	4	.04535	3
.20	.08553	5	.08559	3
.30	.12225	5	.12231	3
.40	.15522	5	.15529	3
.50	.18408	5	.18420	3
.60	.20845	5	.20857	3
.70	.22761	4	.22685	3
.80	.24041	4	.22378	3
.90	.24479	3	.16835	3

Table III shows the results obtained by the nonlinear eigenvalue scheme and by the mechul function scheme. Both calculations were made with $\eta_\infty = 6$, $\Delta\xi = 0.05$ and $\Delta\eta = 0.25$. The convergence test in both cases was $|\beta^{(\nu+1)}(\xi_n) - \beta^{(\nu)}(\xi_n)| < 10^{-4}$. Again quadratic convergence was always observed. For $\xi \geq .80$ the $\beta(\xi)$ values computed by the mechul function scheme are less accurate. This is due to the previously noted weak instability which has propagated fairly close to the plate ($\eta = 0$). It seems likely that a modification in imposing the boundary condition (4.2b) at $\eta = \eta_\infty$ could eliminate the instability.

References

[1] Cebeci, T. and Keller, H. B.: Shooting and Parallel Shooting Methods for solving the Falkner-Skan Boundary-Layer Equation. J. Comp. Phys. v.7, no.2, 1971, pp. 289-300.

[2] Isaacson, E. and Keller, H. B.: Analysis of Numerical Methods, J. Wiley, New York, 1966.

[3] Keller, H. B. and Cebeci, T.: An Inverse Problem in Boundary-Layer Flows; Numerical Determindation of Pressure Gradient for a Given Wall Shear, to appear in J. Comp. Phys. 1973.

[4] Keller, H. B. and Cebeci, T.: Accurate Numerical Methods for Boundary Layer Flow-I. Two-Dimensional Laminar Flows, in Lecture Notes

in Physics, Proceedings of Second International Conference on Numerical Methods in Fluid Dynamics, Springer-Verlag, 1971.

[5] Keller, H. B., A New Difference Scheme for Parabolic Problems in Numerical Solution of Partial Differential Equations, v. II, J. Bramble, Ed. Academic Press, New York, 1970.

[6] Smith, A. M. O.: Improved Solutions of the Falkner-Skan Boundary Layer Equation, Sherman M. Fairchild Fund Paper No. FF-10, 1954.

[7] Stewartson, K.: Further Solutions of the Falkner-Skan Equation, Cambridge Phil. Soc., v. 50, 1954, pp. 454-465.

APPLICATION DE LA METHODE HODOGRAPHIQUE AU TRAITEMENT DES ECOULEMENTS TRANSSONIQUES AVEC ONDE DE CHOC

par Jean-Jacques CHATTOT
Office National d'Etudes et de Recherches Aérospatiales (ONERA) -92320 - Châtillon (France)

et Maurice HOLT
Université de Californie - Berkeley (U.S.A.)

INTRODUCTION

En aérodynamique transsonique, la tendance actuelle est à la résolution directe des équations du mouvement dans le plan physique ou un domaine borné dont il est le transformé. La non-linéarité est prise en compte par les itérations de relaxation ou les itérations en temps, suivant que l'on résout les équations du problème stationnaire ou instationnaire. Les équations linéaires de la méthode hodographique permettent de construire des solutions analytiques fondamentales ou de résoudre numériquement les problèmes et de la façon la moins onéreuse en temps de calcul. En outre, on bénéficie d'une plus grande finesse de description du champ de l'écoulement, en particulier au voisinage du point de rencontre de la ligne sonique et du choc, et du point où celui-ci est évanescent. Les résultats présentés ici concernent le profil losangique dont l'hodographe est particulièrement simple. La méthode numérique employée est la méthode de Télénin qui a été construite pour résoudre les équations mixtes de l'écoulement autour d'un obstacle émoussé, avec choc détaché.

I - LA METHODE HODOGRAPHIQUE

Les caractéristiques fondamentales de la méthode hodographique sont bien connues. Voir par exemple Ferrari et Tricomi (1968). Les écoulements étudiés sont isentropiques, c'est-à-dire dépourvus de chocs ou avec des chocs suffisamment faibles pour que le gradient d'entropie soit négligeable partout. Les singularités de la transformation du plan physique au plan du vecteur vitesse sont caractéristiques du type de problème traité. Dans le cas d'un profil, lorsque le nombre de Mach à l'infini est suffisamment élevé, il apparaît des lignes limites et la solution obtenue est multiforme dans le plan physique.

Les solutions analytiques obtenues par la méthode hodographique ont un intérêt évident. Ce sont des points de repère précis pour les méthodes numériques. Ainsi Murman et Cole (1971) comparent les résultats de leur méthode avec la solution exacte de Nieuwland (1967).

Par combinaison linéaire de solutions élémentaires de l'équation hodographique, on a pu obtenir une grande variété de solutions parmi lesquelles ont été sélectionnées celles qui conduisaient à des écoulements continus, autour d'obstacles pouvant représenter des profils. Il était implicitement compris - depuis que Tollmien (1941) avait démontré que l'on ne pouvait pas simplement remplacer la première ligne limite par un choc, car son intensité serait nulle - que les lignes limites étaient peu souhaitables. Pourtant l'analogie entre les lignes limites et les chocs est grande puisque Lighthill (1947) écrivait que celles-là étaient des "obstacles mathématiques", tandis que celles-ci étaient des "obstacles physiques", à l'existence d'un écoulement continu satisfaisant aux conditions limites. De plus "chaque fois que la méthode hodographique a été employée pour construire des écoulements pour lesquels la présence d'ondes de choc était une évidence expérimentale, des lignes limites apparaissaient à leur place". Il est important de noter que la réciproque est fausse. La présence de lignes limites n'implique pas celle de chocs, et si on étudie l'exemple classique de Von Ringleb (1940), il est clair que l'on ne saurait mettre une onde de choc à la place des lignes limites qui apparaissent dans la partie accélérée de l'écoulement. Le choc ne fonctionne que dans un sens, celui qui fait passer des vitesses amont supersoniques vers des vitesses aval moindres.

Si on analyse la signification des lignes limites, on trouve que dans la partie où l'écoulement est accéléré, elles sont les enveloppes d'ondes de détente, alors que dans la partie retardée, elles représentent les enveloppes des ondes de compression issues de la ligne sonique. Or, comme l'a montré Guderley (1962), les premières n'ont pas de signification physique, puisque contrairement aux secondes, elles n'apparaissent pas sous l'influence de conditions aux limites. Dans ce dernier cas seulement, il est possible d'uniformiser le plan physique en introduisant un choc, qui joue le rôle d'une coupure et élimine les lignes limites.

II - CONSTRUCTION DU CHOC

Lorsque l'on connait une solution multiforme de l'équation de Tchapliguine :

$$(1-q^2)\,q^2\frac{\partial^2\psi}{\partial q^2} + \left(1+\frac{3-\gamma}{\gamma-1}q^2\right)q\frac{\partial\psi}{\partial q} + \left(1-\frac{\gamma+1}{\gamma-1}q^2\right)\frac{\partial^2\psi}{\partial\theta^2} = 0 \qquad (1)$$

on peut construire la courbe de choc dans le plan hodographique. En effet, il existe, le long des images amont et aval du choc une relation différentielle qui exprime que ces deux courbes ont même image dans le plan physique. Soit $tg(\alpha)$, la pente dans le plan (x, y) de la courbe de choc (S) :

$$tg(\alpha) = \left(\frac{dy}{dx}\right)_{(S)} = \frac{q_1\cos\theta_1 - q_2\cos\theta_2}{q_2\sin\theta_2 - q_1\sin\theta_1} \qquad (2)$$

En utilisant les relations de Tchapliguine au point (q_2 , θ_2), on obtient :

$$\left(\frac{\partial\psi}{\partial q}\right)_2\left\{q_1\sin(\theta_1-\theta_2) + \left[q_1\cos(\theta_1-\theta_2) - q_2\right]q_2\left(\frac{d\theta}{dq}\right)_2\right\}$$
$$+\left(\frac{\partial\psi}{\partial\theta}\right)_2\left\{\left[1-\frac{q_1}{q_2}\cos(\theta_1-\theta_2)\right](1-M_2^2) + q_1\sin(\theta_1-\theta_2)\left(\frac{d\theta}{dq}\right)_2\right\} = 0 \qquad (3)$$

Il faut d'autre part utiliser les relations de conservation suivantes :

$$\psi(q_1,\theta_1) = \psi(q_2,\theta_2) \qquad (4)$$

$$\cos^2(\theta_1-\theta_2) - \frac{\gamma+1}{2\gamma}\,\frac{q_1^2+q_2^2+2\frac{\gamma-1}{\gamma+1}}{q_1 q_2}\cos(\theta_1-\theta_2) + \frac{1}{\gamma} + \frac{\gamma-1}{2\gamma}\,\frac{q_1^2+q_2^2}{q_1^2 q_2^2} = 0 \qquad (5)$$

C'est sous forme quadratique en $\cos(\theta_1-\theta_2)$, l'équation de la polaire de choc.

Nous avons analysé en détail la solution de Von Ringleb, $\psi = \frac{\sin\theta}{q}$. L'élimination des lignes limites et l'uniformisation de la solution ont été réalisées par la mise en place du choc (Fig. 1). Les conditions de choc sont bien satisfaites tant que le saut de pression à travers la coupure est suffisamment faible et que l'écoulement potentiel représente bien la solution physique. La position des lignes limites a été indiquée en pointillé. Il faut noter la configuration ligne sonique - onde de choc.

III - APPLICATION

Le contour $\psi = 0$ dans le plan de l'hodographe est connu a priori pour le profil losangique. Il se compose d'un segment radial $\theta = \theta_\ell$, d'un arc de la caractéristique descendante issue du point sonique où $\theta = \theta_\ell$ et d'un segment radial $\theta = -\theta_\ell$. On cherche la solution $\psi(q, \theta)$ analytique à l'intérieur du domaine hodographique prolongé au-delà de la courbe de choc. Puis on construit celle-ci en utilisant les conditions de conservation (3) - (5), ce qui élimine la partie du domaine hodographique qui donne lieu à un recouvrement dans le plan physique (Fig. 2).

La fonction de courant est singulière au point $(q = q_\infty, \theta = 0)$, image du point à l'infini, au voisinage duquel le comportement est connu (Germain - 1962).

La méthode de Télénin a été utilisée par son auteur et ses collaborateurs (1964) pour résoudre le problème mixte de l'écoulement supersonique autour d'un corps émoussé. Plus récemment, elle a été employée par Holt et Ndefo (1970) pour calculer l'écoulement autour d'un cône en incidence lorsque le choc est attaché et l'écoulement transversal est subsonique. Le principe de la méthode de Télénin consiste à chercher une solution approchée sous forme d'un produit de fonctions d'une variable par des polynômes d'interpolation de l'autre variable. On obtient ainsi un système d'équations différentielles ordinaires couplées, que l'on peut intégrer à partir de données du type de Cauchy. Une itération sur une des données initiales permet de satisfaire à la condition aux limites à l'extrémité du domaine d'intégration.

Lorsque l'équation aux dérivées partielles est linéaire, comme c'est le cas ici, le procédé itératif se réduit à une simple combinaison linéaire de solutions indépendantes.

Le domaine hodographique est transformé en un rectangle par le choix d'un système de coordonnées polaires (ν, φ), centrées au point singulier $(q = q_\infty, \theta = 0)$.

On cherche le potentiel ψ sous la forme

$$\psi(\nu, \varphi) = \sum_{i=1}^{N} \sin \frac{i(\pi - \varphi)}{2} \; \psi_i(\nu) \tag{6}$$

Les fonctions d'interpolation sont les polynômes trigonométriques qui s'annulent à $\varphi = \pm \pi$. Les fonctions $\psi_i(\nu)$ sont des combinaisons linéaires des inconnues $\Psi_j(\nu) = \psi(\nu, \varphi_j)$ soit

$$\psi_i(\nu) = \sum_{j=1}^{N} c_{i,j} \, \Psi_j(\nu) \tag{7}$$

L'équation en ψ transformée étant de la forme

$$A \frac{\partial^2 \psi}{\partial \nu^2} + 2B \frac{\partial^2 \psi}{\partial \nu \partial \varphi} + C \frac{\partial^2 \psi}{\partial \varphi^2} + D \frac{\partial \psi}{\partial \nu} + E \frac{\partial \psi}{\partial \varphi} = 0 \tag{8}$$

l'évaluation des dérivées partielles de ψ le long des lignes $\varphi = \varphi_k$ conduit aux expressions suivantes :

$$\left(\frac{\partial \psi}{\partial \nu}\right)_{\varphi_k} = \frac{d \Psi_k(\nu)}{d\nu} \tag{9}$$

$$\left(\frac{\partial \psi}{\partial \varphi}\right)_{\varphi_k} = \sum_{i=1}^{N} -\frac{i}{2} \cos \frac{i(\pi - \varphi_k)}{2} \left(\sum_{j=1}^{N} c_{i,j} \, \Psi_j(\nu) \right) \tag{10}$$

$$\left(\frac{\partial^2 \psi}{\partial \nu^2}\right)_{\varphi_k} = \frac{d^2 \Psi_k(\nu)}{d\nu^2} \tag{11}$$

$$\left(\frac{\partial^2\psi}{\partial\nu\partial\varphi}\right)_{\varphi_k} = \sum_{i=1}^{N} -\frac{i}{2}\cos\frac{i(\pi-\varphi_k)}{2}\left(\sum_{j=1}^{N} c_{i,j}\,\frac{d\Psi_j(\nu)}{d\nu}\right) \qquad (12)$$

$$\left(\frac{\partial^2\psi}{\partial\varphi^2}\right)_{\varphi_k} = \sum_{i=1}^{N} \left(-\frac{i}{2}\right)^2 \sin\frac{i(\pi-\varphi_k)}{2}\left(\sum_{j=1}^{N} c_{i,j}\,\Psi_j(\nu)\right) \qquad (13)$$

Pour rester dans le cadre de l'approximation transsonique, l'application numérique a été faite avec $\theta\ell = 4{,}5°$, $M_\infty = 0{,}89$.

Nous avons placé les lignes à $\varphi = -\pi, -\pi/2, 0, \pi/2, \pi$. Le transfert de la solution dans le plan physique permet d'obtenir la distribution de pression le long du profil (Fig. 3). Les équations de conservation (3) - (5) permettent de construire les courbes de choc dans le plan hodographique. Dans le plan physique on constate (Fig. 4) que la configuration trouvée est en accord avec les études théoriques de Nocilla (1958-59). Le choc prend naissance au sein du domaine supersonique. Son intensité est infiniment faible en ce point car il est tangent à l'onde de compression issue de la ligne sonique. Par contre, au niveau du profil le choc fait passer d'une vitesse supersonique à une vitesse subsonique. Il en résulte l'existence d'un point sonique sur la face aval de l'onde de choc. C'est le point de rencontre de la ligne sonique et du choc. Il faut noter la proximité des deux courbes au voisinage de ce point. Ceci explique qu'il soit aussi difficile de les distinguer expérimentalement ou numériquement, lorsqu'on résout les équations dans le plan physique.

CONCLUSION

La méthode hodographique permet de résoudre des écoulements avec chocs, lorsque ceux-ci sont suffisamment faibles pour que l'hypothèse de mouvement isentropique soit valable. L'application au profil losangique a permis de confirmer les résultats théoriques de Nocilla. Le décollement qui se produit expérimentalement en aval de l'épaulement du profil modifie l'écoulement et rend toute comparaison caduque. On peut espérer que l'application de la présente méthode à un profil présentant une pente continue permettra de lever cette difficulté.

REFERENCES

FERRARI C.,TRICOMI F. — Transonic Aerodynamics, Academic Press (1968)

GERMAIN P. — "Problèmes mathématiques posés par l'application de la méthode hodographique à l'étude des écoulements transsoniques"

Symposium Transsonicum, IUTAM Symposium AACHEN (1962), II Sitzung, p. 24.

GILINSKII S.M., TELENIN G.F., TINYAKOV G.P. "A method for computing supersonic flow around blunt bodies, accompanied by a detached shock wave"
IZV. Akad. Nauk. SSSR, Mekhi i Mash. (1964), 4, 9-28.
NASA Transl. TTF-297.

GUDERLEY K.G. <u>the Theory of Transonic Flow</u>,
Pergamon Press (1962)

HOLT M., NDEFO D.E. "A numerical method for calculating steady unsymmetrical supersonic flows past cones"
Journ. Comp. Phys., <u>5</u>, n° 3 (1970)

LIGHTHILL M.J. "the Hodograph transformation"
<u>Modern Developments in Fluid Dynamics - High Speed Flow</u>, Ed. L. Howarth (1953), Vol. 1, Ch.VII

MURMAN E.M., COLE J.D. "Calculation of plane steady transonic flows"
A.I.A.A. Journal, Vol. 9, n° 1, (1971)

NIEUWLAND G.Y. "Transonic potential flow around a family of quasi-elliptical aerofoil sections"
NLR-TRT 172, (1967)

NOCILLA S. "Flussi transonici attorno a profili alari simmetrici con onda d'urto attaccata ($M_\infty < 1$) - II"
Atti. Accad. Sci. Torino : classe Sci. Fis. Mat. Nat. 93-124-148, (1958-59)

TOLLMIEN N. "Limit lines in adiabatic potential flows"
R.T.P. Transl. 1610, 2, Angew Math. Mech., <u>21</u>, (1941)

Von RINGLEB Z. Angew Math. Mech., <u>20</u>, (1940)

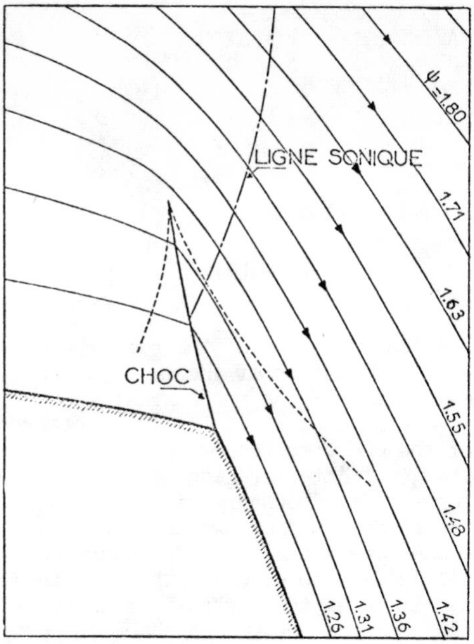

Fig. 1. Solution de Von Ringleb avec choc

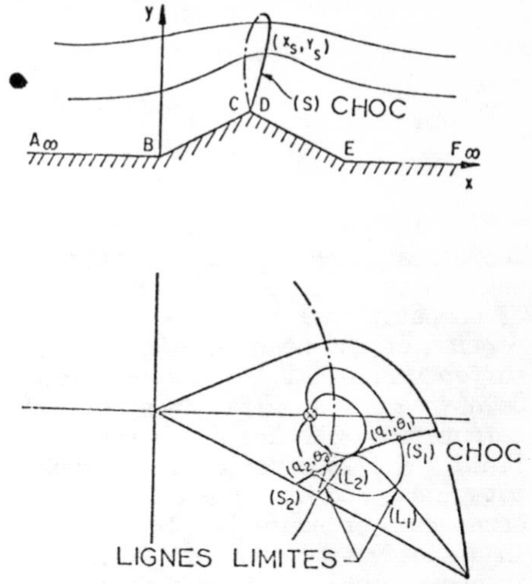

Fig. 2. Ecoulement dans le plan physique et son hodographe

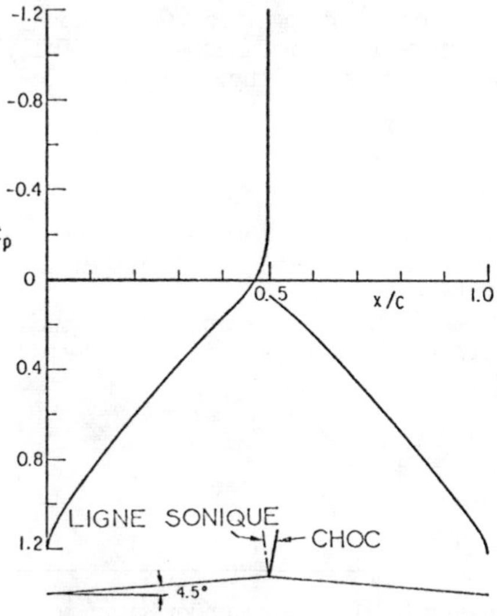

Fig. 3. Coefficient de pression

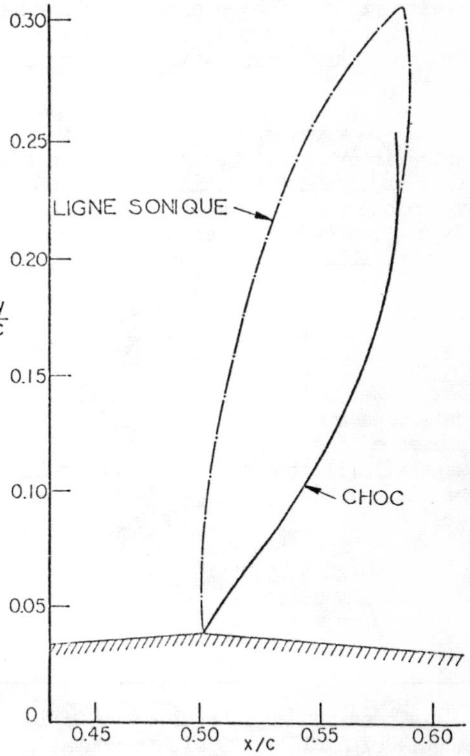

Fig. 4. Ligne sonique et onde de choc

FINITE DIFFERENCE TREATMENT OF STRONG SHOCK OVER A SHARP LEADING EDGE WITH NAVIER-STOKES EQUATIONS†

S. I. Cheng* and J. H. Chen**
Princeton University

INTRODUCTION

In the a posteriori error study of the numerical solution of the planar supersonic near wake, Ross and Cheng found that the small oscillation of the compression wake shock, although confined within 1/4 of a mesh is one of the two largest sources of computational errors. Since the shock is weak and is located in the downstream portion of the near wake, the overall accuracy of the computed results does not suffer seriously. In many supersonic flow problems, our interest generally lies downstream of some fairly strong shock which intersects the boundary of the field of computation. We are apprehensive of the resulting errors in the downstream pressure field. A 1/2% error of shock speed is common in one dimensional time dependent computations, and at a normal Mach number of 10, will lead to an error in the computed downstream pressure field of the same order of magnitude as the upstream pressure. In a steady supersonic flow, the slope of the shock corresponds to the shock speed in one dimensional time dependent flow, and the downstream pressure field is liable to similar computational errors. Moreover, the shock will generally intersect the boundary of the field of computation at some unknown location, where the difference treatment is uncertain. It is simple to demonstrate that inadequate difference treatments of such an extraneous boundary condition leads to significant errors in the shock shape and in the downstream pressure field. The present paper illustrates how such large computational errors downstream of a strong shock may be reduced and controlled.

We treat the strong shock wave emanating from the sharp leading edge as a specific example. We tolerated the additional complications of fluid mechanical (see Hayes and Probstein) and of computational nature since we wish to develop the technique for the computation of many practical problems starting from such a configuration. The difference approximation of the Navier-Stokes equations for a fully compressible fluid was written in terms of the Cheng-Allen algorithm, guided by inferences derived from the model study to be described in the next section.

MODEL ANALYSIS

We shall illustrate how and to what extent the extraneous boundary conditions determine and control the oscillatory components in computed results and why in the presence of such oscillatory components, the truncation errors are not important. We shall illustrate with the Cheng-Allen algorithm as a specific example while much of the results and discussions are applicable to other higher order difference algorithms.

Burgers' equation with both the convective and the viscous terms

$$u_t + uu_x = Re^{-1}\, u_{xx} \tag{1}$$

---

†This research was supported by Office of Naval Research, U.S. Navy under contract N00014067-A-0151-0028 with computations supported by NSF Grants GJ-3157.
*Professor, Department of Aerospace and Mechanical Sciences.
**Currently NSF Postdoctoral Fellow, Geophysical Fluid Dynamics Laboratory, ESSA, Princeton University.

is a useful one dimensional model of the momentum and the energy equation. The difference form, written in terms of Cheng-Allen algorithm is

$$\hat{U}_j^n = U_j^n - \frac{r}{4}[(U_{j+1}^n)^2 - (U_{j-1}^n)^2] + s[U_{j+1}^n - 2\hat{U}_j^n + U_{j-1}^n]$$

$$U_j^{n+1} = U_j^n - \frac{r}{4}[(\hat{U}_{j+1}^n)^2 - (\hat{U}_{j-1}^n)^2] + s[\hat{U}_{j+1}^n - 2U_j^{n+1} + \hat{U}_{j-1}^n] \qquad (2)$$

where $U_j^n = U(n\Delta t, j\Delta x) = r = \Delta t/\Delta x$ and $s = \Delta t/Re(\Delta x^2)$. There are two important properties of this algorithm.

(i) The linear local stability criterion is $r = \Delta t/\Delta x \leq 1$, not limited by the magnitudes of the local Reynolds number Re. In securing this property by treating the viscous terms with the DuFort-Frankel technique, the algorithm is no longer consistent temporarily for finite values of r. It is accordingly useful only for obtaining steady state solutions via time dependent formulations of complicated viscous flow problems as was suggested by Crocco. For both the near wake flow problem and the present problem, we found that the linearized local stability criterion derived from the model is sufficient for securing computational stability.

(ii) This algorithm is a two step scheme formally of second order accuracy $O(\Delta x^2)$ in the steady limit; but like other higher order schemes, is prone to give fairly large oscillations in regions with large gradients. Such large oscillations can, however, be substantially reduced in this algorithm by the following strategem:

The identical form of the predicator step for $\hat{U}_j^n$ and the corrector step for U_j^{n+1} not only facilitates programming but permits the use of identical boundary operators for both steps. Under the circumstances, we shall show that the steady state limits of $\hat{U}_j^n$ and U_j^{n+1} are equal, and the standing oscillations from the extraneous solutions are completely suppressed within the linearized approximation. The linearized Burgers' equation is

$$u_t + u_x = Re^{-1}u_{xx} \qquad (1a)$$

and its difference approximation with Cheng-Allen algorithm is

$$\hat{U}_j^n = U_j^n - \frac{r}{2}(U_{j+1}^n - U_{j-1}^n) + s[U_{j+1}^n - 2\hat{U}_j^n + U_{j-1}^n]$$

$$U_j^{n+1} = U_j^n - \frac{r}{2}(\hat{U}_{j+1}^n - \hat{U}_{j-1}^n) + s[\hat{U}_{j+1}^n - 2U_j^{n+1} + \hat{U}_{j-1}^n] \qquad (2a)$$

where the reference velocity is taken to be unity. In the steady state limit, $(U_j^{n+1} = U_j^n)$, equations (2a) stand as a 4th order linear difference equation whose general solution is: $U_j = C_k[\xi_k(s,r)]^j$ with

$$\xi_1 = 1, \qquad \xi_2 = \frac{s+r/2}{s-r/2} = \frac{1 + Re_{\Delta x/2}}{1 - Re_{\Delta x/2}}, \qquad \xi_{3,4} = \frac{-(1+2s) \pm \sqrt{1+4s+r^2}}{2s-r}$$

where superscript j means "exponent j." Two of the fundamental solutions ξ_1 and ξ_2 are proper: that is, in the limit of $\Delta x \to 0$, they approach respectively the two fundamental solutions of the steady state differential equation 1 and exp (Re x). The other two fundamental solutions ξ_3 and ξ_4 are extraneous one of them is negative and gives a contribution oscillating from mesh to mesh. The four constants c_k are determined by four boundary conditions. Two of these are extraneous, i.e., without counterparts in the differential formulation, and must be imposed somewhat arbitrarily in the difference formulation to effect a solution of the system of difference equations with higher order formal accuracy (Cheng, 1970, p. 2119). The extent which these two extraneous fundamental solutions participate in the computed steady state solution depends on the extraneous boundary conditions.

The errors due to such extraneous fundamental solutions can therefore be controlled by suitable formulation of such extraneous boundary conditions, although inevitably erroneous and often appreciably in practice. For the present difference algorithm with identical predictor and corrector steps, such errors can be suppressed by applying the same extraneous boundary values on both steps. This is because the difference equation for $U_j^{n+1} - \hat{U}_j^n$ is linear and homogeneous, and under zero boundary values, if we ignore the possible existence of eigen solutions, $U_j^{n+1} - \hat{U}_j^n$ must vanish in the steady state ($U_j^{n+1} = \hat{U}_j^n$). Accordingly, the steady state solution will be obtained from either of the second order equations of (2a) with $U_j^{n+1} = \hat{U}_j^n = U_j^n$ and will contain only the proper fundamental solutions. Thus the computed results will be a member of the sequence of solutions which will converge to the "correct" limit as $\Delta x \to 0$, "correct" to the extent that the extraneous boundary conditions approximate the solution of the differential problem on the extraneous boundary. In other words, the primary errors in the computed results ($c_1\xi_1 + c_2\xi_2$) will be in the coefficients c_1 and c_2, as determined by the slightly erroneous extraneous boundary values.

The proper fundamental solution $\xi_2 = (1 + Re\Delta x/2)/(1-Re\Delta x/2)$ can also produce mesh to mesh oscillations when the mesh is so coarse as to have $Re\Delta x > 2$. Under the circumstances, neither the net function as calculated, nor their averages over neighboring mesh points serves as an approximation to the correct fundamental solution exp (Rex) in any sense. Such a situation should be avoided where possible. An important characteristic of such oscillations is that its mesh to mesh growth rate $(1 + 2/Re\Delta x)/(1-2/Re\Delta x)$ appears to decrease with coarser meshes. The improvement in appearance of smaller computed oscillations at coarser mesh size must not be interpreted as providing a "better approximation" but rather a warning that a much finer mesh is needed.

In actual computations we cannot achieve the ideal situation to have all the extraneous solutions suppressed. First, the extraneous boundary conditions cannot be conveniently formulated in terms of approximate boundary values but in terms of some alternative forms of boundary operators. Secondly, for a nonlinear problem we can adopt different models and find different critical values of $Re_{\Delta x}$ such as 2 for model (2a). Such a critical value is determined by the characteristic equation which will vary with the details of the linearized model and of the specific difference algorithm even within the quasi-linear local argument. Anticipating also the results to be given in the next paragraph, we shall be satisfied with the suggestion that there may be such a critical value of $Re_{\Delta x}$ for a nonlinear problem and that we would be warned against exceeding this critical value when a repeated computation at a coarser (finer) mesh should decrease (increase) the amplitudes of oscillations.

For gas dynamic problems, the Burgers' equation models the momentum and the energy equations well with both the convective and the diffusive terms retained. The mass conservation or the continuity equation is quite distinct, however. Let us examine the linearized model system

$$\frac{\partial \rho}{\partial t} + \frac{\partial(\rho u)}{\partial x} = 0 \tag{3}$$

$$\frac{\partial u}{\partial t} + \frac{\partial u}{\partial x} = \frac{1}{Re}\frac{\partial^2 u}{\partial x^2} \tag{1a}$$

The continuity equation (3) is represented in difference form according to the Cheng-Allen algorithm as:

$$\hat{R}_j^n = R_j^n - [(RU)_{j+1}^n - (RU)_{j-1}^n] \tag{4a}$$

$$R_j^{n+1} = R_j^n - [(\hat{R}U)_{j+1}^n - (\hat{R}U)_{j-1}^n] \tag{4b}$$

Here we recognize U_j^n as a known net function obtained from the difference system (2a). In the steady state limit, with $R_j^{n+1} = \hat{R}_j^n$, equations (4) may be written as a 4th order difference equation. When we write $(RU)_j = \eta^j$, we obtain 4 fundamental solutions for η. They are $\eta_{1,2} = \pm 1$ and $\eta_{3,4} = [1 \pm \sqrt{1 + 4U_{j+1}U_{j-1}}\,]/\ 2U_{j+1}$. All three solutions $\eta_{2,3,4}$ are extraneous except $\eta_1 = 1$. While $\eta_{3,4}$ can be suppressed with $\hat{R}_j^n = R_j^{n+1}$ under identical boundary values for (4a) and (4B), the extraneous solution $\eta_2 = -1$ remains. Thus even in the linearized model system (3) and (1a), the use of identical boundary values on the predictor and the corrector steps will not suppress oscillations in the density field although it will do so in the velocity field. The magnitudes of the density oscillations will be determined by the extraneous boundary conditions to be applied on the gas density next to the solid surface and elsewhere.

In computing the full gas dynamic equations it is clear now, that despite all the cautions discussed above, we cannot expect the computed density field to be smooth. We have to be very careful in dealing with the extraneous boundary conditions to minimize the density oscillations from mesh to mesh, which as we shall see, are aggravated by the extreme fluid dynamic situation in the present problem. Since all the gas dynamic equations are coupled, density fluctuations will induce oscillations in the computed velocity and temperature fields. Since we cannot expect the extraneous boundary conditions to be accurate to a couple of percent, any other error sources less than 1% will be insignificant in the presence of oscillatory extraneous solutions, such as the local truncation error that does not accumulate under "conservative difference algorithms."

We shall estimate the local truncation error $E_T(u)$, based on the analytical results given by Cheng (1969), valid for the steady state solutions of the Burgers' equation (1) with a finite velocity at large x:

$$u(x) = -\alpha \tanh\left[\frac{\alpha R_e(x-x_o)}{2} - \theta_o\right] \qquad (5)$$

with $u(x = x_o) = u_o = \alpha \tanh \theta_o$

$$u(x-x_o = -1/2) = \alpha \tanh\left[\frac{\alpha R_e}{4} + \theta_o\right] = 1$$

and $|u(x = \pm \infty| = \alpha = 1/\tanh\left[\frac{\alpha R_e}{4} + \theta_o\right]$

For this solution, the maximum of the local truncation error $E_T(u)$ is given as

$$|E_T(u)|_{max} < |\ 1/2E_3|_{max}(Re_{\Delta x})^2 < 3 \times 10^{-2}(Re_{\Delta x})^2. \qquad (6)$$

for Cheng-Allen algorithm with the simple space centered difference (a = o).

Let us consider that the part of the smooth computed solution with a large gradient is approximated by (5) locally, with x_o designating the point of largest gradient. We can estimate the largest $Re_{\Delta x}$ in this region by

$$Re_{\Delta x} = \frac{2}{\alpha^2-u_o^2}(\Delta u)_{x_o} = \frac{2}{\alpha^2-u_o^2}\left|\frac{\partial u}{\partial x}\right|_{x_o}\Delta x \qquad (7)$$

Now $\alpha > 1$, and the value of u_o where the gradient is large is generally small. Hence we expect $\alpha^2 - u_o^2$ to be $\gtrsim 1$ and we can replace $Re_{\Delta x}$ by $2(\Delta u)$ for the estimate of the truncation error $E_T(u)$ in equation (6). For any computations with a reasonable resolution we should expect Δu over a mesh to be less than 1/3 of the reference velocity. Accordingly we can expect the local truncation errors to be less than 1% except possibly at some singular point such as the leading edge. Consequently, our primary concern in the formulation of the difference approximation for a complicated fluid flow problem is the various treatments of the boundary conditions both proper and extraneous and of the singular points.

DIFFERENCE FORMULATION

The energy equation written in terms of total enthalpy ($H = h + \frac{u^2 + u^2}{2}$) is similar to the momentum equations except for a dissipation term which remains of order of unity even for hypersonic flows. Thus we hope that the results of the model study with Burgers' equation will help to limit the computational errors in momenta and stagnation enthalpy, being computed directly from the conservation relations. For a large Mach number flow, however, the stagnation enthalpy and the gas kinetic energy are nearly equal over most of the flow field. Thus the static temperature (and hence static pressure) calculated from the difference of these two nearly equal quantities becomes much less accurate. These errors are moreover cumulative. Hence, the energy equation is presently written in terms of static energy with its value in the incoming uniform stream as reference. In this form, the energy equation contains dissipation terms like $\gamma(\gamma-1)M_\infty^2 \, Re^{-1} (\frac{\partial u}{\partial y})^2$, which, with $Re^{-1}(\frac{\partial u}{\partial y})^2 \sim O(1)$ near a solid wall are large for high Mach number flows, and lead to large computed results of static energy locally.

With large static energy near the plate surface, the computation of the convective terms in divergence form $\frac{\partial}{\partial x}(\rho ue) + \frac{\partial}{\partial y}(\rho ve)$ involves taking the difference between two large but nearly equal terms of influx and outflux. The net convective flux of energy $\rho u \frac{\partial e}{\partial x} + \rho v \frac{\partial e}{\partial y}$ is altered by the computed residual value of $e\,[\frac{\partial(\rho u)}{\partial x} + \frac{\partial(\rho v)}{\partial y}]$ which should vanish by mass continuity, but remain appreciable because of large e. This error is introduced by the use of the divergence form of the convective terms in generating the difference equations. Hence, we construct our difference approximation for the energy equation with the non-divergence form $\rho u \frac{\partial e}{\partial x} + \rho v \frac{\partial e}{\partial y}$ of the convective energy flux while we use the divergence form $\frac{\partial}{\partial x}(\rho u^2) + \frac{\partial}{\partial y}(\rho uv)$ etc. for the convective momentum flux. With the rectangular cartesian coordinates in the physical space and the centered difference algorithm, the resulting difference formulation is strictly "conservative" in the sense that when summed over neighboring cells, the fluxes crossing the interior cell boundaries cancel identically (Cheng, 1969, 1970). The inviscid Hugoniot relations across a shock wave, if locally valid, will be preserved, although we do not use the divergence form of the flux terms in the energy equation.

The large static energy near a cold plate surface, presents a large gradient normal to the plate which renders the extrapolation of density R from the interior as an extraneous boundary condition highly inaccurate. It induces such large density oscillations near the plate as to sometimes yield negative density (& pressure) on the plate, i.e., the magnitude of the resulting density fluctuation exceeds the local density values. To remedy the situation we implement this extraneous boundary condition as the local mass conservation consideration of the boundary cells on the plate.

$$\hat{R}^n_{j,1} = R^n_{j,1} - \frac{\Delta t}{\Delta x}\left[\frac{3}{8}\{(RU)^n_{j+1,2} - (RU)^n_{j-1,2}\} + \frac{1}{8}\{(RU)^n_{j+1,1} - (RU)^n_{j-1,1}\}\right] - \frac{1}{2}\frac{\Delta t}{\Delta y}\left[(RV)^n_{j,2} - (RV)^n_{j,1}\right] \tag{8}$$

where $R^n_{j,i}$ designates the density R at $(j\Delta x, i\Delta y)$ on the nth time step. We take $(RV)^n_{j,1} \equiv 0$ on the plate $i = 1$, and evaluate $(RU)^n_{j,1}$ from the appropriate velocity slip condition. Equation (8) is written for the control volume centered at (j,1) with the lower half of it (below the plate) taken as the symmetric reflection with respect to $y = 0$.

The conservative difference formulation and the conservative treatment (8) of the boundary cells successfully suppressed the oscillations along the plate except those emerging from the leading edge. If we should calculate the $Re_{\Delta x}$ as $2\Delta E$ for the static energy profile next to the leading edge according to our model, we find it to be fairly large and will likely exceed the critical values for the proper fundamental solution ($\sim \xi_2$) to be oscillatory. Hence, we bring the uniform inflow station as close to the sharp leading edge as is practical (< (Δx)), so as to increase the local value of $Re_{\Delta x} \sim 2\Delta E$ and to reduce the local error.

Now we turn to the extraneous boundary condition where the shock wave intersects the outer edge of the field of computation and where large gradients exist also. We do not know how to implement the local conservation relations here and have to rely on extrapolation procedures. Where the flow is supersonic, however, the direction of extrapolation must lie within the zone of dependence. Moreover to extrapolate a shock along some arbitrary direction is equivalent to imposing a wave entering the field of computation at the point of intersection to produce the required change in the direction of the shock. This "reflected wave" contributes to oscillatory downstream solution. Thus to minimize the wave reflection from the boundary, we prefer to extrapolate in the direction of the shock as if it were locally straight or of some constant curvature or the like. In the present problem along the entire boundary upstream of where the shock intersects the boundary we extrapolate along the "shock direction" since it is immaterial in which direction we extrapolate the upstream uniform flow. The direction of extrapolation is varied linearly from the shock direction at the shock intersection to zero on the plate. The location of the shock intersection is somewhat arbitrary and is taken to be the location of the maximum pressure. In the strongly viscous region near the plate, we have to extrapolate along the plate so as to accommodate the boundary layer-like flow situation, hopefully valid on the downstream outflow boundary.

A practical difficulty is that the direction σ of the leading edge shock when it intersects the boundary cannot be conveniently determined from the computed shock profile. Instead we calculate sin σ from

$$\sin^2 \sigma = \frac{\gamma-1}{2\gamma M^2} \left(\frac{\gamma+1}{\gamma-1} \frac{P_2}{P_1} + 1 \right) \tag{9}$$

although this Hugoniot shock relation may not hold since we cannot identify an inviscid flow region behind it. We select the maximum pressure along the upper and/or the outflow boundary as p_2 in evaluating sin σ from eq. (9). We have thus eliminated noticeable wave reflections from the boundary and the oscillations of the shock wave itself and of its downstream pressure field. We have also checked that a slightly different strategem of choosing p_2 results in no meaningful changes in the computed results.

RESULTS AND CONCLUSIONS

Fig. 1-4 presents the computed results for the hypersonic flow at $M_\infty = 20$ over the sharp leading edge of a flat plate held at a temperature $T_w = 0.1\, T_{o\infty}$, i.e., 10^{-1} of the stagnation temperature of the incoming stream. The specific heat ratio $\gamma = c_p/c_v = 1.40$ and the Prandtl number Pr = 0.75 are chosen to fit the nondissociating diatomic gas. The temperature dependence of the viscosity coefficient is a parabolic fit of the data of Vas and Koppenwallner from Princeton's hypersonic nitrogen tunnel and of Vidal and Bartz from Cornell Aero Lab's hypersonic tunnel both with little dissociation. Bulk viscosity coefficient is taken as zero.

The leading edge is computationally located 1/2 Δx behind the uniform inflow section at j = 1. The field of computation is ($40\Delta x$) x ($30\Delta y$) with $\Delta x = 4\Delta y$. The hypersonic interaction parameter $\bar{\chi} = M_\infty^3 / \sqrt{Re_x}$ is $\geq$ 120 and the corresponding rarefaction parameter $\bar{v} = \bar{\chi}/M_\infty{}^2$ is $\geq$ 0.30. Two stream form slip boundary conditions are adopted on the plate surface with the incoming molecules assuming the average properties of the gas one mean free path λ (assumed to be the same for both momentum and energy) away from the plate and with the out-going molecules totally accommodated on the plate surface and diffusedly emitted. Thus we take

$$\text{velocity slip} \qquad U_g = 1/2\, u(\lambda) \qquad v_g = 0 \tag{10}$$

temperature slip $T_g - T_w = 1/2\ [T(\lambda) - T_w]$ (11)

skin friction $\tau_w = \frac{\rho a}{4}\ u(\lambda)$ (12)

wall heat transfer $q_w = \frac{\rho a}{4}\ [c_p T(\lambda) + c_p T_w + 0.37\ U^2(\lambda)]$ (13)

The linear local stability requirement, obtained from Burgers' model,

$$\Delta t < \Delta y/[\ |u|\ /\beta + |v| + (1 + \beta^2)^{1/2}\ e^{1/2}/\beta M_\infty] \tag{14}$$

with $\beta = \Delta x/\Delta y$, is sufficient to provide computational stability.

The solution presented here satisfies the steady state criterion $\frac{1}{R}\ \frac{\partial R}{\partial t} < 10^{-5}$. There are only minor oscillations in all the computed results emerging from the leading edge. Such oscillations decay fairly rapidly and are negligible 7 cells from the leading edge. On the outflow boundary the velocity slip on the plate is about 5% of free stream velocity. The gas temperature has passed its maximum, but is still substantially above the wall value of 8.1. The gas pressure (wall static pressure) and gas density have, however just reached their peaks. The leading edge shock spreads over the upper third of the outflow-boundary. The profiles across the diffused shock, however, appears reasonable. The downstream boundary is probably close to rather than within the strong shock interaction regime of leading edge hypersonic flow.

We are gratified to have overcome all the major difficulties in computing the static pressure field downstream of a strong shock, crucial in many fluid dynamic problems. We are quite enthusiastic about the general approach. The analysis of selected model equations clarified the nature of the oscillatory computational errors, suggested useful avenues controlling such errors from extraneous solutions, and provided some estimate of the truncation errors of smooth solutions. While we have demonstrated only with the Cheng-Allen algorithm which was after all constructed with all these in prospect, much of the results and discussions are believed to be more widely applicable.

REFERENCES

Allen, J. S. and Cheng, S. I., Physics of Fluids, 13, 37-52 (1970).

Cheng, S. I., Physics of Fluids, Supplement II, 34-41 (1969).

Cheng, S. I., AIAA Journal, 8, 2115-2122 (1970).

Crocco, L., AIAA Journal, 3, 1824-1832 (1965).

DuFort, E. C. and Frankel, S. P., Math. Aids Computation 7, 135 (1953).

Hayes, W. D. and Probstein, R. F. Hypersonic Flow Theory, Academic Press (1959).

Ross, B. and Cheng, S. I., Lecture Notes in Physics, 8, Springer-Verlag, Proc. of Second International Conference on Numerical Methods in Fluid Dynamics, (1970), pp. 164-9.

Vas, I. E. and Koppenwallner, G., AFOSR Report 64-1422, U.S. Airforce and Princeton University Department of Aerospace and Mechanical Sciences Report 690.

Vidal, R. J. and Bartz, J. A., AF-2041-A-2, Cornell Aeronautical Laboratory, 1968.

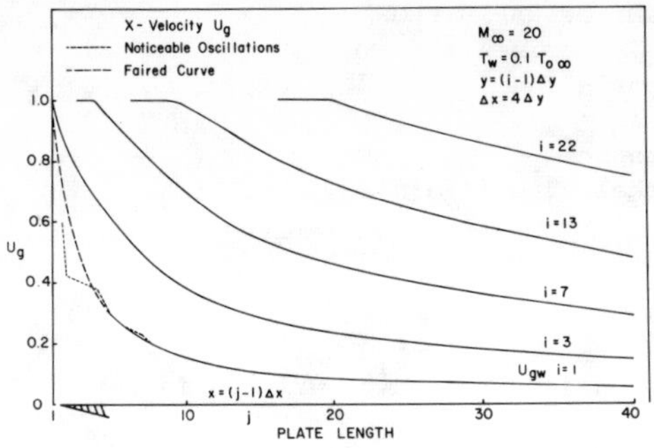

Fig. 1
Velocity Profiles

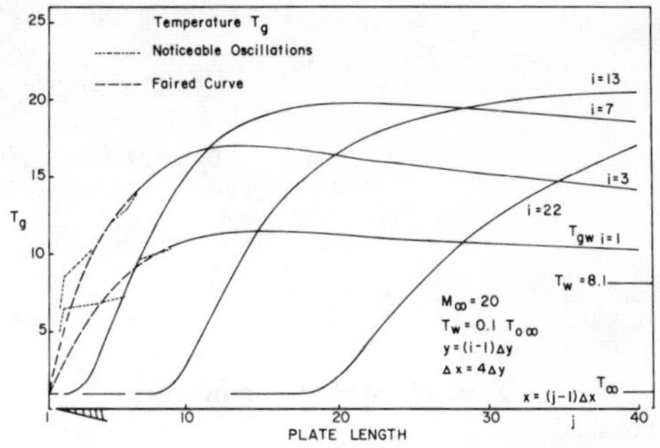

Fig. 2
Temperature Profiles

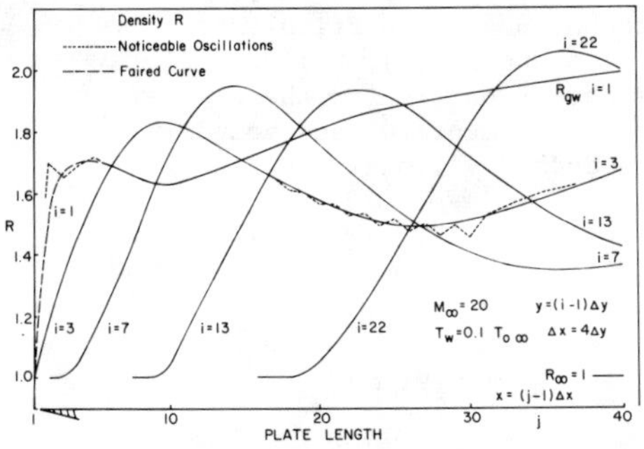

Fig. 3
Density Profiles

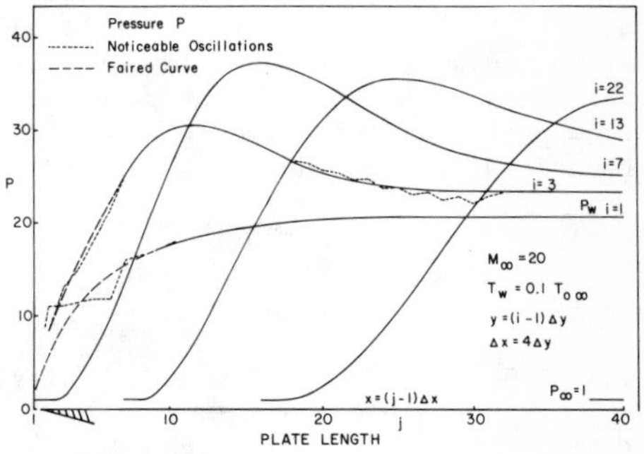

Fig. 4
Pressure Profiles

at various distances y = (i-1)Δy from the plate i = 1 (y = o).

$$\overline{\chi} = M_\infty^{\ 3}/Rex^{1/2} \quad > \quad 120$$

$$\overline{v} = \overline{\chi}/M_\infty^{\ 2} \quad > \quad 0.30$$

Slip conditions on plate surface - two stream form with fully accommodated emitting stream.

A VORTEX METHOD FOR THE STUDY OF RAPID FLOW*

Alexandre Joel Chorin**

Department of Mathematics
University of California, Berkeley, California

INTRODUCTION

The Navier-Stokes equations in two dimensional space can be written in the form

(1a) $$\partial_t \xi + (\underline{u}\cdot\underline{\nabla})\xi = \frac{1}{R}\Delta\xi$$

(1b) $$\Delta\psi = -\xi$$

(1c) $$u = -\partial_y \psi \quad , \quad v = \partial_x \psi$$

where $\underline{u} = (u,v)$ is the velocity, $\underline{r} = (x,y)$ is the position vector, ψ is the stream function, ξ is the vorticity, and R is the Reynolds number. R is assumed to be so large that finite difference techniques are difficult to apply. For a discussion of the relevant circumstances, see Chorin [2], Keller and Takami [4]. In the present paper we shall consider in particular the problem of flow past a cylinder of cross section D', with complement D and boundary ∂D. Equations (1) are thus to be solved in D, with boundary conditions

(2a) $$\underline{u} = 0 \quad \text{on} \quad \partial D$$

(2b) $$\underline{u} \to (U,0) \quad \text{as} \quad r = \sqrt{x^2+y^2} \to \infty \quad ,$$

where U is the velocity at infinity. The initial data are

(3) $$\underline{u}(x,y,t=0) \quad \text{given.}$$

If R were infinite, if the flow were started impulsively from a state of rest, and if the normal component only of (2a) were imposed on ∂D, the resulting flow would be a potential flow. For R finite, however large, both components of (2a) have to be taken into account, and the flow is rotational. For a thorough and lucid discussion of the origin of the vorticity, see Batchelor [1]. The crucial fact is that vorticity is created at the boundary and then diffused into the fluid. The principle of our method is to imitate numerically this process of vorticity generation and dispersal. Before explaining this method, we need some facts about the dynamics of a finite collection of vortices.

DYNAMICS OF A FINITE COLLECTION OF VORTICES

Consider a collection of N point vortices of intensity ξ_i, located at $\underline{r}_i = (x_i,y_i)$, $i = 1,\ldots,N$, in an inviscid fluid. The vorticity field is

$$\xi = \sum_{i=1}^{N} \xi_i \,\delta(\underline{r}-\underline{r}_i) \quad .$$

The stream function is

$$\psi = \frac{1}{2\pi} \sum_{i=1}^{N} \log|r-r_j|$$

---

* Partially supported by the Office of Naval Research under Contract USN N00014-69-A-0200-1052.

**This work was performed while the author was Visiting Miller Research Professor, University of California, Berkeley.

and the equations of motion for the point vortices are (see Batchelor [1])

$$(4a) \qquad (dx_i/dt) = X_i(x_1,\dots,x_N,y_1,\dots,y_N,\xi_1,\dots,\xi_N)$$

$$(4b) \qquad (dy_i/dt) = Y_i(x_1,\dots,x_N,y_1,\dots,y_N,\xi_1,\dots,\xi_N)$$

where

$$X_i = -\frac{1}{2\pi}\sum_{j\neq i}[(y_i-y_j)/r_{ij}]\xi_j$$

$$Y_i = \frac{1}{2\pi}\sum_{j\neq i}[(x_i-x_j)/r_{ij}]\xi_j$$

and

$$r_{ij} = |r_i - r_j| \quad .$$

This set of ordinary differential equations can be solved by an algorithm of the form

$$(5a) \qquad x_i^{n+1} = x_i^n + k\,X_{ik}$$

$$(5b) \qquad y_i^{n+1} = y_i^n + k\,Y_{ik}$$

where k is the time step, $x_i^n \equiv x_i(nk)$, $y_i^n \equiv y_i(nk)$, and X_{ik}, Y_{ik} approximate X_i, Y_i; X_{ik}, Y_{ik} could result from application, e.g., of the Runge-Kutta integration formulas.

Consider next the problem of solving the diffusion equation

$$(6) \qquad \partial_t \xi = \frac{1}{R}\Delta\xi \quad , \quad \xi = \xi(x,y,t)$$

with

$$\xi^\circ \equiv \xi(x,y,0) \text{ given.}$$

A Monte-Carlo solution of this problem can be obtained as follows: Pepper the (x,y) plane with points of masses ξ_i, and locations $\underline{r}_i = (x_i,y_i)$ $i = 1,\dots,N$, N large, in such a way that the mass density approximates ξ°. Then move the points according to the law

$$x_i^{n+1} = x_i^n + \eta_1$$

$$y_i^{n+1} = y_i^n + \eta_2$$

where η_1, η_2 are gaussianly distributed random variables with mean 0 and variance σ, where

$$\sigma = 2k/R \quad .$$

Under these conditions, the mean density after n steps will approximate $\xi(nk)$.

One could therefore attempt to solve equations (1), with an initial vorticity distribution ξ°, and in the absence of boundaries, by placing N vortices on the plane, with vorticity per unit area approximately ξ°, and then moving the vortices according to the rules

$$(7a) \qquad x_i^{n+1} = x_i^n + k\,X_{ik} + \eta_1 \qquad i = 1,\dots,N$$

$$(7b) \qquad y_i^{n+1} = y_i^n + k\,Y_{ik} + \eta_2$$

The density at the nth step should approximate $\xi(nk)$. The exact meaning of this approximation when ξ is smooth will be considered elsewhere. Clearly, some mollification of the vortices will be needed in order to have convergence in a sufficiently strong sense, and the vortices will no longer be point vortices. In particular, it can be seen that vortices with stream function of the form

$$(8) \qquad \psi(x,y) = \begin{cases} \frac{1}{2\pi} \log |r-r_j| & |r-r_j| \geq \sigma \\ \frac{1}{2\pi} \frac{r}{\sigma} & |r-r_j| < \sigma \end{cases}$$

will avoid the qualitative difficulties which can arise from the spurious strong interaction of neighboring vortices. σ is a cut-off length, for which we shall later find a natural value. The form (8) was first suggested by a theory of turbulence (Chorin [3]); the introduction of the cut-off length is essential to the method.

ANALYSIS OF RAPID FLOWS

We are now ready to present the algorithm for solving the problem set in the introduction. For the sake of simplicity, we assume that the flow is started from rest impulsively, and is thus initially irrotational. The appropriate potential flow can be obtained by setting

$$\underline{u}^\circ = \text{flow in the absence of the obstacle},$$

and adding to $\underline{u}^\circ$ a velocity field of the form $\underline{u}' = \text{grad}\ \phi$, with

$$\phi(\underline{r}) = \frac{1}{2\pi} \int_{\partial D} \alpha(q) \log \underline{r}(q)\, ds$$

where q denotes a point on ∂D, $r(q)$ is the distance between $\underline{r}$ and q, ds is an element of ∂D, and $\alpha(q)$ satisfies

$$(9) \qquad \alpha(q) - \int_{\partial D} \frac{\alpha(q(s))}{\pi} (\partial_n \log r(q(s)))ds = -\underline{u}^\circ \cdot \underline{n}$$

where $\underline{n}$ is the unit outward normal to ∂D, and ∂_n denotes differentiation in the direction of $\underline{n}$. (See, e.g., Kellogg [5].) Equation (9) is readily solved numerically (see, e.g., Smith [6]). $\underline{u}^\circ + \underline{\text{grad}}\ \phi$ will fail to satisfy the tangential component of (2a). A vortex sheet will form on ∂D. We now divide ∂D into N segments $S_1, S_2, \ldots, S_N$ of length h, compute the intensities ξ_i of the vortex sheet in the center of each S_i, and create N vortices, of intensity $\xi_i hk$ at those centers, endowing those vortices with the structure (8) where $\sigma = 2\pi hk$. One can see that with this choice of σ, the tangential boundary condition will be satisfied; it is also readily seen that without the introduction of the cut-off, totally unrealistic interaction between boundary and nearby vortices will occur. We then allow these new vortices to move according to the law

$$(10) \qquad \begin{aligned} x_i^{n+1} &= x_i^{n} + \eta_1 \\ y_i^{n+1} &= y_i^{n} + \eta_2 \end{aligned}$$

Those vortices which under the effect of (10) cross into D' are eliminated. This procedure results in the introduction of $N' \leq N$ vortices into the fluid. The next step can then be taken. $\underline{u}^\circ$, the initial velocity field, will now be replaced by the sum of $\underline{u}^\circ$ and the velocity due to the vortices inside the fluid; new vortices will be created, and the old and the new vortices will be moved according to a law of the form

$$(11) \qquad \begin{aligned} x_i^{n+1} &= x_i^{n} + P\, k + X_{ik}\, k + \eta_1 \\ y_i^{n+1} &= y_i^{n} + Q\, k + Y_{ik}\, k + \eta_2 \end{aligned}$$

where the vector (P,Q) will approximate the effect of $\underline{u}^\circ$ and of the potential contribution of the obstacle, and X_{ik}, Y_{ik} will take into account the interaction of the vortices (see equations (7)).

The main attractions of our method are evident: there is no grid in the interior of the fluid, and thus no harmful grid-induced effects, such as numerical diffusion; the applicability of the method depends little on the shape of D, on its size, be it infinite, and most importantly, it depends relatively little on the Reynolds number R.

In summary, at each step we satisfy the equations of motion and the boundary conditions by a flow consisting of a sum of a potential flow and a flow due to vortices created at the boundary and diffused from it; this diffusion process is approximated by a Monte-Carlo method whose variance decreases with increasing R.

NUMERICAL RESULTS

The method described above has been applied to the solution of several problems; in particular, flow past a circular cylinder at various Reynolds numbers R between 1 and 1000 has been studied, with the Reynolds number R based on cylinder radius. For $R \leq 30$ states steady in the mean are reached, and our results regarding drag, vorticity distribution, and length of standing vortices are in good agreement with previous work. At higher Reynolds numbers, there is a dearth of reliable numerical solutions; our calculations are in good agreement with experiment. In particular, vortex shedding and the formation of a vortex street are clearly reproduced. The detailed solutions will be presented elsewhere. An example is presented on the next page. The vortices in the wake of a circular cylinder which has been started from rest are presented. A * is placed for positive vortex and a 0 for negative vortex. The Reynolds number is 200, the dimensionless time is 12, and the computed drag is 1.22, in excellent agreement with experiment.

REFERENCES

[1] Batchelor, G. K., An Introduction to Fluid Mechanics, Cambridge University Press (1967).

[2] Chorin, A. J., Math. Comp., 23, 341 (1969).

[3] Chorin, A. J., Proc. Second International Conference on Numerical Methods in Fluid Mechanics (1970).

[4] Keller, H. B., and Takami, H., in Numerical Solutions of Nonlinear Differential Equations, D. Greenspan (Ed.), Wiley, New York (1966).

[5] Kellogg, O. D., Foundations of Potential Theory, Ungar, New York (1929).

[6] Smith, A. M. O., Proc. Seventh Symposium on Naval Hydrodynamics, ONR, (undated).

Vortices in the Wake of a Circular Cylinder

```
       0 **
      00
    0       **
             *
   00         **
   00*       ***
  00**      *0*
  0 0 *0   0***
 *0 0 * *0* 0 * *
 0 00  0*0* *0*
  0*00   00** *
   0 0**0*   *0
 0 00  0  * 0  *
 0 0  0  *  0**
   0 00      *0
  0 0 00 *    **
   0   0  * **
  0 *0* ** 00*  **
  0 * *    ** 0
   0000 0  **** *
 000 00  *   *
 0   00     * **
  0 0** *  * *
   0000   **
      00 **
*0      * * **
     * 0***
      * *
  0
  0
   *

0 0    *
 *00*
 0 * 00
  00  0
    0


         0

      0   *
          0
```

NUMERICAL STUDIES OF THE HEAT CONDUCTION EQUATION WITH HIGHLY ANISOTROPIC TENSOR CONDUCTIVITY

C.K. Chu, Univ. of Oxford and UKAEA, Culham Laboratory*
K.W. Morton, Univ. of Reading and UKAEA, Culham Laboratory
K.V. Roberts, UKAEA, Culham Laboratory

INTRODUCTION

In practical magnetohydrodynamic calculations [see, e.g. Roberts & Potter, 1970] the transport coefficients long and across the magnetic field differ by several orders of magnitude. If the finite difference grid does t coincide with the direction of the field and its normal, large errors across the field will result from e numerical calculations. Indeed, when the field lines are sufficiently simple, they should be chosen as ordinate directions. But as they are generally complicated, this leads to difficulties and expense. It would e often worthwhile therefore to use fixed coordinates for the difference scheme and devise methods with nproved directional accuracy to treat the anisotropy.

We investigate in this paper several methods for the simple model of a two-dimensional heat equation with ighly anisotropic conductivity $k_\perp/k_\parallel = \epsilon \ll 1$. For application to magnetohydrodynamics we cannot afford to onsider time steps limited by the parallel diffusion: hence we shall only study implicit schemes.

There are two different time scales in this type of problem: that corresponding to diffusion along the ield and that to diffusion across the field. We shall refer to them as the short time scale and the long ime-scale. Our schemes are studies and compared for use in each of these two time scales, and their character-stic features are analysed in detail.

MODEL PROBLEM AND METHODS OF SOLUTION

We consider as the model problem an initial value problem for the anisotropid heat conduction equation

$$u_t = u_{\xi\xi} + \epsilon u_{\eta\eta} \qquad 0 \leq \epsilon < 1 \tag{1}$$

th initial data $u(\xi,\eta,0) = u_o(\xi,\eta)$. The problem is to be solved by finite difference methods with a quare grid of width h in the (x,y) directions, the principal axes (ξ,η) being at an angle φ to the (x,y) es.

In (x,y) coordinates, equation (1) becomes

$$u_t = au_{xx} + 2b\, u_{xy} + cu_{yy} \equiv Lu \tag{2}$$

ere $a = \cos^2\varphi + \epsilon \sin^2\varphi$, $b = (1-\epsilon)\sin\varphi\cos\varphi$, and $c = \sin^2\varphi + \cos^2\varphi$ .. The u_{xx} and u_{yy} terms are ssentially uncoupled and cause no difficulty; the difficulty which is introduced by numerical diffusion in e η direction enters through the u_{xy} term.

We first discuss spatial differencing. Three different schemes are considered in this paper.

Cross derivative scheme

$$u_{xy} \approx [u(x+h,y+h)+u(x-h,y-h)-u(x+h,y-h)-u(x-h,y+h)]/4h^2 \equiv \Delta_{ox}\Delta_{oy}u/h^2$$
$$u_{xx} \approx [u(x+h,y)+u(x-h,y)-2u(x,y)]/h^2 \equiv \delta_x^2 u/h^2 \tag{3}$$

d similarly for u_{yy}. This is the most obvious spatial differencing.

. Diagonal scheme. Assume $b > 0$. Let α be the +45° direction from the x-axis, u_α , $u_{\alpha\alpha}$ the directional rst and second derivatives in the α direction. Since

$$u_{xy} = u_{\alpha\alpha} - \tfrac{1}{2}u_{xx} - \tfrac{1}{2}u_{yy}$$

find

$$Lu = (a-b)\, u_{xx} + (c-b)\, u_{yy} + 2b\, u_{\alpha\alpha}$$
$$\approx [(a-b)\,\delta_x^2 u + (c-b)\,\delta_y^2 u + b\delta_\alpha^2 u]/h^2, \tag{4}$$

ere $\delta_\alpha^2 u$ denotes u (x+h,y+h)+u(x-h,y-h) - 2u(x,y), and $u_{\alpha\alpha} \approx \delta_\alpha^2 u/(\sqrt{2}h)^2$. If $b < 0$, then the -45° rection is chosen as the α direction.

This differencing was studied by Pucci (1958), Bramble and Hubbard (1964), etc. In all these studies, b>0 and c-b>0 are assumed in addition to a,b,c>0 and $b^2 - ac < 0$. With small ϵ, while the latter equalities still obtain, a-b>0 and c-b>0 do not hold simultaneously. Hence we lose the positivity of the lution and further analysis is required.

One could interpret this scheme as an attempt to approximate the diffusion in the ξ direction by that the two nearest grid directions, and consistency enforces some additional diffusion (which may be gative) in a third direction. This suggests the next approximation.

---

*On sabbatical leave from Columbia University; Guggenheim Fellow 1971-72. He also thanks L.C. Woods at xford and his colleagues at Culham for their hospitality.

III. Generalised diagonal scheme. For φ small, $0\leqslant\varphi\leqslant\beta \equiv \arctan 1/2$, we can generalise II to use the grid points $(x+2h,y+h)$ and $(x-2h,y-h)$. Then

$$Lu = (a-2b)\, u_{xx} + (c-\tfrac{1}{2}b)\, u_{yy} + (\tfrac{5}{2}b)\, u_{\beta\beta} \approx [(a-2b)\,\delta_x^2 u + (c-\tfrac{1}{2}b)\,\delta_y^2 u + \tfrac{1}{2}b\delta_\beta{}^2 u]/h^2. \tag{5}$$

For $\beta\leqslant\varphi\leqslant\pi/4$, we approximate Lu by a combination of $u_{\beta\beta}$, $u_{\alpha\alpha}$, and $u_{\alpha'\alpha'}$, α' being the perpendicular direction to α, and so forth.

For time differencing, we employ fractional step fully implicit schemes of first order accuracy in Δt. With the presence of the cross derivative terms, schemes with higher order accuracy in Δt lead to more complicated implicit equations than the simple and practical tri-diagonal systems that we use. In addition, there are two more fundamental reasons for using fully implicit schemes: (a) The correct decay of a mode of wave number k in one time step is

$e^{-k^2\Delta t} = e^{-(kh)^2\tau}$ where $\tau = \Delta t/h^2$. ($\tau\sim 1$ corresponds to the short time scale, while $\epsilon\tau\sim 1$ corresponds to the long time scale). At very large τ, for long time scale calculations, a partly implicit scheme ($\tfrac{1}{2}\leqslant\theta<1$) gives a decay per time step $\sim -(1-\theta)/\theta$, while the fully implicit scheme ($\theta = 1$) gives the much better decay $\sim 1/(kh)^2\tau$. (b) Even in a one dimensional problem, only the fully implicit scheme maintains the positivity of the solution at large τ.

For scheme I, we split as follows [Yanenko, 1971]:

$$\begin{aligned} u^{n+\frac{1}{2}} - u^n &= a\tau\delta_x^2\, u^{n+\frac{1}{2}} + b\tau\Delta_{ox}\Delta_{oy}u^n \\ u^{n+1} - u^{n+\frac{1}{2}} &= c\tau\,\delta_y^2 u^{n+1} + b\tau\Delta_{ox}\Delta_{oy}u^{n+\frac{1}{2}}. \end{aligned} \tag{6}$$

For scheme II, when $a-b>0$ and $c-b<0$,

$$\begin{aligned} u^{n+\frac{1}{2}} - u^n &= (a-b)\tau\,\delta_x^2\, u^{n+\frac{1}{2}} \\ u^{n+1} - u^{n+\frac{1}{2}} &= b\tau\delta_\alpha^2\, u^{n+1} - (b-c)\tau\delta_y^2\, u^{n+\frac{1}{2}}. \end{aligned} \tag{7}$$

If $a-b<0$ and $c-b>0$, we do $\delta_y^2 u$ implicitly in the first half step, and $\delta_x^2 u$ explicitly in the second half step. If both $a-b>0$ and $c-b\geqslant 0$, there will be a mild stability requirement on this scheme, discussed in the next section. Exactly the same kind of splitting is used for scheme III. We observe that the cross term in I is treated explicitly, half in each fractional step, while the negative diffusion term in II and III is treated explicitly in only one of the fractional steps. This is required for stability at all τ, as we shall see in the next section.

STABILITY, ACCURACY, AND NUMERICAL DIFFUSION

Fourier analysis of scheme I leads to the amplification factor

$$r_I = (1-4b\tau s_x c_x s_y c_y)^2/\{(1+4a\tau s_x{}^2)(1+4c\tau s_y{}^2)\} \tag{8}$$

where $s_x = \sin\tfrac{1}{2}k_x h$, $c_x = \cos\tfrac{1}{2}k_x h$, etc. and k_x and k_y are the usual wave numbers in the x and y directions. Unconditional stability follows immediately [Yanenko, 1971] from the ellipticity of L, since then $b^2 s_x{}^2 c_x{}^2 s_y{}^2 c_y{}^2 \leqslant ac s_x{}^2 s_y{}^2$.

For scheme II, we can derive from $b^2 < ac$ the important inequality

$$(a-b)\, s_x{}^2 + (c-b)\, s_y{}^2 + b\, s_\alpha{}^2 \geqslant 0 \tag{9}$$

where $s_\alpha = \sin\tfrac{1}{2}(k_x+k_y)h$. The amplification factor for the splitting scheme (7), which we use when $b>0$, is

$$r_{II} = [1-4(c-b)\tau s_y^2]/\{[1+4(a-b)\tau s_x{}^2][1+4b\tau s_\alpha{}^2]\}. \tag{10}$$

When $b\geqslant c$ (true for all angles φ when $\epsilon = 0$), the inequality (9) therefore ensures unconditional stability. However, when $\epsilon\neq 0$, $b<c$ at angles near $\varphi = 0$ or 45°, and it would be best to treat $\delta_y{}^2 u$ implicitly. As changing schemes in this way is inconvenient in practice, we accept from (10) a mild stability condition $2\tau\epsilon<1$, which is the usual condition corresponding to explicit treatment of the $\epsilon\, u_{\eta\eta}$ diffusion.

For scheme III the amplification factor becomes

$$r_{III} = [1-2(c-b)\tau s_y^2]/\{[1+4(a-2b)\tau s_x{}^2][1+2b\tau s_\beta{}^2]\} \tag{11}$$

with $s_\beta = \sin\tfrac{1}{2}(2k_x+k_y)h$. The analysis is similar to that for scheme II.

The relative accuracy of the schemes may be studied by transforming these expressions to wave numbers k_ξ, k_η in the (ξ,η) directions and expanding in their powers. An important property is the behaviour for $k_\xi = 0$ and $\epsilon = 0$, where each r should equal unity. The lowest order terms for small τ are

$$\begin{aligned} r_I &\sim 1 - \frac{\tau}{16}\sin^2 2\varphi\, k_\eta^4 \\ r_{II} &\sim 1 - \frac{\tau}{16}\sin^2 2\varphi(\cos\varphi-\sin\varphi)^2\, k_\eta^4 \\ r_{III} &\sim 1 - \frac{\tau}{16}\sin^2 2\varphi(\cos\varphi-\sin\varphi)^2(1-2\tan\varphi)k_\eta^4 \end{aligned} \tag{12}$$

Each scheme thus exhibits fourth order numerical diffusion in the η direction. The respective maxima are in the ratios of 1 : 0.15 : 0.06 and occur at the angles 45°, 20.9°, 14.0° respectively.

At large τ, differences in the schemes show up at the level of k_η^2. To this order, the amplification factor for scheme I is identically one for all τ and φ, while that for scheme II has the form $(1 - \gamma k_\eta^2)/[1-\gamma k_\eta^2 \tan\varphi][1-\gamma k_\eta^2(1-\tan\varphi)]$. Clearly, one could reduce this to unity by splitting the explicit y direction diffusion between the two steps of (7) with factors $\tan\varphi$ and $1 - \tan\varphi$; however, this scheme then has a severe stability limit.

Thus, our analysis shows that schemes II and III are decidedly more accurate than I for small τ, while scheme I has this very attractive feature at very large τ. Further properties of these schemes are disclosed by actual computation in the next section.

RESULTS OF COMPUTATION AND DISCUSSION

Short Time Scale The various schemes were tested on an initial Gaussian

$$u(\xi,\eta,0)\ e^{-(\xi^2+\eta^2)/4} \tag{13}$$

in a domain of 50 x 50 grid points ($\Delta x = \Delta y \equiv h = 1$). The boundary condition $u = 0$ was imposed at all edges, but the duration of each run was short enough for the effect of the boundary not to be significant. Various values of τ from $\frac{1}{2}$ to 2 were used, and ϵ from 0 to 0.01. A coarser grid with h = 2 was also tested.

A detailed comparison of the three schemes for $\epsilon=0$, $\tau=2$ at 20 iterations (t=40) is given in fig. 1. By that time, the centre value u(0,0) has dropped to almost 15% of the initial value. Calculations were made for φ = 0°, 15°, 30°, 45° for schemes I and II, and φ = 0° and 15° and 26.6° (=β) for scheme III. The cross section profiles $u(\eta)$, i.e. across the diffusion direction, are given at ξ= 0, 5, 10. Ideally, these profiles should be independent of φ.

The profiles at φ = 0° are the same in all cases, since in each case we then have a one-dimensional calculation in a grid direction. In addition, they agree with the exact solution for equation (1) with $\epsilon = 0$ and initial condition (13),

$$u(\xi,\eta) = (1 + t)^{-\frac{1}{2}} e^{-[\eta^2+\xi^2/(1+t)]/4},$$

to within 2% at the parameters and time chosen. In the diagonal scheme II, φ = 45° corresponds again to a calculation along a grid direction, so that it agrees with φ= 0° extremely closely. The same holds true for the angle φ = 26.6° for the generalised diagonal scheme III.

As is evident from the curves of fig. 1, the cross derivative scheme I is the least accurate, and the generalised diagonal scheme III is the most accurate, with respect to changes in the angle φ. This agrees with the analysis of the previous section on the relative magnitude of the numerical diffusion in the η direction.

In the small τ regime, smaller τ and greater ϵ both improve the accuracy of all these schemes. A coarser grid not only results in less accuracy, but also accentuates the difference in accuracy of the three schemes.

Long Time Scale The various schemes were also tested for large τ (τ = 40). In addition to the symmetric Gaussian initial data (13), we used an elongated Gaussian

$$u(\xi,\eta,0) = e^{-(\eta^2+\xi^2/25)/4} \tag{14}$$

This is a closer approximation to the situation where an equilibrium almost exists in the ξ direction, and we are only interested in the behaviour in the η direction.

The results for the symmetric Gaussian are shown in fig. 2. The most noticeable and surprising fact is that scheme I now shows inaccuracy associated with under-damping at off-angles $\varphi \neq 0$, in contrast to the increased diffusion at off-angles for small τ. In addition, large negative values of u appear. Schemes II and III still show essentially increased diffusion at off-angles, although at the first few iterations, there also appeared some underdamping at off-angles.

This off-angle underdamping of I can be seen from expanding the amplification factor (8) to second order in k. For $\epsilon = 0$ and large τk^2, all modes with $k_x \neq 0$ and $k_y \neq 0$ damp like $(1 + \tau k^2)^{-1}$ for $\varphi = 0$, but like $1 - O(1/\tau k^2)$ for $\varphi \neq 0$. Hence, at off angles, all these modes are virtually undamped. On the other hand, for the diagonal scheme, (9) shows that all modes damp like $1/\tau k^2$, except those with $k_x = 0$, which damp like $1 - \tan\varphi$, and those with $k_x = -k_y$, which damp like $\tan\varphi$. (For scheme III, these modes are $k_x = 0$ and $k_x = -\frac{1}{2}k_y$). In addition, this gives the diagonal schemes the desirable side effect of very small negative values of u.

The results for the elongated Gaussian (fig.3) show some improvement of Scheme I over the symmetric Gaussian results (fig. 2). This is because the small k_ξ modes are now more important: for $\epsilon = 0$ the $k_\xi = 0$ mode should be undamped and the analysis described above gives $r_I = 1 - O(k_\eta^4)$ for this mode. Unfortunately, all other modes with $k_x \neq 0$ and $k_y \neq 0$ are also underdamped. By contrast, schemes II and III damp this $k_\xi = 0$ mode, but still keep underdamped the two exceptional modes referred to above.

The effect of $\epsilon > 0$ is seen strongly at $\varphi = 0$ for all schemes (and at $\varphi = 45^o$ and 26.6^o for II and III respectively in addition), but the underdamping or overdamping discussed above mask the $\epsilon u_{\eta\eta}$ diffusion at off angles. As a result, scheme I becomes even more sensitive to the angle φ, while schemes II and III become less sensitive.

CONCLUSIONS

For short time scale calculations our studies have shown the diagonal schemes to be much more accurate than the cross derivative scheme. If the angle φ between the (x,y) axes and the (ξ, η) axes varies little over the region, they are also very practical. This covers many cases of interest: but if φ varies greatly, probably only scheme II will be useful.

Cases where the long time scale calculations are of interest are usually where τk_ξ^2 and $\epsilon\tau k_\eta^2$ are small while τ and τk_η^2 are large. We have shown that the diagonal schemes also have advantages here over the cross derivative scheme, but more development is necessary before a successful scheme could be claimed.

REFERENCES

Bramble, J.H. and Hubbard, B.E., Contributions to Diff. Eq. 3, 339 (1964).

Pucci, C. Some Topics in Parabolic and Elliptic Equations. U. of Md. Inst. Fluid Dynamics and Appl. Math. Lect. Series, 36, (1958).

Roberts, K.V. and Potter, D.E., in Methods of Computational Physics, Vol. 9 (ed. Alder et al), p. 339, Academic Press (1970).

Yanenko, N.N. Methods of Fractional Steps. Springer Verlag 1971.

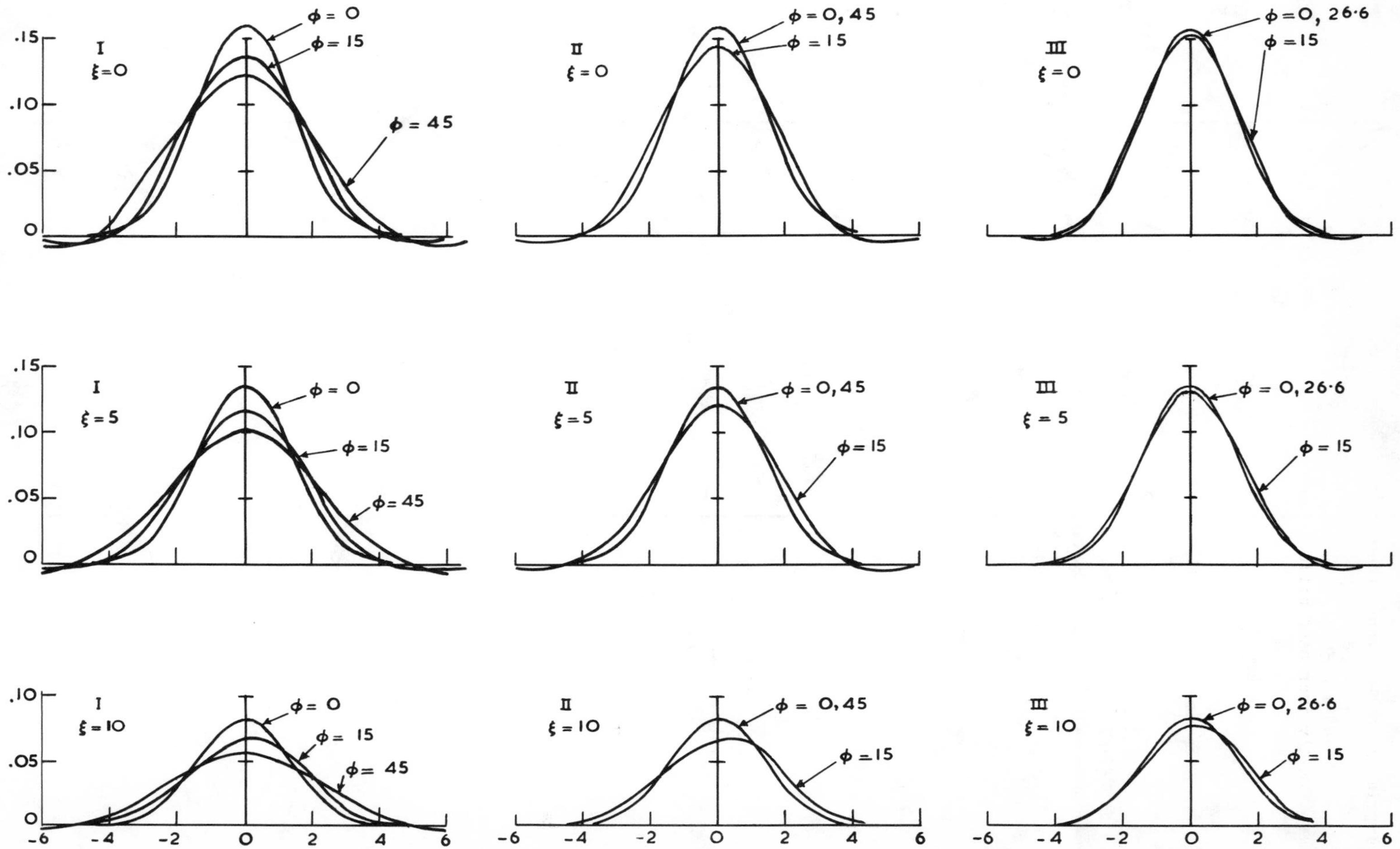

FIGURE 1. Solutions plotted against η for schemes I II III and for $\xi = 0, 5, 10$ with $\epsilon = 0, \tau = 2, t = 40$

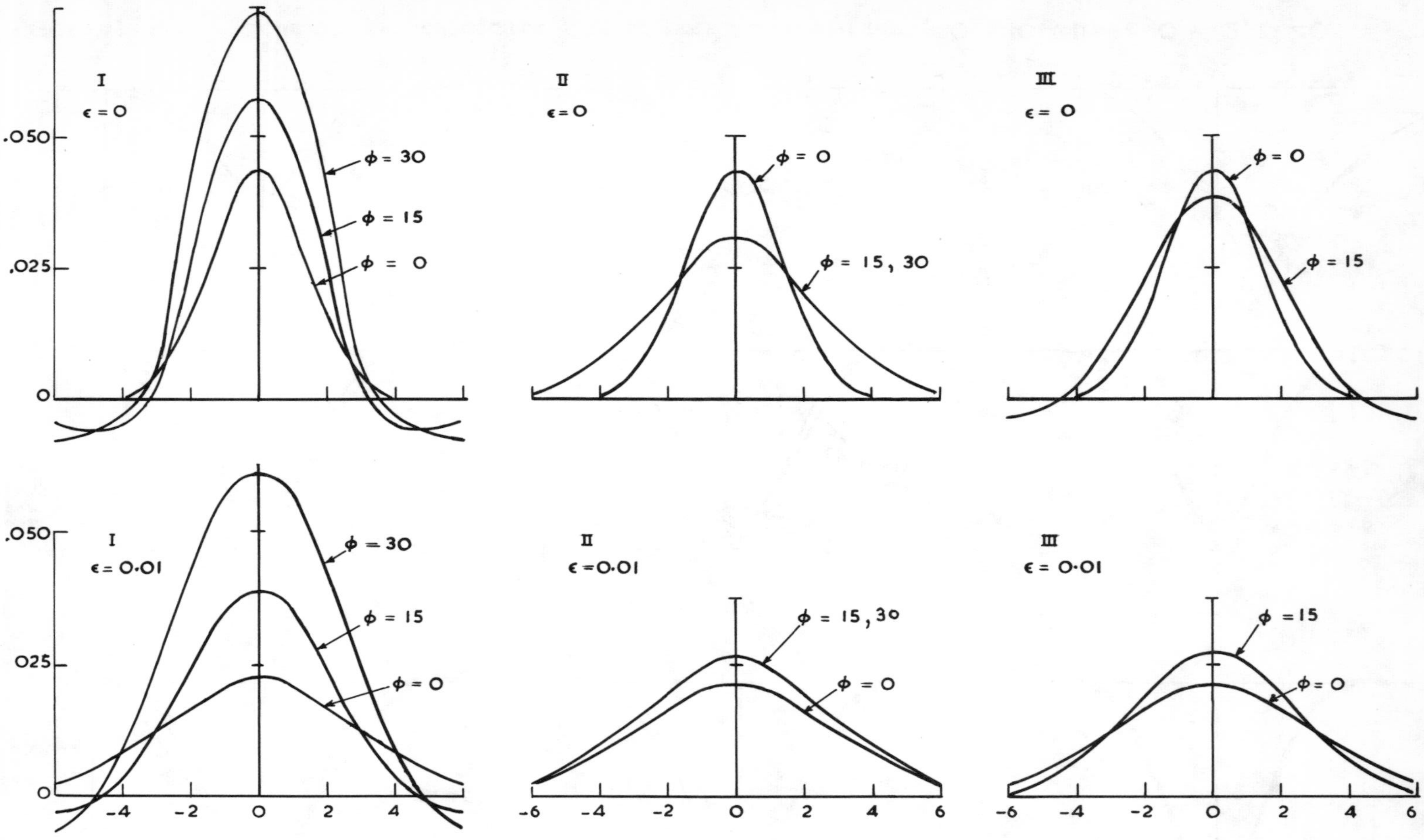

FIGURE 2. Result for symmetric Gaussian and $\epsilon = 0, 0{\cdot}01$ at $\xi = 0$, $t = 320$ with $\tau = 40$

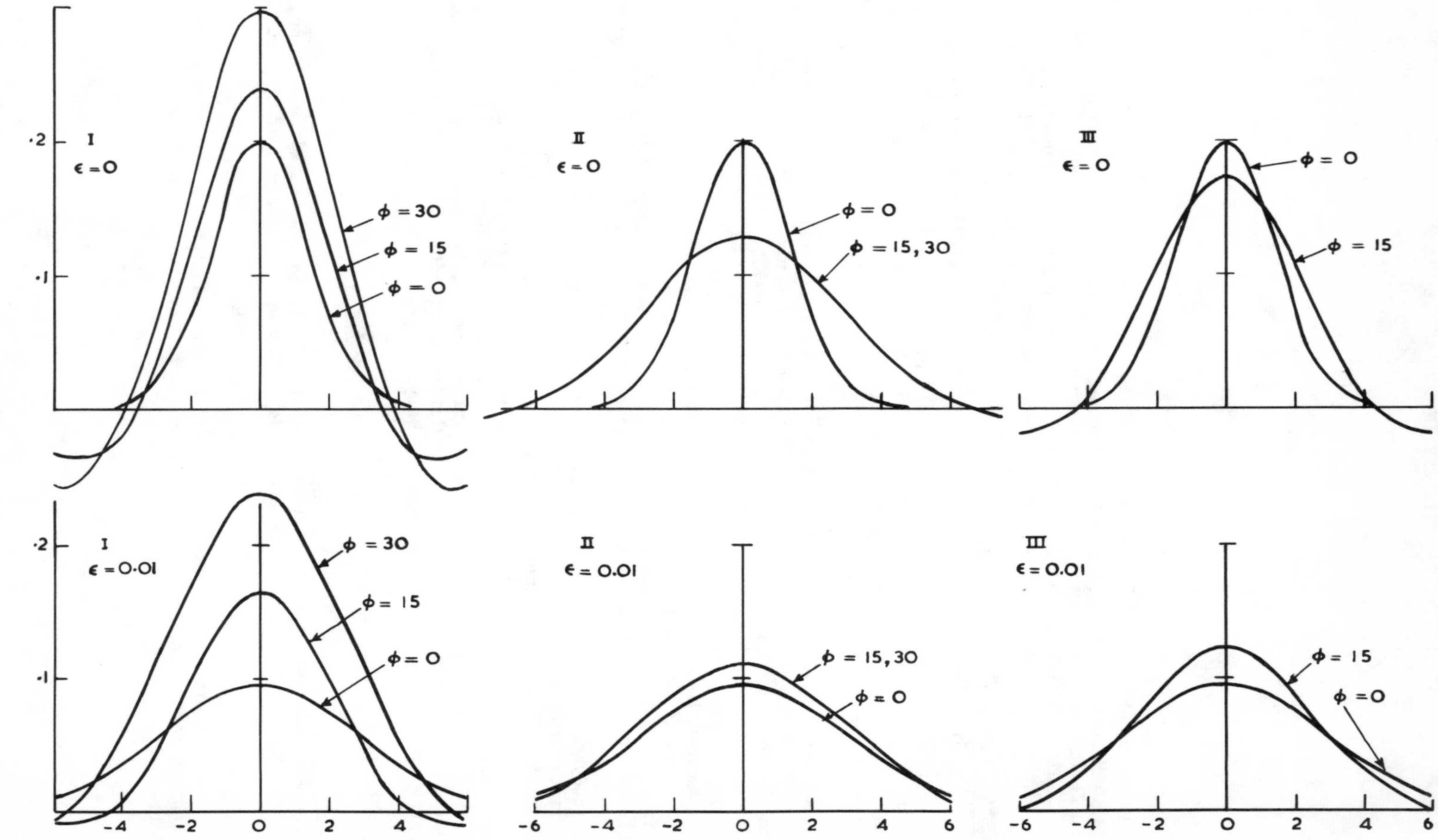

FIGURE 3. Results for elongated Gaussian and $\epsilon = 0, 0{\cdot}1$ at $\xi = 0$, $t = 320$ with $\tau = 40$

TURBULENCE TRANSITIONS IN CONVECTIVE FLOW*

Bart J. Daly
Los Alamos Scientific Laboratory
University of California
Los Alamos, New Mexico

INTRODUCTION

In 1969 Harlow and Hirt proposed a transport equation method for representing turbulent fluid flow. Beginning with the fluctuating Navier-Stokes equations (in which the fluid velocity is composed of mean and fluctuating contributions), they derived equations to describe the rate of change with time of the turbulence Reynolds stress, $R_{ij} = \overline{u_i'u_j'}$, and the rate of decay of the turbulence energy $D = \frac{1}{2}\overline{(\partial u_i'/\partial x_k)^2}$. (The overbar indicates an ensemble average.) Closure for this system of equations was achieved through the use of various flux approximations to evaluate the higher order moments that appear in the derivation. If these equations are to have wide applicability, then the coefficients that are introduced by the flux approximations must be universal.

This turbulence representation has been used to study numerically several turbulent flow problems for which experimental comparisons could be made, in order to determine whether a universal set of coefficients does exist. In the first two studies of this kind, that of turbulence distortion in a nonuniform tunnel, reported by Harlow and Romero (1969), and of turbulent channel flow by Daly and Harlow (1970), good comparison with experiment was obtained using the same set of coefficients. In each of these papers a discussion is given of the results obtained as a result of changes in the value of the various coefficients. It was discovered that the calculated results were not extremely sensitive to small changes in parameter values.

The study reported in this paper affords a further test of the turbulence transport equations by applying them to a problem in which the driving force is buoyancy rather than mean flow shear, thereby allowing us to test the appropriateness of the body force terms that appear in the R_{ij} and D equations. This problem also marks the first application in which the turbulence equations have been coupled with the Navier-Stokes and heat equations to obtain time dependent, two dimensional solutions. The details of these equations and the calculational procedure used to solve them will be presented in a forthcoming paper by Daly.

THE TURBULENT BENARD PROBLEM

When a fluid is confined between two horizontal plates and heated from below, various regimes of fluid flow are observed, depending upon the values of the fluid

*This work was performed under the auspices of the United States Atomic Energy Commission and was partially supported by the Office of Naval Research, Government Order NAonr-13-72.

Rayleigh number, $R = \beta g h^3 \Delta T/\lambda\nu$, and Prandtl number, $Pr = \nu/\lambda$. Here β = volumetric coefficient of expansion, g = gravitational acceleration magnitude, h = plate separation distance, ΔT = temperature difference between the plates, λ = heat conduction coefficient, and ν = kinematic viscosity coefficient. For $R < 1708$ the fluid is static and heat transfer between the plates occurs only through conduction. For $R \gtrsim 1708$ flow circulations develop in the form of two dimensional rolls with axes parallel to the plates, resulting in increased heat transfer through convection. The nature of the flow for $R > 1708$ depends upon the Prandtl number of the fluid, but for some sufficiently high Rayleigh number the flow becomes turbulent. We examine this initial transition to turbulence, as well as transitions to other turbulent regimes at higher Rayleigh numbers, for the case of air.

In experiments with air, Willis, Deardorff and Somerville (1971) observe a rapid increase in the horizontal wavelength of the two dimensional rolls from 2.016h at $R = 1708$ to 3.7h at $R \approx 8000$, followed by a more gradual growth rate at larger R. Willis and Deardorff (1970) report that the rolls develop sinuous oscillations along their axes at $R \approx 5800$, and that the amplitude of these oscillations increases rapidly until $R \approx 8000$, after which little further increase is observed. In view of the ordered structure of these oscillations, one should not view them as evidence of the onset of turbulence. However, these authors indicate the existence of short-lived wave crests that develop and travel along a roll edge by $R \approx 12{,}000$. At $R \approx 30{,}000$ these processes have produced a sufficiently disordered structure that the flow may be classified as turbulent.

In the numerical study we do not attempt to resolve the details of this transition to turbulence in the mean flow calculation. To do this would require a three dimensional calculation mesh of large extent and fine resolution. Instead we represent the gross motion of the fluid, the rolls, by <u>two dimensional</u> solutions of the Navier-Stokes equations in a fixed reference frame. This ignores the experimentally observed sinuous oscillations of the rolls, their lateral drift with time, and presence of rolls that extend only part of the way across the experimental region. The fine scale structure of the flow is represented in an average sense by the turbulence transport equations, which are coupled with the mean flow equations.

The first phase of this study neglects the observed increase of roll wavelength with R. In this phase, calculations were performed at Rayleigh numbers,

$$R = 5000 \times 2^n, \quad n = 0, 1, \ldots, 7$$

in which the roll wavelength was fixed at 2h by the presence of reflective boundaries. These calculations predict a steady state heat flux that is large compared to experiment, the disparity increasing with R. Other investigators have observed this disagreement in two dimensional numerical studies in the laminar regime of this problem. However, in two of these studies [Willis, Deardorff and Somerville (1971), Lipps and Somerville (1971)] it is shown that better agreement can be obtained between calculation and experiment by forcing the calculation wavelength to correspond to the experimental wavelength.

This procedure is followed in the second phase of this study. A variable width calculation mesh is used, such that at each Rayleigh number the circulation wavelength is forced to correspond to the estimated experimental wavelength. A plot of these estimated experimental wavelengths of the large scale motions in air for $5000 < R < 640{,}000$ is shown in Fig. 1. For $R < 3 \times 10^4$ the wavelength estimates in Fig. 1 are probably reliable since they were obtained from direct measurement in three dimensional flow by Willis, Deardorff and Somerville (1971). However, for $R > 3 \times 10^4$ the estimated wevelengths may be in error, since very little experimental information is available at these higher Rayleigh numbers. Some guidance can be obtained from a measurement by Deardorff and Willis (1965) at $R = 1.6 \times 10^5$, but it is quite possible that their value is influenced by two dimensional constraints on the experimental apparatus. Some additional guidance is given by the spectral analyses of fluctuating quantities presented by Deardorff and Willis (1967).

Thus, in the second phase, the calculations are performed relative to a two dimensional, rectangular mesh of uniform height but of width that increases with R. The horizontal boundaries are rigid, with the lower boundary maintained at a temperature higher than that of the upper boundary. The vertical boundaries are reflective, indicating the presence of adjoining flows of opposite circulation.

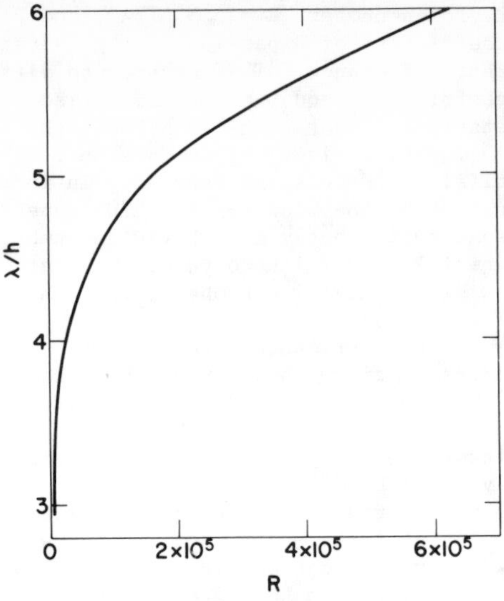

Fig. 1. Wavelength of the large scale motions in air, estimated from experimental data

The fluid is assumed to be incompressible with the effect of temperature variations introduced through the Boussinesq approximation. Turbulence contributes to the diffusion of momentum and temperature, and the variation of these quantities provides the driving force for the creation of turbulence. Solutions are obtained by integrating this coupled system of equations through time to steady state. The turbulence equations and the parameters that appear therein are unchanged from those given by Daly and Harlow (1970), except for the addition of one component interchange term in the R_{ij} equation.

This addition appears as a contribution to Φ_{ij} in Eq. (19) of that reference, of the form,

$$- \Omega \left(R_{ik} \frac{\partial u_k}{\partial x_j} + R_{jk} \frac{\partial u_k}{\partial x_i} - R_{ik} \frac{\partial u_j}{\partial x_k} - R_{jk} \frac{\partial u_i}{\partial x_k} \right) ,$$

where we have taken $\Omega = 0.2$ (constant). This term, which has zero contraction, serves to distribute energy from components that receive contributions from mean flow shear to those that do not. It is shown by Nakayama (1972) that an energy redistribution of this type is crucial to the attainment of realistic results in the calculation of a turbulent plane free jet. This new term would not alter the results obtained by Harlow and Romero (1969), but could have an effect in the channel flow study [reported by Daly and Harlow (1970)] and should be examined in that context.

Two parameter values, dealing with the effect of buoyancy forces in the turbulence equations, were not specified by Daly and Harlow (1970). The results of the present study indicate that these parameters should be evaluated as follows:

$$\tau = 0.8 \text{ (constant)},$$

$$f(\xi) = \begin{cases} 1.7 & \xi \geqslant 5 \\ 1.0 & \xi < 5 \end{cases} ,$$

where ξ is the turbulence Reynolds number.

THE TURBULENCE TRANSITIONS

In experimental studies [Malkus (1954), Willis and Deardorff (1967), Krishnamurti (1970)] of Benard convection it is possible to obtain detailed plots of the variation with R of the heat flux between the plates. These plots show several distinct discontinuities in slope between linear segments of the heat flux curve, plotted as a product of R and Nu, where Nu, the Nusselt number, is a dimensionless measure of the horizontally averaged heat flux between the plates. While there is considerable discrepancy among experimenters (and even between different experiments in the same study) regarding the details of these slope discontinuities, there is little doubt that they exist and exhibit a certain basic consistency for a wide variety of fluids. However, the physical processes that produce these transitions are not well understood, especially in the turbulent regime.

Because of the large number of calculations involved, it is not practical in numerical studies to obtain the detailed heat flux information needed to delineate the linear segments of the NuR versus R curve. Numerical studies can contribute most to the understanding of these transitions by examining the problem at a sequence of Rayleigh numbers for evidence of changes in the structure of the anisotropic turbulent flow. This is the procedure followed in the present study. In order that these results provide a useful complement to the experimental findings, it is important that results of the two investigative techniques be compared wherever possible.

Figure 2 shows a comparison of experimental and calculated Nusselt numbers as a function of R. The two numerical curves show results obtained using fixed and variable width calculation meshes. The use of the variable width mesh produces better agreement with experiment than was obtained with the fixed width mesh, but a discrepancy remains. This discrepancy may be attributable to the inaccuracy of our estimate of the wavelength of the large scale motions in Fig. 1, or it may be indicative of the neglected three dimensional effects. The experimental data, given by Mull and Reiher (1930), is very limited, but this is the only reliable data available for air at these values of R. Much more extensive information is available for water [Rossby (1966), Silveston (1958)], and these curves tend to confirm Mull and Reiher's data, lying about midway between his points and the lower numerical curve in Fig. 2. Unfortunately, calculations appropriate to water would have required about ten times as much computer time as these calculations, in order to satisfy numerical stability conditions.

The arrows along the abcissa in Fig. 2 mark the mean positions of the experimentally observed transitions. Of the five transitions, there is least agreement among experimenters regarding the one shown at 1.8×10^4. However, this value corresponds approximately to the R for which transition to turbulence is reported in air by Willis and Deardorff (1970) and in water by Malkus (1954). This point is also in close agreement with the R for which transition to turbulence occurs in this numerical study. Thus, in the calculation at $R = 10^4$ the initial seed turbulence decayed away and the flow reverted to the laminar steady state, but for $R \geqslant 2 \times 10^4$ the turbulence survived and contributed to an increased heat flux. The fact that this transition to turbulence could be calculated in reasonable agreement with experiment marks an important accomplishment of this study.

To investigate any further comparisons between the numerical results and the higher Rayleigh number transitions, one needs to consider the calculated flow structure in more detail. Figures 3-5 show velocity vector and contour plots of steady state results obtained from calculations at Rayleigh numbers of 4×10^4, 1.6×10^5 and 6.4×10^5, respectively. The velocity plots are formed by drawing vectors from the center of each calculation cell, of magnitude and direction appropriate to the local mean-flow cell velocity. The magnitudes of the vectors are normalized through division by $\delta x/2\delta t$, where δx and δt are the space and time increments. In the contour plots, the contour interval is always a power of two (the power is denoted by CI), thus facilitating comparisons between plots. The minimum contour line is marked by asterisks. Notice the symmetry that exists about the center of the calculation mesh in these plots.

The three sets of results shown in Figs. 3-5 are representative of three regimes of turbulent flow observed in this study. The results at $R = 4 \times 10^4$ are typical of a low intensity turbulent flow that was also observed at $R = 2 \times 10^4$ and 8×10^4. The characteristics of this turbulence regime are:

1. A temperature field that differs little from a laminar temperature distribution, indicating that the effect of turbulent diffusion is small.

2. A turbulence integral scale, or scale of the energy carrying turbulent eddies, that is directly proportional to the microscale, which is the scale of the smallest eddy to survive the dissipation of turbulence energy to heat. Thus the integral scale is small (compare the contour interval with that of Figs. 4 and 5) and decreases rapidly toward the rigid walls. The maximum turbulence scale is attained midway between the walls.

3. A turbulence energy whose creation is primarily attributable to buoyancy rather than shear effects. Thus the zz component of R_{ij}, which is created mostly through buoyancy, has a much greater magnitude than the xx component, which is produced by mean flow shear. The total turbulence energy, $Q = (R_{xx} + R_{yy} + R_{zz})/2$, reflects the dominance of R_{zz} by exhibiting a very similar appearance. The maximum intensity in both of these plots occurs at the upper left and lower right corners where the temperature gradient, and therefore the buoyancy force, is largest. R_{xx} reaches its maximum near the top and bottom walls where the shear is greatest.

Figure 4 is typical of a flow in which the turbulence is beginning to enter the high intensity regime. In this calculation low intensity turbulence persists in those regions of the flow where neither the turbulence scale nor the turbulence energy are large. The nature of this flow is such that:

1. The temperature field shows the effects of turbulent diffusion, especially in the central region away from the plates. Large temperature gradients exist very near the plates where viscous diffusion prevents a buildup of turbulence energy.

2. There is a sharp dichotomy of turbulence scales in the problem. A very large eddy with little turbulence energy exists in the central part of the mesh, while the energy containing eddies near the walls have wavelengths that are on the order of the microscale.

3. The contribution of shear creation to turbulence energy is approaching that of buoyancy creation. Thus Q shows almost as much contribution from R_{xx} as it does from R_{zz}. The peak of R_{xx} has been convected slightly downstream from the region of greatest mean flow shear, but the R_{zz} peak remains in the area of the greatest temperature gradients.

Figure 5 shows a flow in which the turbulence is high intensity throughout. This flow regime is characterized by a strong turbulent diffusion of the mean flow and turbulent quantities such that:

1. The turbulent diffusion of the temperature field in the midplate region is more pronounced than in Fig. 4 and strong temperature gradients exist only near the walls.

2. The concentration of large scale turbulence in the center of the mesh and fine scale structure elsewhere is not as evident as it was in Fig. 4. In that problem, 90% of the turbulence energy was concentrated in scales of size less than 0.2h, whereas, in Fig. 5, 90% of the energy is in scales of size 0.45h or less.

3. The turbulence energy is diffused more uniformly throughout the mesh, with sharp gradients existing only near the walls. The increased importance of shear relative to buoyancy as the driving force for turbulence is evident in that the contour interval for R_{xx} is now greater than that for R_{zz}.

To visualize the effect of turbulence on the diffusion of temperature between the plates it is useful to think of this as being accomplished by the vertical motion of eddies through regions of strong temperature gradient. The heat flux due to turbulence will therefore depend, in our model, on the distribution in space as well as the magnitudes of s, R_{zz} and $\partial T/\partial z$. These are the important points to consider when analyzing the effect of changes in flow structure observed in Figs. 3-5 for correlation with the experimental transitions in the NuR versus R curve.

The outstanding difference between the turbulent structure at $R = 4 \times 10^4$ and that at $R = 1.6 \times 10^5$ is the remarkable change in distribution of the turbulence scale. In particular, the uniformly small scale near the wall at $R = 1.6 \times 10^5$ results in a strong coupling among eddies in this region but little contact with the flow midway between the plates. Thus there is little diffusion of the turbulence energy but much interchange among its components near the wall, so that the peak R_{zz} is simultaneously sharper and more elongated than at $R = 4 \times 10^4$. The combination of this effect with the steepened temperature gradient and uniform scale near the wall at $R = 1.6 \times 10^5$ results in a considerably different mode of turbulent heat flux from that at $R = 4 \times 10^4$.

The scale structure near the wall changes again at $R = 6.4 \times 10^5$, becoming larger and more variable than at $R = 1.6 \times 10^5$. This increases the turbulent diffusion near the wall, tending to move the R_{zz} concentration away from the wall. Thus there is less overlap of the R_{zz} peak with the steep temperature gradient, but more overlap with the large scale region. This produces a transition in mode of turbulent heat flux from that at the smaller R.

Another mechanism for the turbulent transport of heat between the plates is the turbulent shear stress. The shear stress, R_{xz}, is not plotted in Figs. 3-5, but its appearance is similar to that of R_{xx}. This source of heat transport was not important at the lower Rayleigh numbers but, because of the increased importance of mean flow shear as a driving force at $R = 6.4 \times 10^5$, it makes an important contribution at this Rayleigh number. This, too, causes a transition in the heat flux curve.

REFERENCES

Daly, B. J., J. Fluid Mech., submitted for publication.

Daly, B. J. and Harlow, F. H., Phys. Fluids 13, 2634 (1970).

Deardorff, J. W. and Willis, G. E., J. Fluid Mech. 23, 337 (1965).

Deardorff, J. W. and Willis, G. E., J. Fluid Mech. 28, 675 (1967).

Harlow, F. H. and Hirt, C. W., University of California, Los Alamos Scientific Laboratory Report LA-4086 (1969).

Harlow, F. H. and Romero, N. C., University of California, Los Alamos Scientific Laboratory Report LA-4247 (1969).

Krishnamurti, R., J. Fluid Mech. 42, 295, 309 (1970).

Lipps, F. B. and Somerville, R. C. J., Phys. Fluids 14, 759 (1971).

Malkus, W. V. R., Proc. Roy. Soc. London 225, 185 (1954).

Mull, W. and Reiher, H., Beih. Gesundh. Ingr., Ser. 1, 28, 1 (1930).

Nakayama, P. I., Ph.D. Thesis, Purdue University (1972).

Rossby, H. T., Ph.D. Thesis, Massachusetts Institute of Technology (1966).

Silveston, P. L., Forschung auf dem Gebiete des Ingenieurwesens 24B, 29, 59 (1958).

Willis, G. E. and Deardorff, J. W., Phys. Fluids 10, 1861 (1967).

Willis, G. E. and Deardorff, J. W., J. Fluid Mech. 44, 661 (1970).

Willis, G. E., Deardorff, J. W. and Somerville, R. C. J., NCAR Manuscript No. 71-154 (1971).

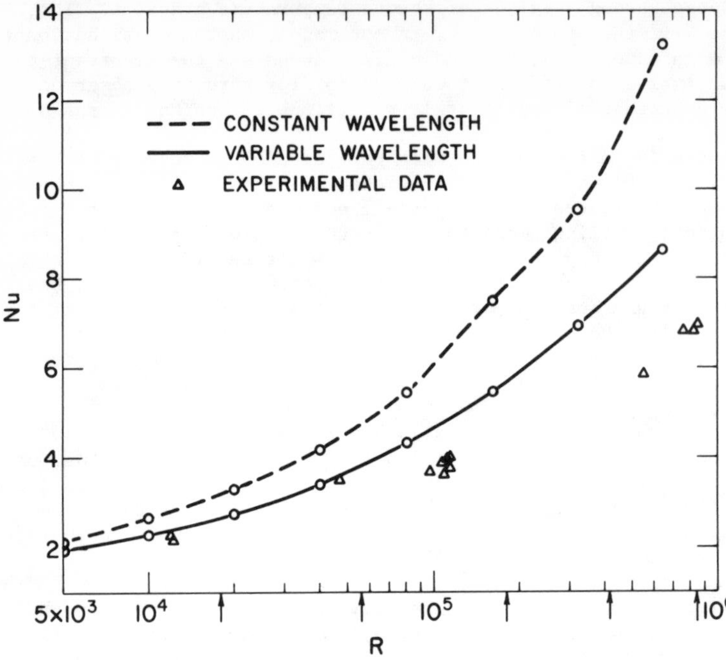

Fig. 2. Comparison of calculated Nusselt number, obtained with constant and variable wavelength, with experimental measurements by Mull and Reiher (1930)

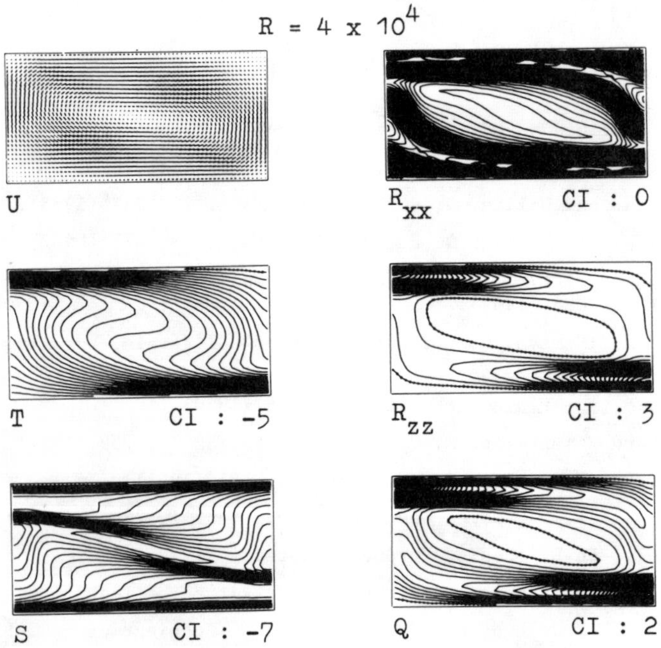

Fig. 3. Velocity vector and contour plots at $R = 4 \times 10^4$. T = temperature, s = integral scale, $Q = (R_{xx} + R_{yy} + R_{zz})/2$ = turbulence energy. Contour interval = 2^{CI}

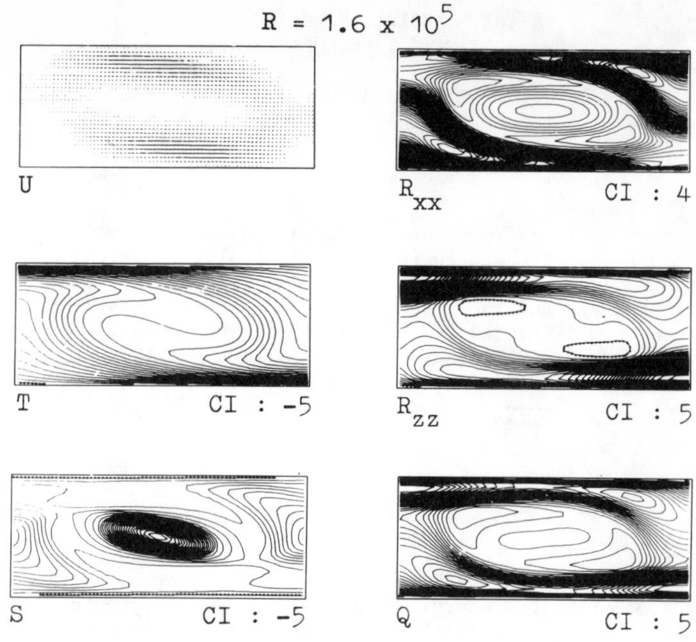

Fig. 4. Same as Fig. 3. Rayleigh number is 1.6×10^5

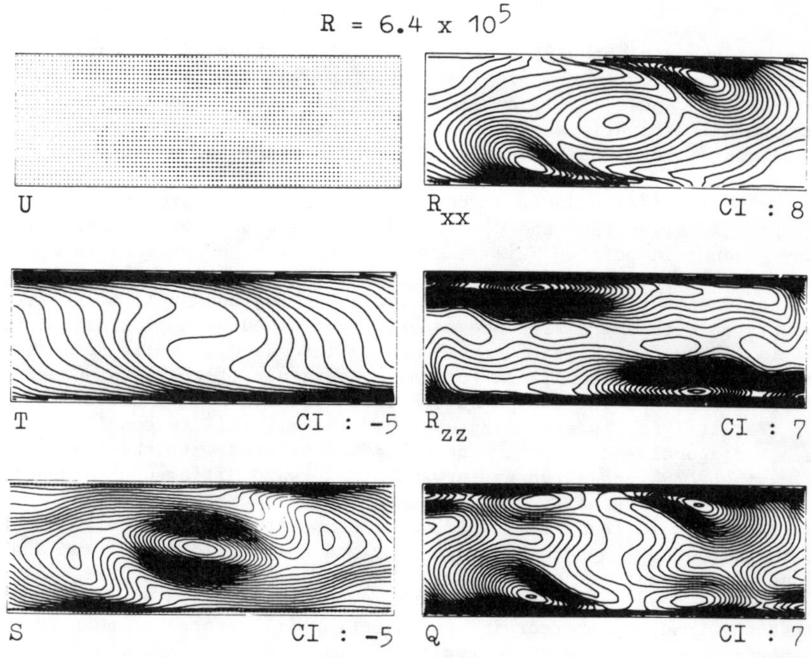

Fig. 5. Same as Fig. 3. Rayleigh number is 6.4×10^5

THE NUMERICAL SOLUTION OF THE VORTICITY TRANSPORT EQUATION

S.C.R. Dennis
CERN, Geneva, Switzerland*)

INTRODUCTION

For simplicity we shall restrict consideration to the vorticity transport equation in two dimensions, although there is no difficulty in extending the method to be described to other equations of similar type, and in more than two dimensions. For the steady flow of a viscous incompressible fluid in two dimensions, the vorticity vector $\underset{\sim}{\omega}$ is given by $\underset{\sim}{\omega} = (0,0,\zeta)$, where ζ satisfies the equation

$$\frac{\partial^2\zeta}{\partial\xi^2} + \frac{\partial^2\zeta}{\partial\eta^2} + 2\lambda\frac{\partial\zeta}{\partial\xi} + 2\mu\frac{\partial\zeta}{\partial\eta} = 0 \ , \tag{1}$$

with

$$\lambda = -\frac{R}{4}\frac{\partial\psi}{\partial\eta} \ , \qquad \mu = \frac{R}{4}\frac{\partial\psi}{\partial\xi} \ .$$

Here ψ is the stream function and R is the Reynolds number. The variables (ξ,η) are coordinates in some orthogonal curvilinear system, usually chosen so that the domain of the solution of (1) is a rectangle. In obtaining numerical solutions to problems governed by the two-dimensional Navier-Stokes equations, it is quite customary to solve equation (1) numerically in conjunction with the equation

$$\frac{\partial^2\psi}{\partial\xi^2} + \frac{\partial^2\psi}{\partial\eta^2} + \zeta/H^2 = 0 \ , \tag{2}$$

where $H(\xi,\eta)$ is a function which depends upon the particular coordinate system used.

In most of the earlier work on the numerical solution of the Navier-Stokes equations using finite-difference approximations, for example the work of Thom (1933), Kawaguti (1953) and Apelt (1961) on flow past a circular cylinder, it has been usual to approximate all derivatives of ζ in (1) by central-difference formulae. More recently Spalding (1967) and Greenspan (1968) have suggested a method in which, although the second derivatives of ζ in (1) are approximated as usual by central differences, the terms $\partial\zeta/\partial\xi$ and $\partial\zeta/\partial\eta$ are approximated by forward or backward differences depending upon the signs of λ and μ, i.e., depending upon the local direction of the flow at any given grid point. This method will briefly be reviewed in the next section, but basically the object is as follows. If λ and μ become large compared with the grid size in some regions of the flow field, as may happen if R is large, the matrix associated with the difference equations obtained by approximating all derivatives of ζ in (1) by central differences may fail to be diagonally dominant. Diagonal dominance is a sufficient condition for the convergence of the point Gauss-Seidel iterative procedure, and for the point SOR procedure for a well-defined range of the SOR factor (Varga, 1962), whereas these procedures may fail to converge for matrices which are not diagonally dominant. The matrix associated with the equations obtained by approximating $\partial\zeta/\partial\xi$ and $\partial\zeta/\partial\eta$ by forward or backward differences depending on the direction of flow is diagonally dominant under all circumstances. This method of approximation therefore leads to a successful computational procedure. It has been discussed in some detail by Runchal, Spalding and Wolfshtein (1968) and a number of applications of this and methods of similar type have been given, for example, by Kawaguti (1969), Thoman and Szewczyk (1969). The disadvantage of methods of this type is that the truncation error in approximating (1) is greater than that when all derivatives of ζ are replaced by central differences.

*) On leave from the University of Western Ontario, London, Ontario, Canada.

The object of the present paper is to discuss a method of approximating (1) in which the matrix associated with the difference equations is diagonally dominant and, further, the truncation error is the same as that of the fully central-difference approximation. The origin of the method is in a paper by Allen and Southwell (1955) and it was subsequently considered by Dennis (1960) and by Allen (1962). The basic properties of the approximation were not, however, recognized at the time and they are presented here with some numerical examples which indicate that the method may be fruitful in solving problems governed by the Navier-Stokes equations and in boundary-layer theory. In particular, the method of h^2-extrapolation, where h is the grid size, can be used for solutions obtained using this approximation as it can for solutions obtained using central differences. Only h-extrapolation is appropriate for the method of forward and backward differences. An illustration from boundary-layer theory is given by calculating the viscous stagnation point flow at the nose of a cylinder. Some new solutions of the Navier-Stokes equations are obtained for symmetrical flow past a flat plate of finite length for R = 40 and 100, where R is the Reynolds number based on the length of the plate.

BASIC APPROXIMATIONS

We assume that the region of integration in the (ξ,η) plane is rectangular and covered by a rectangular grid of elements whose sides are of lengths h and k parallel to the ξ and η directions, respectively. The Southwell notation is used, in which the subscript 0 denotes a quantity at a typical grid point (ξ_0,η_0) and the subscripts 1, 2, 3, 4 denote quantities at the points $(\xi_0 + h, \eta_0)$, $(\xi_0, \eta_0 + k)$, $(\xi_0 - h, \eta_0)$ and $(\xi_0, \eta_0 - k)$, respectively. A discretized approximation to (1) at (ξ_0,η_0) defines a typical member of a set of equations spanning all the internal grid points of the rectangle. We can write this as

$$d_1\zeta_1 + d_2\zeta_2 + d_3\zeta_3 + d_4\zeta_4 - d_0\zeta_0 = 0 , \qquad (3)$$

and this approximation results from neglecting some discretization error E on the right side of (3). The associated matrix is diagonally dominant (Varga, 1962) provided

$$|d_1| + |d_2| + |d_3| + |d_4| \le |d_0| \qquad (4)$$

and is strictly diagonally dominant if the inequality holds at least at one point (ξ_0,η_0). If the equality holds generally at all internal grid points and Dirichlet boundary conditions are assumed on the boundary, the strict condition is ensured by consideration of points adjacent to the boundary.

If $\gamma = h^2/k^2$, the result of replacing all derivatives of ζ in (1) by central differences gives $d_n = a_n$ in (3), where

$$a_1 = 1 + \lambda_0 h , \quad a_2 = \gamma(1 + \mu_0 k) ,$$
$$a_3 = 1 - \lambda_0 h , \quad a_4 = \gamma(1 - \mu_0 k) , \quad a_0 = 2(1 + \gamma) . \qquad (5)$$

The associated error term on the right of (3) is

$$E_1 = O(h^4) + O(\gamma k^4) . \qquad (6)$$

Diagonal dominance is assured if the coefficients $a_n (n = 1, 2, 3, 4)$ are positive at every grid point, but if the Reynolds number is large the absolute values of $\lambda_0 h$ and $\mu_0 k$ may well exceed unity in some regions of the flow field. The method of Spalding (1967) and Greenspan (1968) obviates this difficulty by replacing the term $\partial\zeta/\partial\xi$ in (1) by a forward difference at points where $\lambda > 0$ and by a backward difference at points where $\lambda < 0$, with a similar treatment of the term $\partial\zeta/\partial\eta$ depending on the sign of μ. In this case we have $d_n = b_n$ in (3), where

$$
\begin{array}{llll}
b_1 = 1 + 2\lambda_0 h , & b_3 = 1 , & \text{if } \lambda_0 > 0 ; & \\
b_1 = 1 , & b_3 = 1 - 2\lambda_0 h , & \text{if } \lambda_0 < 0 ; & \\
b_2 = \gamma(1 + 2\mu_0 k) , & b_4 = \gamma , & \text{if } \mu_0 > 0 ; & \quad (7) \\
b_2 = \gamma , & b_4 = \gamma(1 - 2\mu_0 k) , & \text{if } \mu_0 < 0 ; & \\
b_0 = 2\{1 + |\lambda_0|h + \gamma(1 + |\mu_0|k)\} . & & &
\end{array}
$$

The associated matrix is now diagonally dominant under all circumstances, but the associated truncation error is now

$$E_2 = O(h^3) + O(\gamma k^3) , \tag{8}$$

which must be the main objection to the method.

The approximation of Allen and Southwell (1955) is obtained by writing (1) in the form of two equations

$$\frac{\partial^2\zeta}{\partial\xi^2} + 2\lambda \frac{\partial\zeta}{\partial\xi} = A , \qquad \frac{\partial^2\zeta}{\partial\eta^2} + 2\mu \frac{\partial\zeta}{\partial\eta} = -A \tag{9}$$

and assuming that in the neighbourhood of (ξ_0,η_0) the functions λ, μ, and A can be approximated by their local values λ_0, μ_0, and A_0. With these approximations, the first of the equations (9) can be solved along the grid line $\eta = \eta_0$ as an ordinary differential equation with constant coefficients. The two constants of integration which appear are eliminated in terms of values of ζ at $\xi = \xi_0 - h$, ξ_0, $\xi_0 + h$, and this finally gives an expression for A_0 in terms of ζ_0, ζ_1 and ζ_3. The second of (9) is likewise solved along the grid line $\xi = \xi_0$, giving an expression for A_0 in terms of ζ_0, ζ_2 and ζ_4. Elimination of A_0 then gives an equation of the form (3) with $d_n = c_n$, where

$$c_1 = \exp(\lambda_0 h) , \quad c_2 = \alpha \exp(\mu_0 k) , \quad c_3 = \exp(-\lambda_0 h) , \quad c_4 = \alpha \exp(-\mu_0 k) ,$$
$$c_0 = 2\{\cosh(\lambda_0 h) + \alpha \cosh(\mu_0 k)\} \tag{10}$$

and

$$\alpha = \frac{\mu_0 h \sinh(\lambda_0 h)}{\lambda_0 k \sinh(\mu_0 k)} .$$

It is difficult to identify the error term E_3 on the right side of (3) directly using this procedure, and no attempt was made by Allen and Southwell to do so. The ideas were intuitive, the object being to simulate the local behaviour of the transport of vorticity, and in this sense there is some accord with the basic ideas of the method of forward and backward differences as set out by Spalding (1967).

PROPERTIES OF THE ALLEN AND SOUTHWELL APPROXIMATION

The most obvious property is that the coefficients c_n in (10) are all positive and the associated matrix is diagonally dominant. It is less obvious that the error term E_3 is of the same order as that of the central-difference approximation (5) and we shall show this indirectly. We are concerned with the behaviour as h, k $\rightarrow$ 0. If we first expand the exponentials in the expression

$$c_1\zeta_1 + c_3\zeta_3 - 2\zeta_0 \cosh(\lambda_0 h)$$

in powers of h it is seen that these terms are equivalent to

$$(1 + \lambda_0 h)\zeta_1 + (1 - \lambda_0 h)\zeta_3 - 2\zeta_0 + \frac{1}{2}\lambda_0^2 h^2(\zeta_1 + \zeta_3 - 2\zeta_0) + \frac{1}{6}\lambda_0^3 h^3(\zeta_1 - \zeta_3) + \cdots .$$

Thus they are equivalent to

$$(1 + \lambda_0 h)\zeta_1 + (1 - \lambda_0 h)\zeta_3 - 2\zeta_0 + O(h^4) . \tag{11}$$

It is easily shown that

$$\alpha = \gamma + O(h^2) + O(\gamma h^2)$$

and we may therefore show that the terms

$$c_2\zeta_2 + c_4\zeta_4 - 2\alpha\zeta_0 \cosh(\mu_0 k)$$

are equivalent to

$$\gamma\{(1 + \mu_0 k)\zeta_2 + (1 - \mu_0 k)\zeta_4 - 2\zeta_0\} + O(h^4) + O(\gamma k^4) . \tag{12}$$

It follows that the difference between the left side of (3) with $d_n = a_n$ and the left side with $d_n = c_n$ amounts only to terms $O(h^4) + O(\gamma k^4)$ and since it is already known that E_1 is given by (6), then also

$$E_3 = O(h^4) + O(\gamma k^4) . \tag{13}$$

In particular, if $h = k$ both E_1 and E_3 are $O(h^4)$ and we can make use of the Richardson h^2-extrapolation procedure.

One further property of the approximation of interest in solving the Navier-Stokes equations is that it appears to be capable of adequately describing the nature of boundary-layer flow in regions of the flow field where boundary-layer theory applies. This will now be illustrated in the case of flow near the nose region of a cylinder. We can show that, provided the grid sizes are adjusted in accordance with boundary-layer theory, the approximation deals correctly with the situation. It is convenient to identify the nose of the cylinder with the point $\xi = 0$, $\eta = \eta^*$. We suppose the coordinates to be chosen so that $H(\xi,\eta) = 1$ in equation (2) and introduce new variables defined by

$$\xi = R^{-1/2}x , \quad \eta = \eta^* - \theta , \quad \psi = 2R^{-1/2}\Psi , \quad \zeta = 2R^{1/2}\phi . \tag{14}$$

As $R \to \infty$, equations (1) and (2) tend to the boundary-layer equations

$$\frac{\partial^2\phi}{\partial x^2} + 2F\frac{\partial\phi}{\partial x} + 2G\frac{\partial\phi}{\partial\theta} = 0 , \tag{15}$$

$$\frac{\partial^2\Psi}{\partial x^2} + \phi = 0 , \tag{16}$$

where

$$F = \frac{1}{2}\frac{\partial\Psi}{\partial\theta} , \quad G = -\frac{1}{2}\frac{\partial\Psi}{\partial x} .$$

Consider now the effect of putting $h = R^{-1/2}h_1$ in (10), where h_1 is the grid size in the x coordinate, and keeping k fixed. Then $\lambda_0 h = F_0 h_1 = O(1)$ and $\mu_0 k = -RG_0 k = O(R)$. Moreover, since no separation takes place in the boundary layer, $\partial\Psi/\partial x$ is positive at all points in the flow field and hence $G_0 < 0$ at all points. It is now found that as $R \to \infty$ the limits of the coefficients c_n in (10) are (where here and in the remaining equations of this section, we have interchanged the subscripts 2 and 4 so that the grid line 402 is in the direction of increasing θ)

$$c_1 = \exp(F_0 h_1)\,, \quad c_2 = 0\,, \quad c_3 = \exp(-F_0 h_1)\,, \quad c_4 = -\frac{2G_0 h_1}{F_0 k}\sinh\left(F_0 h_1\right)\,,$$
$$c_0 = 2\left\{\cosh\left(F_0 h_1\right) - \frac{G_0 h_1}{F_0 k}\sinh\left(F_0 h_1\right)\right\}. \tag{17}$$

The fact that $c_2 \to 0$ as $R \to \infty$ reflects the fact that equation (15) is parabolic and the solution for ϕ can be advanced step by step from known conditions at $\theta = 0$. The appropriate equations in the limit as $R \to \infty$ are obtained by substituting the c_n in (17) for the d_n in (3) and writing ϕ in place of ζ. There is then no forward influence in (3) and they can be solved along a line of constant θ simultaneously with some approximation to the equation (16) to give an implicit step-by-step procedure for the solution of (15) and (16). The important inference to be drawn from this limiting behaviour is that when the Reynolds number is moderately large, but not infinite and the equation (1) is not parabolic, but weakly elliptic, the Allen and Southwell approximation will automatically adapt itself to the nearly parabolic situation. This will not occur in the case of the central-difference approximation, since the approximation of the derivative $\partial\zeta/\partial\eta$ in (1) by a central difference automatically commits the difference equations to forward influence which will not diminish as $R \to \infty$. The method of forward and backward differences would, however, take account of the parabolic nature of the problem as $R \to \infty$. In the limit this approximation becomes, since $F_0 > 0$ and $G_0 < 0$ at all points,

$$\left(1 + 2h_1F_0\right)\phi_1 + \phi_3 - 2\left(1 + h_1F_0 - \frac{h_1^2}{k}G_0\right)\phi_0 - \frac{2h_1^2}{k}G_0\,\phi_4 = 0\,. \tag{18}$$

The only way of applying central differences to (1) to ensure that the parabolic nature of the problem is properly taken account of as $R \to \infty$ is to express the ξ derivatives of ζ in terms of central differences but the term $\partial\zeta/\partial\eta$ in terms of a forward difference (i.e. in terms of a backward difference of $\partial\phi/\partial\theta$). This gives, in the limit as $R \to \infty$ and in terms of the variables used in equation (15),

$$\left(1 + h_1F_0\right)\phi_1 + \left(1 - h_1F_0\right)\phi_3 - 2\left(1 - \frac{h_1^2}{k}G_0\right)\phi_0 - \frac{2h_1^2}{k}G_0\,\phi_4 = 0\,. \tag{19}$$

CALCULATION OF STAGNATION POINT FLOW

For symmetrical flow near a front stagnation point we put $\Psi = f(x)\theta$, $\phi = -g(x)\theta$ in (15) and (16) which gives the two equations

$$g'' + fg' - f'g = 0\,, \tag{20}$$

$$f'' = g\,, \tag{21}$$

where the prime denotes differentiation with regard to x. If the external flow is taken to be the potential flow, the boundary conditions are

$$f(0) = f'(0) = 0\,, \qquad f'(\infty) = 1\,, \qquad g(\infty) = 0\,. \tag{22}$$

It is of course possible to eliminate g between (20) and (21) and then integrate once to obtain the Hiemenz equation for stagnation point flow (Schlichting, 1960)

$$f''' + ff'' - f'^2 = 1\,, \tag{23}$$

but the object of the present approach is to test the methods of approximating the solution of (20). In the case of the Allen and Southwell approximation the appropriate equations are obtained as follows. We first put $F_0 = \frac{1}{2}f_0$, $G_0 = -\frac{1}{2}f_0'k$ in the coefficients (17). These are then substituted for d_n in (3), ζ_4 is put equal to zero, and g is written in place of ζ. Approximations can also be obtained corresponding to (18) and (19) by the same substitutions for F_0 and G_0, putting $\phi_4 = 0$, and writing g for ϕ. Any one of these sets of equations can then be solved in conjunction with some method of approximating (21) to give a numerical approximation to the functions $f(x)$ and $g(x)$.

Numerical solutions have been obtained in all three cases according to the following procedure. It may be seen from the equation (21) and the conditions (22) that

$$\int_0^\infty g\,dx = 1 \tag{24}$$

and $g(x)$ must also satisfy $g(\infty) = 0$. The equations (20) and (21) are solved by obtaining an iterative sequence of successive approximations to $f(x)$ and $g(x)$. Corresponding to an approximation to $f(x)$ at some stage of the process, substitution of f and f' in (20) gives a linear equation for g. An approximation to a particular solution of this equation, say $g^*(x)$, can be found to satisfy the boundary conditions

$$g^*(0) = 1\ , \qquad g^*(x_m) = 0\ , \tag{25}$$

where x_m is some sufficiently large value of x. This approximation is found by solving the appropriate set of difference equations by the SOR procedure. It now follows that $g(x) = Ag^*(x)$ is a solution and an approximation to the constant A is found from the equation

$$A = \frac{1}{\int_0^\infty g^*(x)\,dx}\ , \tag{26}$$

which follows from (24). The upper limit of this integral is replaced by x_m and numerical integration is used. This determines completely an approximation to $g(x)$. Approximations to $f(x)$ and $f'(x)$ are then found by integrating (21) twice by a step-by-step procedure subject to $f(0) = f'(0) = 0$. The value of $f'(x_m)$ should be unity and this gives a check on the procedure. The sequence is then repeated until convergence, which is very rapid.

Two solutions with $h_1 = 0.1$, $h_1 = 0.05$ were obtained for each of the three methods of approximation. The value $x_m = 15$ was taken in each case. This is much larger than necessary since $1 - f'(x) < 10^{-7}$ at about $x = 5.6$. It is of interest to note that it is always possible to make the SOR procedure diverge for the central-difference approximation (19) by taking x_m large enough, whereas the same procedure remains convergent with the other two approximations. For comparison purposes we shall use two quantities, firstly the value of $A = g(0) = f''(0)$ and secondly the constant B in the asymptotic expression

$$f(x) \sim x - B\ , \qquad \text{as } x \to \infty\ .$$

These are measures of the skin friction and displacement thickness near the nose and they have been calculated very accurately by Tifford (1954). Tifford's values, rounded to five decimals, are

$$A = 1.23259\ , \qquad B = 0.64790\ . \tag{27}$$

Results from the present calculations are given in Table I. In this instance the central-difference approximation denotes that obtained from equation (19).

TABLE I

Calculated values of A and B

APPROXIMATION	FORWARD - BACKWARD		CENTRAL DIFFERENCE		ALLEN & SOUTHWELL	
h_1	0.1	0.05	0.1	0.05	0.1	0.05
$A(h_1)$	1.20776	1.21975	1.23396	1.23293	1.23339	1.23279
$B(h_1)$	0.66569	0.65694	0.64710	0.64770	0.64757	0.64781

In order to illustrate the truncation errors inherent in the approximations we shall use the Richardson extrapolation procedure to estimate A and B. For either of the central-difference or Allen and Southwell approximations we can write

$$A(h_1) \sim A + Kh_1^2 , \quad \text{as } h_1 \to 0 \tag{28}$$

which gives an estimate

$$3A = 4A(0.05) - A(0.1)$$

from each pair of values in Table I, with a similar estimate from each pair of values of $B(h_1)$. It can be verified that both estimates of A agree with (27) to all the figures quoted and that the estimates of B differ by not more than one unit in the fifth decimal from the value given in (27). For the forward and backward difference approximation, the extrapolation formula is similar to (28) but with h_1^2 replaced by h_1. Application of this formula to the results of Table I gives estimates A = 1.23174 B = 0.64819.

SKIN FRICTION ON A FLAT PLATE

The second problem considered is that of flow past a flat plate of finite length at zero incidence to a uniform stream. The governing equations are (1) and (2), where (ξ,η) are elliptic coordinates defined in terms of dimensionless Cartesian coordinates (X,Y) by

$$X = \cosh \xi \cos \eta , \qquad Y = \sinh \xi \sin \eta . \tag{29}$$

The quantity H in (2) is given by

$$H^2 = (\partial\xi/\partial X)^2 + (\partial\xi/\partial Y)^2 = 2/(\cosh 2\xi - \cos 2\eta) . \tag{30}$$

The leading and trailing edges of the plate are situated at the points (-1,0) and (1,0) of the (X,Y) plane and the plate therefore occupies the position $\xi = 0$ of the (ξ,η) plane with leading edge at $\eta = \pi$ and trailing edge at $\eta = 0$. The Reynolds number in (1) is given by $R = UL/\nu$, where L is the length of the plate, U is the velocity of the steady stream and ν is the coefficient of kinematic viscosity.

There is a substantial literature of both theoretical and numerical investigations on this problem which will not be cited here because it is not relevant to the calculations to be presented. Our present object is to give some results which illustrate the application of the various methods of approximation to the solution of equation (1) for this particular problem, which provides a good test of these methods. The method of solution used is essentially that described by Dennis and Chang (1969a, 1969b) and we only briefly note the major points relevant to the solution of (1). The flow is symmetrical about the X-axis and the computational domain is the semi-infinite rectangle $\xi \geq 0$, $0 \leq \eta \leq \pi$, with the boundary conditions

$$\zeta = 0 \quad \text{when } \eta = 0 , \ \eta = \pi .$$

The condition for ζ as $\xi \to \infty$ is that $\zeta \to 0$ for all values of η although, in practice, a more satisfactory condition to be applied on a boundary $\xi = \xi_m$ has been given by Dennis and Chang (1969b). This is based on the known asymptotic nature of the flow as $\xi \to \infty$. One of the features of the present work is that this boundary condition has been improved by taking account of higher terms in the asymptotic expansion, although this will not be reported here.

Finally, the boundary condition for ζ at $\xi = 0$ is actually calculated from values of ζ in the flow field. It has been shown by Dennis and Chang (1969b) that $\zeta(0,\eta)$ may be calculated from the expression

$$\zeta(\xi,\eta) = -H^2 \sum_{n=1}^{\infty} r_n(\xi) \sin n\eta , \tag{31}$$

where the functions $r_n(\xi)$ must satisfy the conditions

$$\int_0^{\infty} e^{-n\xi} r_n(\xi)\, d\xi = \begin{cases} 1 & (n = 1) \\ 0 & (n = 2,3,4, \ldots) . \end{cases} \tag{32}$$

It may be noticed that the set of conditions (32) is rather similar to the condition (24) and the method of satisfaction of (32) is also similar, but more complicated. From a given approximation to ζ the values of $r_n(\xi)$ for $\xi \neq 0$ are calculated from the Fourier integral

$$r_n(\xi) = -\frac{2}{\pi} \int_0^{\pi} (\zeta/H^2) \sin n\eta \, d\eta , \tag{33}$$

which corresponds to (31), and then the set of conditions (32) is used to calculate values of $r_n(0)$ using numerical integration with the upper limit in the integral replaced by ξ_m, the finite value of ξ which limits the field. A contribution for the part of the integral from $\xi = \xi_m$ to $\xi = \infty$ is added using known asymptotic theory for the behaviour of $r_n(\xi)$ as $\xi \to \infty$. The estimation of this contribution has been substantially improved in the present calculations by utilizing higher terms in the asymptotic expansion. The values of $r_n(0)$ calculated to satisfy (32) are then introduced into (31) to give an estimation of $\zeta(0,\eta)$.

The method of calculating $\psi(\xi,\eta)$ and the iterative procedures of solving the simultaneous equations (1) and (2) are essentially as reported by Dennis and Chang (1969a, 1969b). Calculations have been carried out for two Reynolds numbers, $R = 40$ and 100, with two sets of calculations in each case. In the first the calculations of Dennis and Chang (1969a) using the central-difference approximation (5) to equation (1) in the manner already reported have been repeated, but using two smaller grid sizes ($h = k = \pi/60$, $h = k = \pi/80$). In the second, this same set of calculations has been repeated using the Allen and Southwell approximation (10). In both sets the number of terms taken to approximate the infinite sum in (31) was 50 and the value $\xi = \xi_m$ which limits the field in the ξ direction was taken as $\xi_m = \pi$. By reasonably careful checks on the solutions it is believed that these values of the parameters are adequate to give accurate approximations.

There are a number of calculated properties to be compared with theory in this problem, which is as yet still not completely understood, particularly in the trailing edge region of the plate. A detailed analysis and comparison of the results will be reported elsewhere and we shall here use only one property to illustrate typical results obtained using the central-difference and Allen and Southwell approximations. The coefficient of total frictional drag on the plate, C_f, is defined by $C_f = D_f/(\rho U^2 L)$, where D_f is the total frictional drag on the plate and ρ is the density. It may be expressed as

$$C_f = -\frac{2}{R} \int_0^{\pi} \zeta(0,\eta) \sin \eta \, d\eta$$

and hence, using (31), as

$$C_f = \frac{2\pi}{R} \sum_{n=1}^{\infty} r_{2n-1}(0) \ . \tag{34}$$

Calculated results for C_f are shown in Table II, where the sum in (34) has been approximated using computed values of $r_n(0)$ from n = 1 to n = 50.

TABLE II

Calculated values of the total drag coefficient C_f

R	h = k	CENTRAL DIFFERENCES	ALLEN AND SOUTHWELL
40	π/60	0.31516	0.30485
40	π/80	0.31330	0.30740
100	π/60	0.18342	0.16895
100	π/80	0.18186	0.17346

The most interesting feature of these results, and this applies to other properties of the solutions, is that the values obtained from the central-difference approximation are converging downwards as the grid size is decreased while those from the Allen and Southwell approximation are converging upwards. It seems probable that we can therefore deduce that

$$0.3074 < C_f < 0.3133 \ , \qquad \text{for } R = 40 \ ;$$
$$0.1735 < C_f < 0.1819 \ , \qquad \text{for } R = 100 \ ,$$

where C_f is the correct frictional drag. The results of applying h^2-extrapolation for the two grid sizes to each of the four pairs of values in Table II are shown in Table III

TABLE III

Extrapolated values of C_f

R	CENTRAL DIFFERENCES	ALLEN AND SOUTHWELL
40	0.31091	0.31068
100	0.17985	0.17926

In view of these results, the true values of C_f would appear to be not greatly different from $C_f = 0.311$ at R = 40, and $C_f = 0.180$ at R = 100. We can, therefore, present a more detailed analysis of some of the properties, for example the behaviour of the singularity in the skin friction at the trailing edge, with reasonable expectations of accuracy.

A solution was also obtained for the case R = 40, h = k = π/80, using the method of forward and backward differences, but the calculated value of C_f was found to be some 10% lower than the corresponding value obtained from the central-difference approximation, with a similar order of discrepancy in other properties. It was not considered to be worthwhile to investigate this approximation in this problem.

ACKNOWLEDGMENT

It is well known that computing accurate solutions of the Navier-Stokes equations involves substantial amounts of computing time. The calculations presented here were carried out on the CDC 6500 and 6600 machines at CERN. Grateful acknowledgment is made of the opportunity to use these facilities.

REFERENCES

Allen, D.N. de G., and Southwell, R.V. Quart. J. Mech. Appl. Math. 8, 129 (1955).

Allen, D.N. de G. Quart. J. Mech. Appl. Math. 15, 11 (1962).

Apelt, C.J. Aero. Res. Counc. R. and M. No. 3175 (1961).

Dennis, S.C.R. Quart. J. Mech. Appl. Math. 13, 487 (1960).

Dennis, S.C.R., and Chang, G.Z. Phys. Fluids Suppl. II 12, II-88 (1969a).

Dennis, S.C.R., and Chang, G.Z. Mathematics Research Center, University of Wisconsin, Technical Summary Report No. 859 (1969b).

Greenspan, D. in Lectures on the Numerical Solution of Linear, Singular and Non-linear Differential Equations, Prentice-Hall, Englewood Cliffs, New Jersey, 1968.

Kawaguti, M. J. Phys. Soc. Japan 8, 747 (1953).

Kawaguti, M. Phys. Fluids Suppl. II 12, II-101 (1969).

Runchal, A.K., Spalding, D.B., and Wolfshtein, M. Imperial College Report No. SF/TN/14 (1968).

Schlichting, H. Boundary Layer Theory, McGraw-Hill, New York, 1960.

Spalding, D.B. in Numerical Methods for Viscous Flows, AGARD Conference Proceedings No. 60 (1967).

Tifford, A.N. Wright Air. Dev. Center Tech. Report 53-288 (1954).

Thom, A. Proc. Roy. Soc. A141, 651 (1933).

Thoman, D.C., and Szewczyk, A.A. Phys. Fluids Suppl. II 12, II-76 (1969).

Varga, R.S. Matrix Iterative Analysis, Prentice-Hall, Englewood Cliffs, New Jersey, 1962.

REFLEXIONS D'ONDES DE CHOC SUR L'AXE EN ECOULEMENT PERMANENT DE REVOLUTION

par P. DIRINGER

Office National d'Etudes et de Recherches Aérospatiales (ONERA) - 92 Châtillon (France)

INTRODUCTION

La connaissance de l'écoulement à la sortie d'un éjecteur supersonique est d'un grand intérêt pratique mais sa détermination devient particulièrement difficile lorsqu'apparaissent des ondes de choc qui se réfléchissent au voisinage de l'axe en formant une petite onde de choc quasi-normale à l'axe appelée "disque de MACH". Le mécanisme de la formation d'une telle onde de choc a pour origine deux tendances antagonistes puisque les ondes de forte détente situées au voisinage de l'axe et issues de la sortie de l'éjecteur se heurtent aux ondes de compression situées en aval du choc engendré par réflexion des premières sur la ligne de jet. Le saut de pression entre ces deux zones est tel que la courbure prise par l'onde de choc incidente induit des déflexions des lignes de courant incompatibles avec celles d'un choc attaché. Si le calcul progresse par les caractéristiques montantes (fig. 1), en allant de l'axe vers la ligne de jet, la poursuite du calcul, illicite d'ailleurs à partir d'un point P_o , montre de plus que les caractéristiques montantes focalisent en aval du choc, indiquant par conséquent la présence d'une onde de choc réfléchie plus tôt que ne le laissait prévoir une impossibilité de calcul au voisinage immédiat de l'onde de choc.

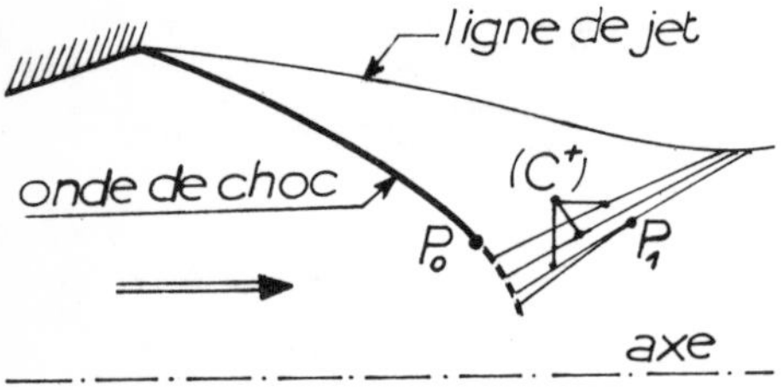

Fig. 1

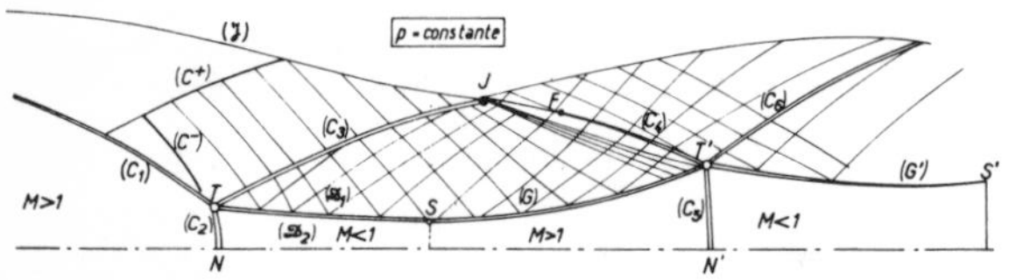

Fig. 17 - Schéma de l'écoulement en aval d'une réflexion de Mach sur l'axe.

(C^+) caractéristique montante
(C^-) caractéristique descendante
(C_1), (C_4) ondes de choc incidentes
(C_2), (C_5) disques de Mach
(C_3), (C_6) ondes de choc réfléchies
(G), (G') lignes de glissement
($\mathcal{J}$) ligne de jet

Fig.2

Soit alors T (fig.2) le point situé à la confluence des trois chocs.

La ligne de glissement (G) issue de ce point sépare l'écoulement en deux régions, la région ($\mathcal{D}_1$) au-dessus de (G) où l'écoulement est entièrement supersonique et la région ($\mathcal{D}_2$) comprise entre (G) et l'axe où l'écoulement, d'abord subsonique jusqu'au voisinage du point S, peut continuer à s'accélérer pour devenir supersonique si les conditions aval le permettent. Le tube de courant limité par (G) peut alors être considéré comme le contour d'un convergent - divergent.

L'écoulement est donc régi par des équations aux dérivées partielles du type mixte. Le point T étant supposé connu et les conditions de RANKINE-HUGONIOT satisfaites le long de TN, le calcul des écoulements à l'intérieur et à l'extérieur du tube peut être effectué de proche en proche en respectant la condition de symétrie sur l'axe et la continuité des pressions à la traversée de (G). La condition de fermeture du domaine subsonique implique alors que la section du tube passe par un minimum lorsque la vitesse devient sonique.

MEL'NIKOV a formulé correctement le problème en utilisant la méthode des relations intégrales pour la partie subsonique mais n'a pas fourni de résultats. ASHRATOV a supposé le tube de courant monodimensionnel et fourni des résultats sur un cas précis en donnant la variation de la position du point T en fonction de la pression extérieure. Plus récemment ABBETT a montré que son résultat, obtenu en faisant également la même hypothèse pour l'écoulement subsonique, était en accord avec l'expérience de LOVE, mais assez différent de ceux que l'on pouvait déduire d'hypothèses émises sur la position du point T - disque normal à (G) en T (BOWYER et alias) ou disque rectiligne normal à l'axe (KAWAMURA).

Nous présentons ici des calculs effectués en utilisant le théorème de la quantité de mouvement et celui de la conservation du débit pour l'écoulement subsonique ; on a pu, dans un cas, franchir le domaine transsonique et poursuivre le calcul au-delà du second disque de MACH jusqu'à l'approche du second col sonique. Les compléments relatifs à cette étude se trouvent dans la note citée en référence.

I - CALCUL DES ECOULEMENTS.

I.1 - Ecoulement extérieur. (Région $\mathcal{D}_1$)

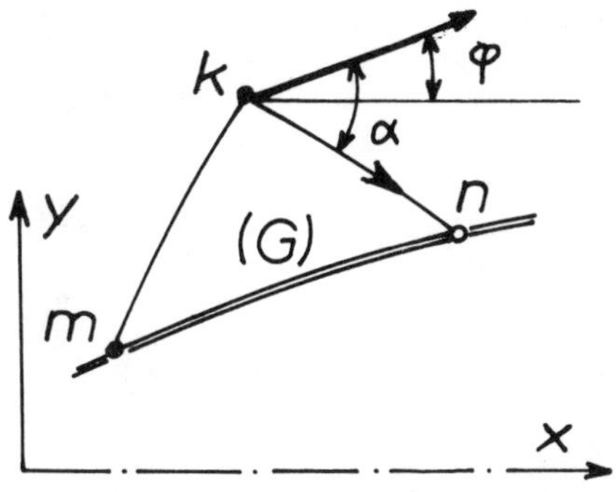

Fig. 3

- On utilise la méthode des caractéristiques en progressant par caractéristiques descendantes.

La position du point n est déterminée (figure 3) à partir de celle du point k par la relation

(1) $$y_n = y_k + (x_n - x_k)\,\mathrm{tg}(\varphi - \alpha)_{kn}$$

Sur cette même caractéristique la pression p et l'angle φ de la vitesse sont liées par la relation :

(2) $$\frac{\sin(2\alpha)_{kn}}{2\gamma}\log\frac{p_n}{p_k} - \varphi_n + \varphi_k = -\left(\frac{\sin\alpha\sin\varphi}{y}\right)_{kn}\overline{kn}$$

l'indice double indiquant la valeur moyenne des quantités qui en sont affectées.

Les points situés sur les ondes de choc sont calculés également par itérations en utilisant les relations de RANKINE-HUGONIOT sous forme finie et l'autre relation de compatibilité analogue à (2) dans le cas des chocs montants.

I.2 - Détermination des paramètres au point triple.

Fig. 4

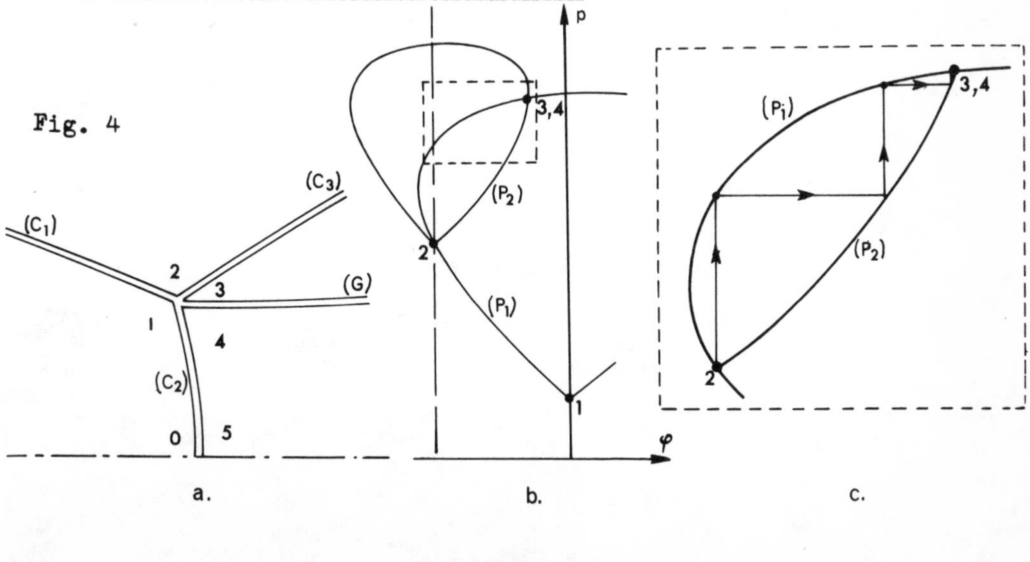

Le choc incident étant calculé ainsi qu'une partie du champ en aval, et le point T étant "choisi" sur ce choc, il s'agit tout d'abord de déterminer les paramètres caractérisant l'écoulement dans les différentes régions séparées par les 3 ondes de choc. Les notations étant celles de la Fig. (4, a), le calcul se fait par itérations dans le plan (φ, p). Les paramètres relatifs aux régions 1 et 2 étant connus, ceux relatifs aux régions 3 et 4 sont déterminés par l'intersection des polaires (2, 3) et (1, 4) fig (4, b) suivant le schéma indiqué par les flèches de la figure (4, c).

Le disque de MACH étant assimilé à la parabole d'axe OX tangente en T au choc (1, 4), la position du point N s'en déduit ; la pression en o est extrapolée sur l'axe suivant une loi parabolique à l'aide des 3 points les plus proches ; les paramètres en 5 se déduisent des équations relatives au choc droit.

Selon que φ_4 est supérieur ou inférieur à φ_1, l'angle $\sigma_{1,4}$ du choc (C_2) avec la direction φ_1 est positif ou négatif, l'inclinaison du disque de MACH en T étant bien entendu égale à ($\varphi_1 + \sigma_{1,4}$).

I.3 - Ecoulement intérieur.

ρ désigne la masse volumique, u et v les composantes de la vitesse q suivant les axes Ox et Oy. Nous sommes intéressé plus spécialement aux valeurs des paramètres à la frontière du tube de courant. D'après l'équation de BERNOULLI valable dans tout le champ, on peut remarquer que les paramètres sur l'axe ne dépendent que de p_o tandis que ceux caractérisant la ligne de glissement peuvent s'exprimer à l'aide de p_1 et φ_1, l'entropie ayant une valeur constante sur chacune de ces lignes de courant.

Les inconnues principales sont alors au nombre de 3 : p_o, p_1, φ_1. L'équation (2) dans laquelle n désigne un point de la ligne de glissement fournit une première relation entre p_1 et φ_1. Les autres sont obtenues à l'aide de l'équation de continuité et de l'équation d'EULER suivant Ox. Celles-ci, intégrées par rapport à y, depuis l'axe (y = o) jusqu'à la ligne (G) ($y = y_1$), traduisent respectivement la conservation du débit et le théorème de la quantité de mouvement :

$$(3) \quad \frac{d}{dx}\int_0^{y_1^2} \rho u \, d(y^2) = 0 \quad \text{et (4)} \quad \frac{d}{dx}\int_0^{y_1^2} (p + \rho u^2)\, d(y^2) = p_1 \frac{dy_1^2}{dx}.$$

. Hypothèse A - En supposant pour des raisons de parité que les paramètres thermodynamiques varient linéairement en fonction de y^2, les intégrales se calculent. Une nouvelle quadrature, par rapport à x, entre deux sections voisines E (entrée) et S (sortie) fournit les équations cherchées :

$$(5) \quad y_{1S}^2 (\rho_o q_o + \rho_1 u_1)_S - y_{1E}^2 (\rho_o q_o + \rho_1 u_1)_E = 0$$

$$(6) \quad y_{1S}^2 \left[(p_o + \rho_o q_o^2 + \rho_1 u_1^2)_S - p_{1E} \right] - y_{1E}^2 \left[(p_o + \rho_o q_o^2 + \rho_1 u_1^2)_E - p_{1S} \right] = 0$$

. Hypothèse B - Dans le cas où l'on suppose ρu indépendant de y l'équation (3) devient :

$$(7) \quad \rho_1 q_1 \cos\varphi_1 \, y_1^2 = \text{constante} = \rho_T q_T \cos\varphi_T \, y_T^2 .$$

(Cette équation est préférable à l'équation purement monodimensionnelle

$\rho_1 q_1 y_1^2$ = constante, lorsque φ_1 est supérieur à 10 degrés environ).

Seuls subsistent alors les inconnues p_1 et φ_1.

De toute manière, que les paramètres thermodynamiques soient indépendants de y ou fonction linéaire de y^2, les équations (3) et (4) sont simultanément satisfaites lorsque les conditions écrites ci-dessous le sont :

(8) $\varphi_1 = 0 \ , \ M_0 = 1 \ , \ M_1 = 1 \ .$

Ceci a pour conséquence que le seul réglage de la position du point T suffit à réaliser (8).

I.4 - Calcul couplé des écoulements sur la ligne de glissement.

Le calcul se fait par itérations. Le point n est déterminé géométriquement par intersection de la caractéristique descendante avec la ligne de glissement : celle-ci est assimilée à la parabole d'équation

$$y - y_m = a(x - x_m)^2 + (x - x_m)\, tg\, \varphi_m$$

tangente à (G) au dernier point calculé m. Pour chaque valeur du paramètre a l'ordonnée y_1 et l'angle φ_1 sont donc connus. Les pressions en n calculées indépendamment selon qu'elles appartiennent à $(\mathcal{D}_1)$ ou $(\mathcal{D}_2)$ sont comparées. A chaque itération on modifie le paramètre a jusqu'à ce que soit réalisée l'égalité des pressions en n.

II - CALCUL AUTOMATIQUE DE LA POSITION DU POINT T.

Le calcul de la position du point T a pu être rendu automatique lorsque le champ extérieur est régulier et en particulier lorsque les caractéristiques issues du faisceau de détente formé à l'intersection de l'onde de choc réfléchie (fig. 1) avec la ligne de jet n'atteignent la ligne de glissement qu'en aval du point sonique S.
Selon que le point T est choisi trop près ou trop loin de l'axe sur l'onde de choc incidente, la ligne de glissement obtenue tend à s'incurver vers l'axe (famille $\mathcal{F}_1$) avec un fort gradient du nombre de MACH ou à présenter une tangente horizontale (famille $\mathcal{F}_2$) correspondant à un nombre de MACH maximum inférieur à 1.

L'ordonnée optimale y_T^* satisfaisant (8) est comprise entre deux valeurs y_{min} et y_{max}; on cherche à rendre leur différence aussi petite que le permet le nombre de chiffres significatifs de l'ordinateur. On est alors amené à introduire 2 paramètres : un angle φ_{min} négatif (de l'ordre de quelques degrés) et une abscisse x_{test} à partir de laquelle s'effectue un test :

- si pour $x > x_{test}$ on a $\varphi \leqslant \varphi_{min}$ alors $(G) \in \mathcal{F}_1$ et $y_{min} = y_T$
- si pour $x > x_{test}$ on a $\varphi \geqslant 0$ alors $(G) \in \mathcal{F}_2$ et $y_{max} = y_T$.

III - DOMAINE TRANSSONIQUE.

Dans le plan (φ , M) de la figure 6, où φ et le nombre de Mach M se rapportent à la ligne de glissement, on peut poser au voisinage de M = 1, $M = 1 + \tau z$, où $z = tg\,\varphi$ et $\tau = (M_0 - 1)/z_0$, l'indice o se rapportant au point de partage P_0 des dernières courbes appartenant respectivement à $\mathcal{F}_1$ et $\mathcal{F}_2$.

Compte tenu de (7) on déduit les coordonnées paramétriques de (G) dans le domaine transsonique

$$y = y_s \frac{(1+z^2)^{\frac{1}{4}} \left(1 + 2\frac{\gamma-1}{\gamma+1}\tau z + \frac{\gamma-1}{\gamma+1}\tau^2 z^2\right)^{\frac{\gamma+1}{4(\gamma-1)}}}{(1+\tau z)^{\frac{1}{2}}} \qquad (9)$$

puis, du fait que $dx = \dfrac{dy}{dz}\,\dfrac{dz}{z}$

$$(10)\quad x = x_o + \frac{y_s}{2}\int_{z_o}^{z} \frac{\left(1+2\frac{\gamma-1}{\gamma+1}\tau z + \frac{\gamma-1}{\gamma+1}\tau^2 z^2\right)^{\frac{5-3\gamma}{4(\gamma-1)}}}{(1+z^2)^{\frac{3}{4}}\,(1+\tau z)^{\frac{3}{2}}}\left[(1+\tau z)\left(1+2\frac{\gamma-1}{\gamma+1}\tau z+\frac{\gamma-1}{\gamma+1}\tau^2 z^2\right)+\frac{2\tau^2}{\gamma+1}(2+\tau z)(1+z^2)\right]dz$$

La quadrature s'effectue numériquement.

La poursuite du calcul lorsque le domaine ($\mathcal{D}_2$) devient supersonique ne présente pas de difficulté, seul est modifié le calcul du nombre de Mach à partir duquel est déduite la pression.

IV - <u>RESULTATS</u>.

Les calculs ont été effectués sur l'ordinateur IBM 360/50 de l'ONERA.

IV - 1 - Jet supersonique (fig. 5)

Le calcul concerne l'écoulement à la sortie d'une tuyère conique divergente de 9°468 de demi-angle d'ouverture en gaz parfait (γ = 1,4). Le nombre de Mach M_s est égal à 8,15 sur la calotte sphérique s'appuyant sur le cercle de la section de sortie de 72 mm de rayon. La pression extérieure p_e est égale à 4,8276 la pression p_s.

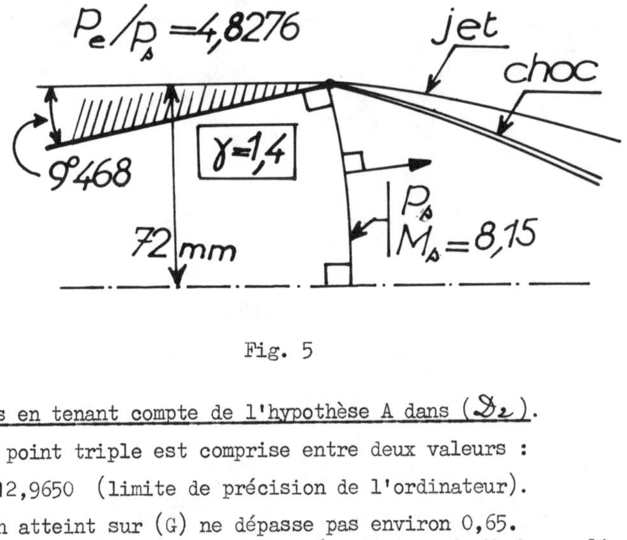

Fig. 5

. Calculs effectués en tenant compte de l'hypothèse A dans ($\mathcal{D}_2$).

L'abscisse x_T du point triple est comprise entre deux valeurs :

$312{,}9648 < x_T < 312{,}9650$ (limite de précision de l'ordinateur).

Le nombre de Mach atteint sur (G) ne dépasse pas environ 0,65.
Il existe très peu de différence entre les répartitions de Mach sur l'axe et sur (G).

. Calculs effectués en tenant compte de l'hypothèse B dans ($\mathcal{D}_2$).

Les résultats obtenus avec l'hypothèse A autorisent l'utilisation de l'hypothèse B. x_T étant grand devant y_T, le point T est désormais repéré par son ordonnée y_T. Les calculs étant un peu plus rapides on peut opérer en double-précision et progresser ainsi jusqu'au domaine transsonique. Le calcul a pu être poursuivi jusqu'à épuisement des chiffres significatifs ; on trouve donc :

$5{,}884\ 624\ 494\ 464\ 176 < y_T < 5{,}884\ 624\ 494\ 464\ 180.$

Dans le plan (φ, M) (fig. 6) sont représentées les principales étapes du calcul, les courbes se divisant en deux familles. Le point de divergence P_o des dernières courbes y_{min} et y_{max} correspond à M = 0,925 environ.

Le tableau ci-dessous donne les valeurs de y_T correspondant à chacune des courbes.

Fig. 6	y_T
10	5,884 65
8	5,884 625
6	5,884 624 495
4	5,884 624 494 47
2	5,884 624 494 464 180
	5,884 624 494 464 178
1	5,884 624 494 464 176
3	5,884 624 494 45
5	5,884 624 493
7	5,884 623 437 5
9	5,884 60

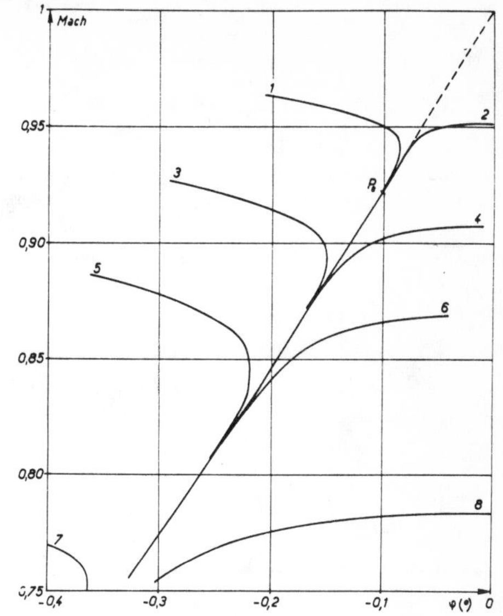

Fig. 6 - Jet supersonique. **Hodographes** pour différentes valeurs de y_T.

Le domaine transsonique s'étend de M = 0,925 à M = 1,25 (fig. 7)

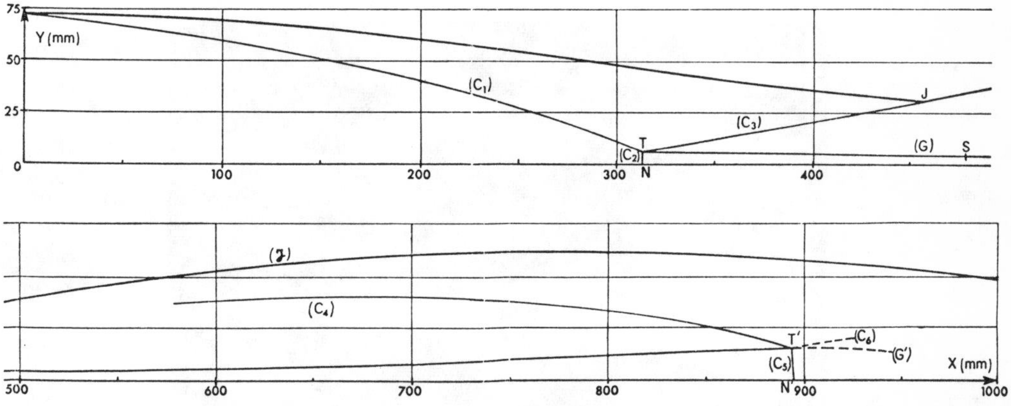

Fig. 7 - Réflexion d'une onde de choc sur l'axe dans le cas d'un jet supersonique (M = 8,15).

La reprise des calculs dans ($\mathcal{D}_1$) s'est effectuée sans difficulté ce qui prouve que l'extrapolation de la ligne de glissement à partir de résultats obtenus en subsonique était compatible avec la régularité de l'écoulement dans le domaine supersonique déjà acquis.

Une autre onde de choc incidente s'est formée par confluence des caractéristiques. Elle rencontre la ligne de glissement en T'. La détermination des conditions aval s'effectue sans difficulté suivant un procédé analogue à celui décrit au § I.2. Le calcul poursuivi comme précédemment à partir du point T' a montré que la ligne de glissement s'incurvait vers l'axe alors que le nombre de MACH valait 0,843. Il n'y a d'ailleurs pas de raison physique pour que le second col se place systématiquement à l'intersection de la ligne de glissement avec le choc incident.

Pour chacun des points T et T' les angles φ_1 , $\sigma_{1,4}$ et l'angle du choc avec Ox ont pour valeur :

point T $\varphi_1 = 0°46$, $\sigma_{1,4} = 89°90$, $\varphi_1 + \sigma_{1,4} = 90°36$
point T' $\varphi_1 = 2°32$, $\sigma_{1,4} = 89°72$, $\varphi_1 + \sigma_{1,4} = -87°40$
} concavité vers l'amont.

Une confrontation avec l'expérience a été tentée en respectant le saut de pression p_e/p_4 mais dans le cas de l'expérience la température extérieure étant bien supérieure à celle régnant dans le plan de sortie de la tuyère, les effets de couche limite ont modifié la configuration de l'écoulement.

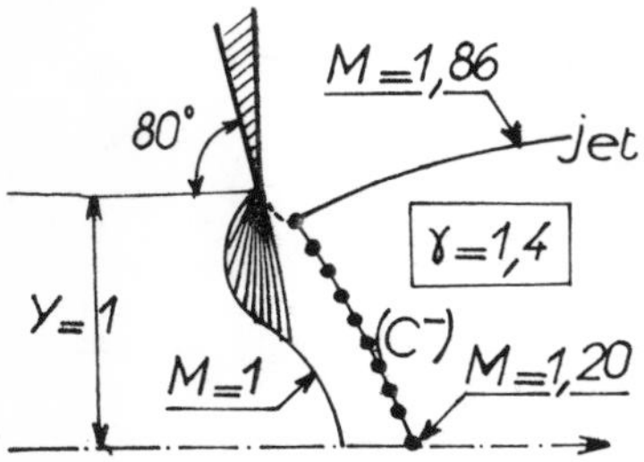

Fig. 8

V.II - Jet sous-détendu (fig. 8).

Il s'agit d'un convergent conique fortement incliné sur l'axe (80°) fonctionnant en régime sous-détendu (la dernière onde du faisceau de détente rencontre la ligne sonique). La caractéristique de départ a été obtenue par SOLIGNAC à partir de résultats d'interférogramme.

Le choc incident se forme par focalisation des caractéristiques (fig. 9).

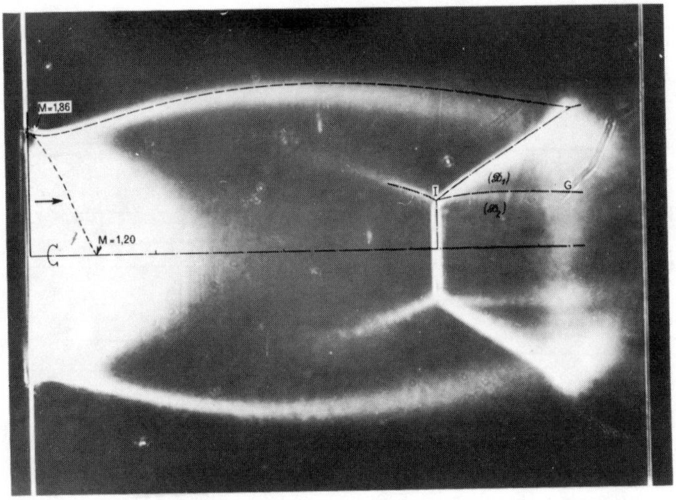

Fig. 9 - Réflexion d'une onde de choc sur l'axe dans le cas d'un jet subcritique.

Les ondes de détente issues du point d'intersection du choc réfléchi avec la ligne de jet influencent la ligne de glissement en amont du point sonique (fig. 10), provoquant une discontinuité de pente dans le diagramme (φ ,M). Bien que le calcul (effectué d'après l'hypothèse B) n'ait pas été poursuivi au-delà de $M \simeq 0,6$, la position du disque ($0,375 < y_T < 0,377$) recoupe exactement celle obtenue en lumière diffractée (SOLIGNAC).

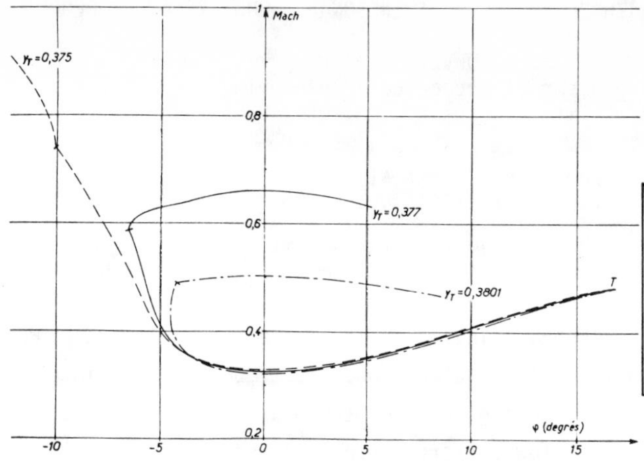

Fig. 10 - Hodographes du jet subcritique pour différentes valeurs de y_T .

ce qui par ailleurs, prouve l'exactitude des données de départ. Les angles φ_1 , $\sigma_{1,4}$ et l'angle du choc avec l'axe ont pour chaque y_T les valeurs suivantes

$y_T = 0,375$	$y_T = 0,377$
$\varphi_1 = 3°64$	$\varphi_1 = 3°66$
$\sigma_{1,4} = 85°62$	$\sigma_{1,4} = 85°54$
$\varphi_1 + \sigma_{1,4} = 89°26$	$\varphi_1 + \sigma_{1,4} = 89°20$

La concavité du disque est tournée vers l'aval.

- CONCLUSION -

La méthode décrite permet de prédire avec précision la position des différentes lignes de discontinuité. Les hypothèses utilisées s'avèrent suffisantes et adaptées à la poursuite du calcul au-delà du disque de MACH. L'approche du domaine sonique demande une précision extraordinaire mais en revanche les perturbations en aval du point triple affectent peu la position du disque. Les valeurs obtenues pour chacun d'eux montrent que leur concavité est tournée suivant les cas vers l'amont ou l'aval.

- REFERENCES -

ABBETT M. — AIAA Paper 70-231 January 19-21 1970

ASHRATOV E.A. — MEKHANIKA ZHIDKOSTI I GAZA, vol. 1 pp. 158-161 (1966)

BOWYER J.) D'ATTORRE L. and) YOSHIHARA H.) — dans "Supersonic Flow, chemical Processes and Radiative Transfer", Pergamon Press p. 177 à 200 (1964).

DIRINGER P. — N.T. ONERA N° 183 (1971).

KAWAMURA — Tokyo University, Institute of Science and Technology, Reports V6, n° 3, 1952.

MEL'NIKOV D.A. — IZV. AN SSSR, OTN, Mekhanika i Mashinostroyanye, n° 3, 1962.

SOLIGNAC J.L. — N.T. ONERA (à paraître) 1972.

COMPUTATIONAL PROBLEMS IN THREE AND FOUR DIMENSIONAL BOUNDARY LAYER THEORY

H. A. Dwyer
Associate Professor of Mechanical Engineering
University of California, Davis, CA, USA 95616

W. J. McCroskey
Research Scientist
Ames Research Center, Moffett Field, CA, USA 94035

INTRODUCTION

Over the past decade there has been an increasing interest in the computation of boundary layer problems involving time, three dimensionality or both. This interest has been due largely to the fact that the digital computer has made it possible to obtain the solutions to these complex flow patterns in a reasonable amount of time. However, even with the increasing capabilities of the digital computer and many research papers (References 3,6,7,8,9 and 11), exact solutions of the multi-dimensional boundary layer problems are at the present time of only limited usefulness to the design engineer. It is the purpose of this paper to show the many difficulties which remain to be solved, and to exhibit some solutions to these problems (only laminar boundary layer problems will be studied, since turbulent flow would involve too many uncertainties due to the inadequacies of eddy viscosity formulations).

Many unsolved problems remain in all parts of a time dependent or three dimensional boundary layer problem. The main things involved in numerically solving a problem are: (1) obtaining proper and consistent initial and boundary conditions for the equations; (2) developing a stable and unique numerical scheme to solve the equations; and (3) calculating the flow up to the separate line. In general it can be said that each physical problem must be treated individually concerning the above requirements, and there does not exist today a general and unique way for treating a problem which will carry out all of the necessary parts of the solution. In the present paper each of the above three parts is solved for a particular problem, and some satisfactory results are presented. The physical problems which are used to illustrate the results have a wide application, and are the following: (1) oscillating flows over cylinders with both the wall and the inviscid flow oscillating, and (2) time-dependent and three-demensional flow over a rotating airfoil in forward flight.

ANALYSIS OF THE BASIC EQUATIONS

The laminar boundary layer equations for three-dimensional and time-dependent flow can be written in terms of cartesion coordinates as

Continuity $$\frac{\partial u}{\partial x} + \frac{\partial v}{\partial y} + \frac{\partial w}{\partial z} = 0 \tag{1}$$

X-Momentum $$\rho\left\{\frac{\partial u}{\partial t} + \frac{\partial u}{\partial x} + \frac{v\partial u}{\partial y} + \frac{w\partial u}{\partial z}\right\} = \frac{-\partial p}{\partial x} + \frac{\partial}{\partial y}\left(\frac{\mu\partial u}{\partial y}\right) \tag{2}$$

Z-Momentum $$\rho\left\{\frac{\partial w}{\partial t} + \frac{u\partial w}{\partial x} + \frac{v\partial w}{\partial y} + \frac{w\partial w}{\partial z}\right\} = \frac{\partial p}{\partial z} + \frac{\partial}{\partial y}\left(\frac{\mu\partial w}{\partial y}\right) \tag{3}$$

where u, v, and w are the boundary layer velocities in the x, y and z directions, respectively; t is the time coordinate; μ , viscosity; ρ , density; and p , pressure. In these equations the x and z coordinates play a "time-like" role since no diffusion occurs along them. Also, because of the lack of diffusion along these coordinates in the boundary layer approximation, initial value problems are posed for x and z as well as time. The "time-like" behavior can also be recognized by observing that the x and z derivative operators, $\frac{u\partial}{\partial x}$ and $\frac{w\partial}{\partial z}$, have the dimensions of an inverse time. Along the y coordinate where viscous diffusion is important, a boundary value problem is posed with the boundary conditions given at the wall and at the edge of the inviscid flow. Therefore, the system of equations given above has one diffusion and three "time-like" coordinates. The physical consequence of the above analysis is that information is diffused along the y coordinate in the boundary layer and convected along the other three coordinates.

The above equations have been written in terms of physical coordinates, but these coordinates are very inefficient for numerically solving boundary layer problems. It is much better to perform a transformation of coordinates which will remove many of the difficulties associated with boundary layer problems. The transformations which exist for three-dimensional and unsteady flow are not as highly developed as those for steady two-dimensional flow, but they are very valuable. The ones used most frequently are direct extensions of the two-dimensional ones, Reference (2) and succeed in locating the edge of the boundary layer, in removing leading edge singularities, and in stretching the coordinates so that very accurate finite difference approximations to derivatives can be made with relatively large grid sizes. The transformation used for the individual problems will be presented with the results of the calculations carried out on that problem.

NUMERICAL METHOD

For the solution of boundary layer problems it is generally best to use an implicit finite difference scheme, References (4,9) because of stability problems that can occur with explicit schemes at leading edges, solid walls and separation lines. The implicit schemes used for time-dependent and three-dimensional flows retain the tridiagonal form of the simultaneous equations, since there is only one diffusion coordinate. The real difference in the time and three-dimensional cases is the treatment of the marching or convective coordinates. A rather simple way of looking at the basic marching scheme is to consider the differential operator given below:

$$\frac{\partial}{\partial s} = \frac{u\partial}{\partial x} + \frac{w\partial}{\partial z} + \frac{\partial}{\partial t}$$

The operator is essentially a local time operator. In terms of it, the equations (1) through (3) have a form very similar to the two dimensional boundary layer equations, Reference (2). The stability and convergence properties of the two-dimensional equations are well documented, and they have been shown to be unconditionally stable for a positive $\vec{\Delta} s$. The problem that occurs in multi-dimensional boundary layer theory is non-uniqueness and not stability. When all of the "time-like" coordinates are positive (positive u and w) there are no problems with uniqueness, however, when one of the coordinates becomes negative (negative u and w) non-unique results may occur.

Another finite difference scheme which may be employed is to evaluate all y-derivatives at the unknown station, and employ backward differences for the x-, z- and t-directions. This method has the same stability behavior as the previous one, and generally exhibits the same uniqueness characteristics. Both methods were employed in the present paper and lead to the same results. The choice of which method should be employed is purely a matter of convenience and will depend on the problem being solved.

RESULTS OF THE NUMERICAL CALCULATIONS

A. Oscillating Time-Dependent Flows

An interesting type of time-dependent boundary layer flow is oscillating flow over a cylinder where it is possible to have either the wall or the inviscid flow oscillate. For the oscillating wall case there is no change in the inviscid flow, if the oscillations have small amplitudes; however, the initial conditions at the stagnation point and the boundary conditions at the wall are different than the steady flow. The initial conditions at the stagnation point have been worked out previously by Rott (Reference 10), but Rott's calculations are not complete enough for numerical solution in certain dimensionless frequency (k) ranges. Rott gave the velocity in the vicinity of the stagnation point as

$$u = axf'(n) + be^{i\omega t}g(n)$$

where $g(n)$ has both real and imaginary parts, and $g(n)$ is a function of $k = \omega/a$. Exact numerical solutions to Rott's equations for $k = 1.0$ and 2.0 are given in Table 1, and were used for initial conditions in the present investigation. The wall boundary conditions were given as a sinusoidally oscillating wall with $k = 0.25$ and $.5$ while the amplitude of the change in angular position was plus or minus 8°.

The condition $k = .025$ corresponds roughly to the oscillating wall condition that a helicopter rotor would experience in forward flight due solely to changing angle of attack, while the condition $k = .5$ corresponds to a high frequency oscillation. Figures (1) and (2) exhibit some of the results of our calculations. Figure (1) shows the steady separation line for the ideal inviscid flow

$$Ue = 2U_\infty \sin\theta$$

and the unsteady flow reversal line for $k = .5$. The flow reversal line is the point in the boundary layer where a substantial portion of the velocity in the boundary layer becomes reversed and the numerical calculations become unstable. The numerical calculations can accommodate small amounts of reversed flow, if the finite difference operator $(\frac{1}{\Delta t} + \frac{u}{\Delta x})$ is kept positive. The case of $k = .025$ essentially was the same as the steady case and the difference between steady separation and unsteady flow reversal was negligible. For $k = .5$ there was a significant difference between the flow reversal and separation lines and the flow reversal line was almost exactly in phase with the oscillating wall velocity. Figure (2) shows some typical velocity profiles near the flow reversal line for maximum wall velocity Uw. A comparison with a local Stokes solution for an oscillating plate shows that the oscillating wall acts very much like a superposition of the Stokes flow near the wall. There does not seem to be a strong interaction between the basic flow over the cylinder and the oscillating wall.

A very interesting problem is the flow over a cylinder where the inviscid flow itself is oscillating. This problem occurs naturally quite frequently over a wide Reynold's number range, and is quite famous at relatively low Reynold's numbers where the name Karmon vortex streets is associated with the phenomenon. For the case to be investigated in the present paper a relatively high Reynold's number case of $1.06 \cdot 10^5$ was studied. In order to carry out the boundary layer calcualtions measurements had to be made of the inviscid flow over the forward portion of the cylinder where a boundary layer exists. The results of the measurement of the inviscid flow with hot wire anemometers is shown in Figure (3). Also shown in Figure (3) are the measured limits of the zero wall shear location, which were obtained with hot surface film probes.

An important aspect of the data in Figure (3) is that the stagnation point θ_0 moves plus or minus 3.7°. This makes the boundary layer calculation very unusual, since one should always begin from the stagnation point where the flow attaches to the body. Therefore, the appropriate coordinates for the problem are those attached to the stagnation point, and moving with respect to the body in time. In this coordinate system the wall is moving; however, there are entirely different pressure gradients acting on the flow, and the initial conditions developed by Rott are not completely suitable. An analysis of the initial value problem showed that the Rott solution was a reasonable approximation to the initial conditions over a major portion of the oscillation (iteration and small steps were taken for the first few steps to relax the errors in the Rott initial conditions).

The complete calculation of the boundary layer showed a very much different behavior than the oscillating wall case, and some of the results are shown in Figures (4) and (5). Figure (4) exhibits the motion of the stagnation point and the wall zero shear stress line as a function of time. Also included are the measured limits of the zero shear stress line, and average measurements of Achenbach, Reference (1). As can be seen from the figure the numerical calculations agree almost exactly with both the average and time dependent measurements. A fascinating part of Figure (4) is the secondary zero shear stress line shown. During some parts of the cycle the shear stress at the wall went slightly negative, then became positive, then negative again. The secondary shear stress line is the location of the second reversal in shear stress. A more detailed picture of this phenomenon can be obtained from the velocity profiles shown in Figure (5). These profiles were calculated for the dimensionless time $\tau = 3/10\pi$, and are given for various angular positions around the cylinder. At $\theta = 78°$ the shear reversal has just occurred, while at $\theta = 82°$ the shear has again become positive. For an angular position of 90° the shear is very negative and the velocity profile is similar to a steady separated one. Clearly, however, it cannot be said that separation has occurred at $\theta = 78°$.

B. Flow Over a Rotating Plate in Forward Flight

As a final example the calculation of the flow over a rotating flat plate in forward flight will be presented, Figure (6). This problem is both time-dependent and three-dimensional and is of central importance to helicopter technology. The boundary conditions or inviscid flow for the problem are given in Reference (6) and represent an exact solution to the inviscid flow. Initial conditions must be given for the time coordinate t and the two "time-like" coordinates x and z. Since the problem is cyclic, the specification of the initial conditions in time is not difficult. This is the result of the fact that the problem can be started from some arbitrary condition, and the calculations carried out until each cycle is repeatable. In practice a good initial starting condition will considerably reduce the amount of calculation time, and a good initial condition is the purely rotating flat plate solution given in Reference (6). The initial conditions for the x coordinate are a local time-dependent Blasius solution, because of the singularity which exists at the sharp leading edge. In almost all problems with sharp leading edges the singularity at the leading edge usually dominates over time-dependent and three-dimensional influences. Initial conditions for the crossflow coordinate z are very difficult to obtain and approximate techniques have to be employed. The technique used in the present investigation was to employ the small crossflow approximation far out on the blade and calculate inward.

The results of our calculations are shown in Figure (7) where the primary flow relative shear at the wall is given at various dimensionless distances from the hub. The ratio of the forward flight velocity to the rotational velocity at a given position on the blade (x/z) is given by the parameter $V_1/\Omega Z$, and the primary flow wall shear has been made dimensionless with a local Blasius value. Clearly the local Blasius value is not a good approximation, since it is twenty times less than the true value of the shear.

For this particular problem the non-uniqueness which was found when integrating against the crossflow in Reference (4) did not appear. The only explanation is that the problem does not have a strong convective history. Also, it should be mentioned that this problem required only fifteen minutes of computer time on an IBM 7044 digital computer.

SUMMARY

As can be seen from the results presented in this paper, three-dimensional and time-dependent boundary layer flows can be calculated in a reasonable amount of time with the current generation of computers. However, serious individual problems with the initial conditions and uniqueness of the numerical methods still exist, and of course there is the ever-present problem of very few time-dependent and three-dimensional inviscid flow solutions.

REFERENCES

1. Achenbach, E. A., J. Fluid Mech., V. 34, Part 4, 1968, pp. 625-639.
2. Blottner, F. B., AIAA J., Vol. 8, No. 2, Feb. 1970, pp 193-205.
3. Der, J., AIAA J., V. 9, No. 7, July 1971, pp. 1204-1302.
4. Dwyer, H. A., AIAA Paper No. 71-57, AIAA 9th Aerospace Sciences Meeting, Jan 1971.
5. Dwyer, H. A., AIAA J., V. 9, No. 2, Feb. 1971, pp. 277-284.
6. Dwyer, H. A., and McCroskey, AIAA J., V. 9, No. 8, August 1971, pp. 1498-1505.
7. Dwyer, H. A., AIAA Paper No. 72-109, AIAA 10th Aerospace Sciences Meeting, Jan. 1972.
8. Fannelop, T. K., AIAA J., V. 6, No. 6, June 1968, pp. 1075-1084.
9. Krause, E. and Hirschel, E., Proceedings 2nd International Conference on Numerical Methods in Fluid Dynamics, Holt, M., Ed., Berkeley, Calif. pp. 132-137, 1970.
10. Rott, N., Quart. Appl. Math 13, 444-51, 1956.
11. Wang, K. C., J. Fluid Mech., V. 43, Part 1, 1970, pp. 187-209.

K	1.0		2.0	
n	g_i	g_r	g_i	g_r
0	0	1.0	.0	1.0
.2	-.07446	.8229	-.12832	.78848
.4	-.11640	.6566	-.19424	.59466
.6	-.13350	.5085	-.21493	.42890
.8	-.13290	.3821	-.20582	.29529
1.0	-.12100	.2785	-.17969	.19334
1.2	-.10310	.1968	-.14629	.11967
1.4	-.08316	.1347	-.11234	.06932
1.6	-.06388	.08927	-.08191	.03693
1.8	-.04692	.05721	-.05691	.01747
2.0	-.03303	.03542	-.03778	.00670
2.4	-.01451	.01219	-.01458	-.00085
2.8	-.00544	.00361	-.00473	-.00135
3.2	-.00175	.00091	-.00129	-.00068
4.0	-.00011	.00003	-.00005	-.00006
5.0	.0	.0	0	0
6.0	.0	.0	0	0

K	g'_{iw}	g'_{rw}
.1	-.049305	-.81226
.2	-.098426	-.81509
.5	-.24292	-.83429
1.0	-.46619	-.89594
2.0	-.82713	-1.0748
5.0	-1.5001	-1.5974
10.0	-2.1938	-2.2414

$u = axf'(n) + be^{i\omega t} g(n)$

$g = g_r + ig_i$

$n = y \frac{a}{\nu}^{1/2}$

Table I. Addition Solutions to Rott's Equations

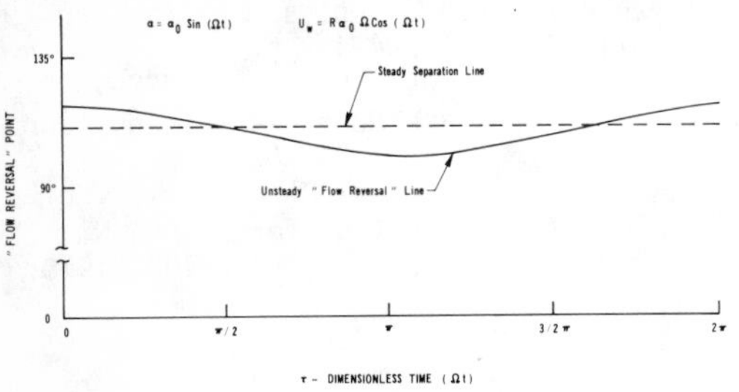

Figure 1. Calculation of Flow Reversal Line in the Flow over an Oscillating Wall

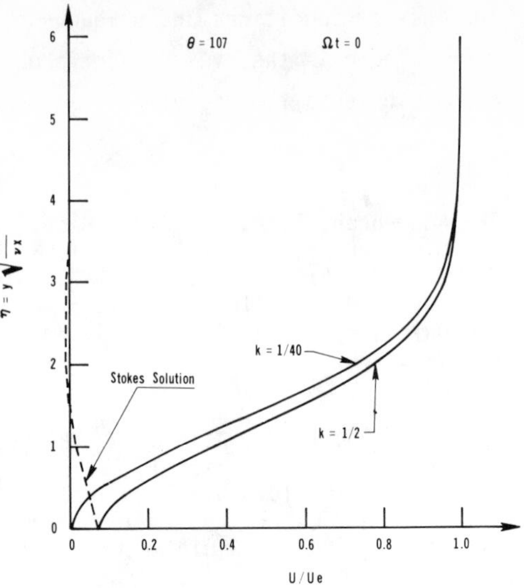

Figure 2. Velocity Profiles for Flow over an Oscillating Wall

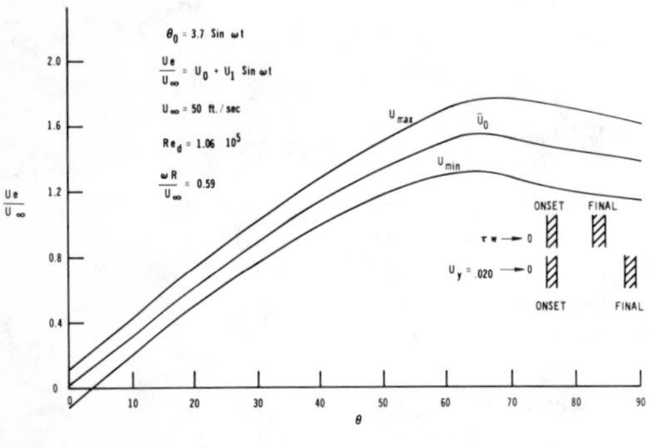

Figure 3. Inviscid Flow for Oscillations over a Cylinder

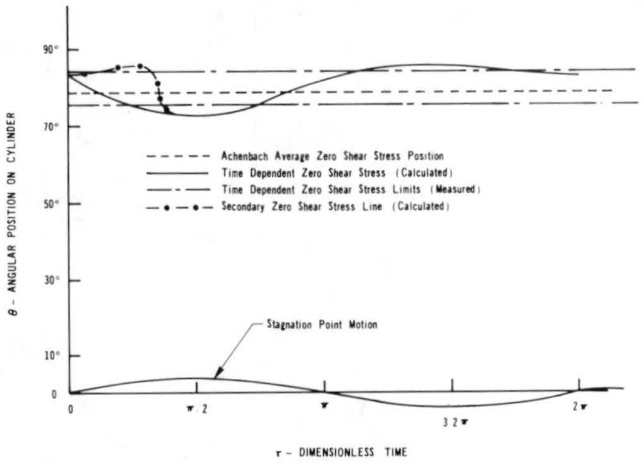

Figure 4. Calculation of Flow Reversal Line in an Oscillating Flow over a Cylinder

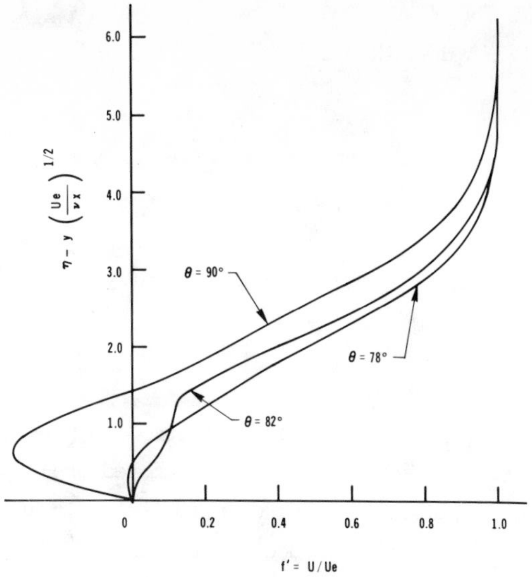

Figure 5. Velocity Profiles for Oscillating Flow over a Cylinder

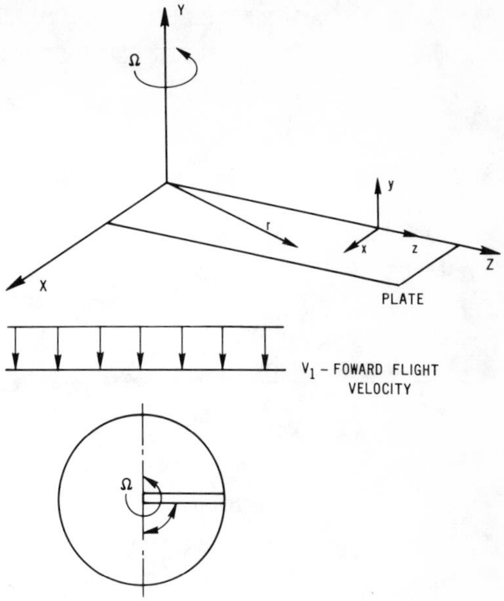

Figure 6. Geometry of the Rotating Plate in Forward Flight

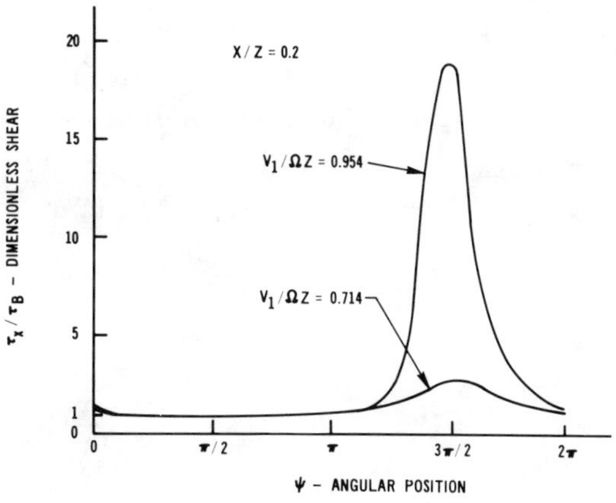

Figure 7. Primary Flow Relative Shear for a Rotating Plate in Forward Flight

A DIRECT METHOD FOR COMPUTING THE STEADY FLOW AT MACH NUMBER ONE PAST A GIVEN WING AIRFOIL

D. Euvrard and G. Tournemine

Département de Mathématiques et Informatique - Université de Rennes (France)

1. *Flow pattern at* $M_\infty = 1$

The steady flow pattern over a wing airfoil with free-stream Mach Number One is given by figure 1. Four successive regions can be distinguished. Region 1 is purely subsonic and spreads from infinity upstream down to the sonic lines. Region 2 includes two narrow supersonic strips bounded upstream by the sonic lines and downstream by the so-called "limiting characteristics", i.e. the only two characteristics stretching to infinity. Region 3 is a set of two supersonic flow fields, the downstream boundaries of which are shock waves starting from the trailing edge. Region 4 lies beyond the shock-waves and exhibits a rather intricate nature. Regions 1 et 2, the union of which is often called the "transonic range", are essentially connected one with the other and must be computed together. Once the transonic range is known, region 3 corresponds to a classical problem of pure hyperbolic type that we can easily solve thanks to the method of characteristics, in order to find the whole velocity distribution over the airfoil. Fortunately, the difficult region 4 has no practical interest. In fact, the only problem consists of computing the transonic range. This is a non-linear problem of mixed elliptic-hyperbolic type ; both sonic lines and limiting characteristics are unknown, and the general direction of characteristics is such that the perturbations due to the airfoil travel very far from the latter, which makes the problem still more difficult.

2. *General outline of the method*

This method is devoted to the computation of the flow at Mach number onte past any convex airfoil. However we shall henceforth restrict ourselves to the case of symmetrical flows. The fluid is assumed to be inviscid.

2.1. We look for the shape of the limiting characteristic and the velocity distribution over the part of the airfoil imbedded in the transonic range in two families of functions depending on a small number of parameters $p_1, \ldots, p_n$. The predicted velocity distribution over the airfoil, together with the given shape of the latter, defines Cauchy data, while the predicted shape of the limiting characteristic and the symmetry condition constitute boundary data. The integration of the mixed elliptic-hyperbolic system of non-linear equations is performed towards infinity thanks to a finite-difference scheme involving some numerical viscosity. At a certain distance from the airfoil we stop computing and we compare the numerical result with an asymptotic expansion. From this comparison there results the value of a distance named J. We can consider J as a function of the n parameters $p_1,\dots,p_n$, the calculation of which needs the integration of the differential equations of the motion. The problem is to minimize J $(p_1,\dots,p_n)$.

2.2. The asymptotic expansion of the perturbation velocity potential φ^P can be written in the form

$$\varphi^P = r^{2/5} f(s) + r^o f^* (s) + r^{-2/5} f^{**}(s) + O(r^{-4/5}), \qquad (1)$$

r and s being respectively some generalized distance and polar angle with respect to a well-fit origin [4].

We must point out that the perturbations due to the airfoil mainly travel in the direction normal to the flow and have still an influence far away from the body, so that it would be quite incorrect to do as though the uniform sonic flow was to be reached at a finite distance from the airfoil. This difficulty is quite analogous to the one encountered in wind-tunnel experiments. On the contrary, the expansion (1) is still valuable at a rather short distance from the body. This idea has also been retained in a recent work by Berndt and Sedin [1].

Equation (1) involves two unknown coefficients. When integration from airfoil towards infinity has been performed, these coefficients are computed by fitting the last numerical and asymptotical values of the velocity on the axis of symmetry and of the position of the limiting characteristic.

2.3. Our method is an aiming one, like Chushkin's [3]. The minimization of the final distance J is performed thanks to Hooke and Jeeves' classical method [7], still valuable when J cannot be expressed in analytical form, and whose running times are approximately proportional to the number of parameters. In fact, two of these parameters, which are connected with the shape of the limiting characteristic, play a limited part and can be computed with the help of the last values of the two coefficients involved in (1). In this way, minimization by means of Hooke and Jeeves' method is restricted to n-2 parameters, while the precision is the one of n parameters ; this mixed iteration-minimization process has been found to converge quickly.

3. *Cauchy problems with boundary conditions for a mixed elliptic-hyperbolic system of non-linear equations*

3.1. We integrate the full equations of inviscid fluid flow ; we have chosen a perfect gas with constant specific heats. The airfoil is explicitely assumed to have a blunt nose. As we want to write with great care the initial data on the airfoil and the boundary conditions on the limiting characteristic, we introduce curved coordinates named ξ and η so that $\xi = 0$ should be the airfoil, $\xi = \xi_\infty$ the outer edge of the domain of integration, $\eta = 0$ the axis of symmetry and $\eta = 1$ the limiting characteristic [figure 2]. In this system of coordinates, the equations of the motion can be written in the general form

$$\frac{\partial u_i}{\partial \xi} = \sum_{j=1}^{2} D_{ij} \frac{\partial u_j}{\partial \eta} , \quad i = 1,2, \tag{2}$$

$u_i(\eta, \xi)$, $i = 1,2$, being the two unknown velocity components and D_{ij}, $i = 1,2$, $j = 1,2$, four known functions of the independant variables ξ, η and of the unknowns u_1, u_2. We can notice that the shape of the limiting characteristic changes from one iteration (or integration) to the other, so that such a system of coordinates is a floating one. In the transformed (η, ξ) plane, the domain of integration is the rectangle $[0,1] \times [0, \xi_\infty]$, with initial data u_1, u_2 on $\xi = 0$, $0 \leq \eta \leq 1$, boundary data u_1, u_2 on $\eta = 1$, $0 \leq \xi \leq \xi_\infty$ and symmetry condition on $\eta = 0$, $0 \leq \xi \leq \xi_\infty$.

3.2. System (2) is quasi-linear and of mixed type. It is a well-known fact that a Cauchy problem for such a system is "badly-posed" because it does not satisfy the third Hadamard's condition. If the data are regular enough, the analytical continuation exists very likely and is probably unique, but does not depend continuously upon these data ; specifically, the oscillations of short wave-lenght are rapidly amplified, which can cause catastrophic numerical consequences. However one can exhibit neighbouring problems by making the system (2) slightly parabolic, according to the idea of the method of quasi-reversibility [8]; we can interpret this technique by saying that we introduce a numerical viscosity which kills the oscillations of short wave-length. To realise this technique, we have chosen to use Lax scheme. It is a well known fact that this scheme is stable in supersonic regions whenever Courant-Friedrichs-Lewy condition is realised. The possibility of applying the same scheme to subsonic flows has been pointed out by Bratos, Burnat and Prosnak [2]. Exhaustive numerical experiments on problems whose exact solutions were known have been performed in Rennes University ; they have confirmed and completed the results given in [2], together with a study of the matrix of amplification. Mathematically speaking, Lax scheme is not "stable" in subsonic regions. But, thanks to certain conditions, it can be considered as "pseudo-stable", i.e. it damps a little number of the possible oscillations of shorter wave-length ; our precise studies have shown that this was quite sufficient for our practical purpose. Lax scheme is a first-order one, but extrapolation procedures provide second-order accuracy. In fact, as far as we want to use an orthogonal grid in the (η, ξ) plane, the numerical viscosity involved in the scheme may be just sufficient in some regions to ensure stability or "pseudo-stability", while it may be superabundant in some others. This is why, instead of the Lax scheme S_1 itself, we use a convex combination $S = (1 - \alpha) S_1 + \alpha S_2$ of S_1 and of another first-order but unstable scheme named S_2, with α $[0,1[$. To give an example, the viscosity term corresponding to Laplace equation is

$$\varepsilon \approx \frac{1}{2} \left[\Delta \xi + (1 - \alpha) \frac{(\Delta\eta)^2}{\Delta \xi}\right] \frac{\partial^2 u_i}{\partial \eta^2} \quad , \; i = 1,2, \qquad (3)$$

so that ε can be controlled with the help of α ; α is not constant, but depends on the mesh point. Figure 3, which is related to supersonic regions, shows another example of the usefulness of controlling the numerical viscosity : α will be taken equal to zero in A_3 because of Courant-Friedrichs-Lewy condition, while, in A_1 and A_2, α will have some values belonging to $]0,1[$. We have exhibited precise conditions for stability and pseudo-stability of S scheme.

4. *Numerical results*

Figure 4 shows the computed transonic range corresponding to classical NACA 0012 airfoil with zero angle of attack, and gives a little number of our numerical results. In fact, the whole domain $0 \leq \xi \leq \xi_\infty$ has been computed ; for $\xi > \xi_\infty$, one can describe the flow thanks to equ. (1), the coefficients of which have been calculated by our program for NACA 0012. Figure 5 gives the Mach distribution over the airfoil as a function of the abscissa, the origin being the leading edge and the length unit the airfoil chord. Figure 6 presents the same Mach distribution as a function of the velocity direction over the airfoil.

Very nice experimental measurements were achieved for us in Lille Institute for Fluid Mechanics (I.M.F.L.) for the same NACA 0012 airfoil (G. Gontier, A. Dyment, P. Gryson). Their results are dotted on figures 5 and 6. The agreement between numerical and experimental data are fairly good.

Il must be pointed out that our numerical Mach distribution over the airfoil is given by the analytical formulas we spoke of in § 2.1. whose parameters are available as soon as minimization indicated in § 2.3. has been achieved.

Our first approximate results were given in [5] ; our last and precise results are to appear in [6]. A detailed paper is to be published in a few months in "Journal de Mécanique".

References

[1] S.B. Berndt, Y. Sedin, "A numerical method for transonic flow fields", I.C.A.S. 7, Roma, sept. 1970.

[2] M. Bratos, M. Burnat, W.J. Prosnak, IXth Symposium of Fluid Dynamics, Kazimierz, Poland, sept. 1969.

[3] P.I. Chushkin, Dokl. Akad. Nauk SSSR 113, pp. 517-519 (1957).

[4] D. Euvrard, "Journal de Mécanique", Vol. 6, n° 4, déc. 1967, pp. 547-592.

[5] D. Euvrard, G. Tournemine, C.R. Acad. Sc. Paris, t. 274, série A, 27 mars 1972, pp. 1071-1074.

[6] D. Euvrard, G. Tournemine, C.R. Acad. Sc. Paris, Juillet 1972 (to appear).

[7] R. Hooke, T.A. Jeeves, ""Direct Search" Solution of Numerical and Statistical Problems", J. Assoc. Comp. Mech., 8, 2 (April 1961), pp. 212-229.

[8] R. Lattès, J.L. Lions, "Méthode de Quasi-Réversibilité et Applications", Dunod, Paris, 1967.

This research has been developped under contract of Direction des Recherches et Moyens d'Essai, French Ministry of Defence.

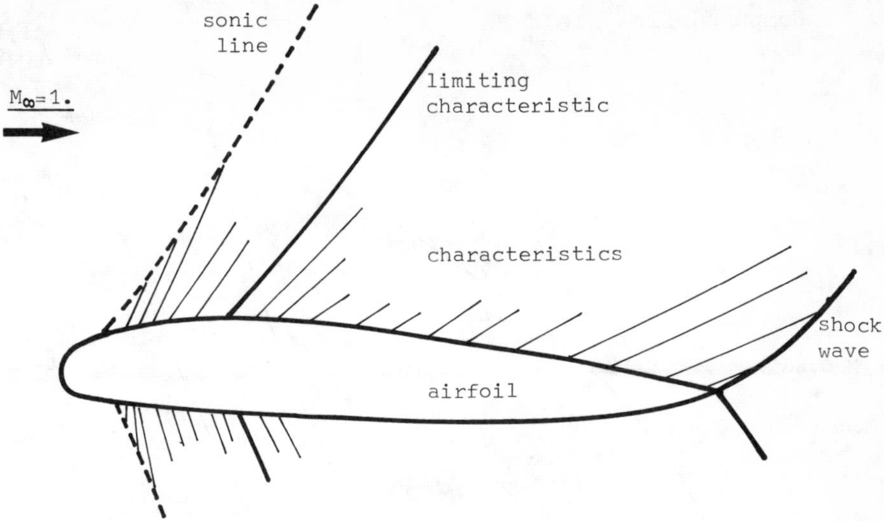

Fig. 1. Flow pattern at $M_\infty = 1$

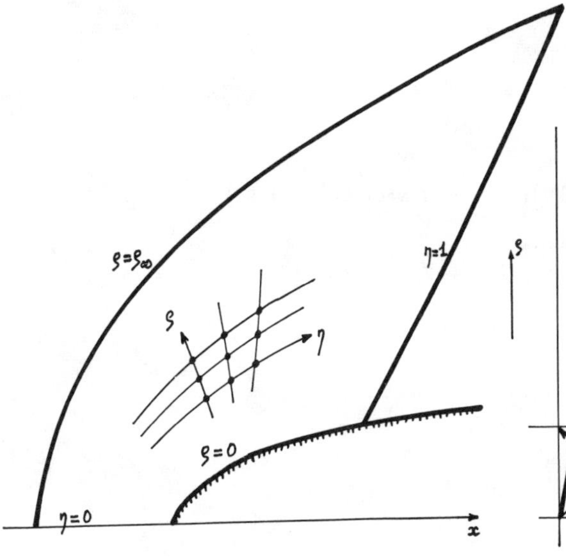

Fig. 2. Curved coordinate system

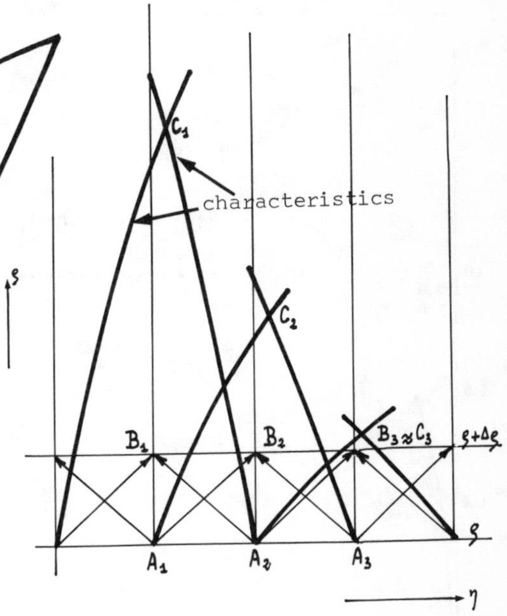

Fig. 3. Courant-friedrichs-lewy condition in supersonic regions

Fig. 4. Computed flow field

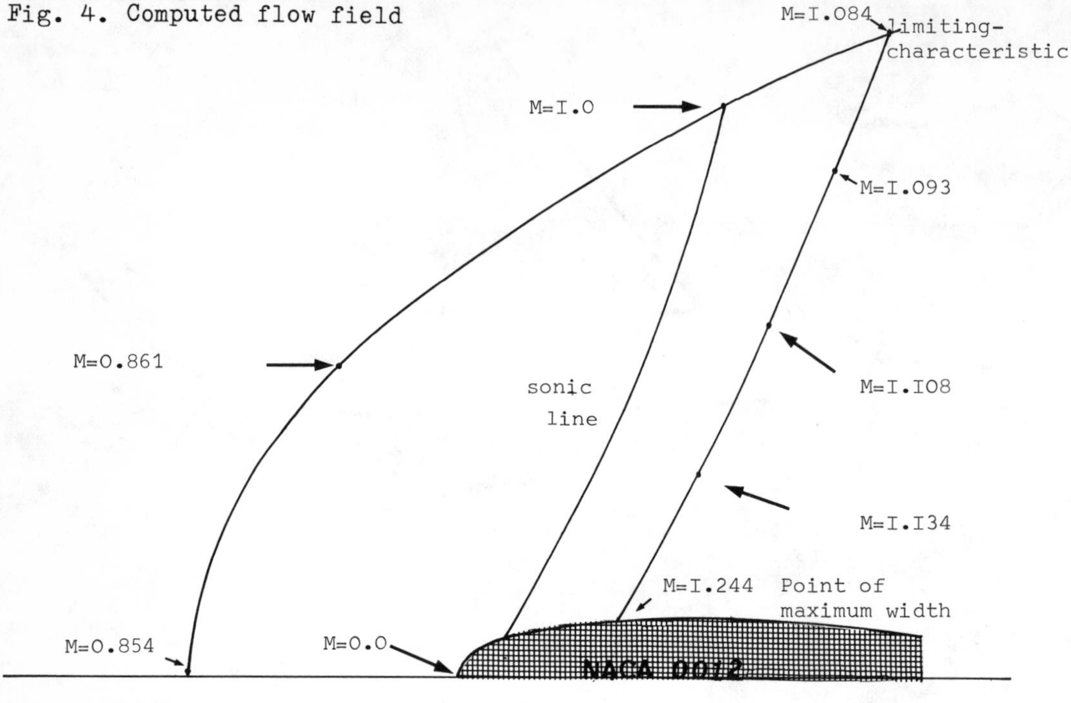

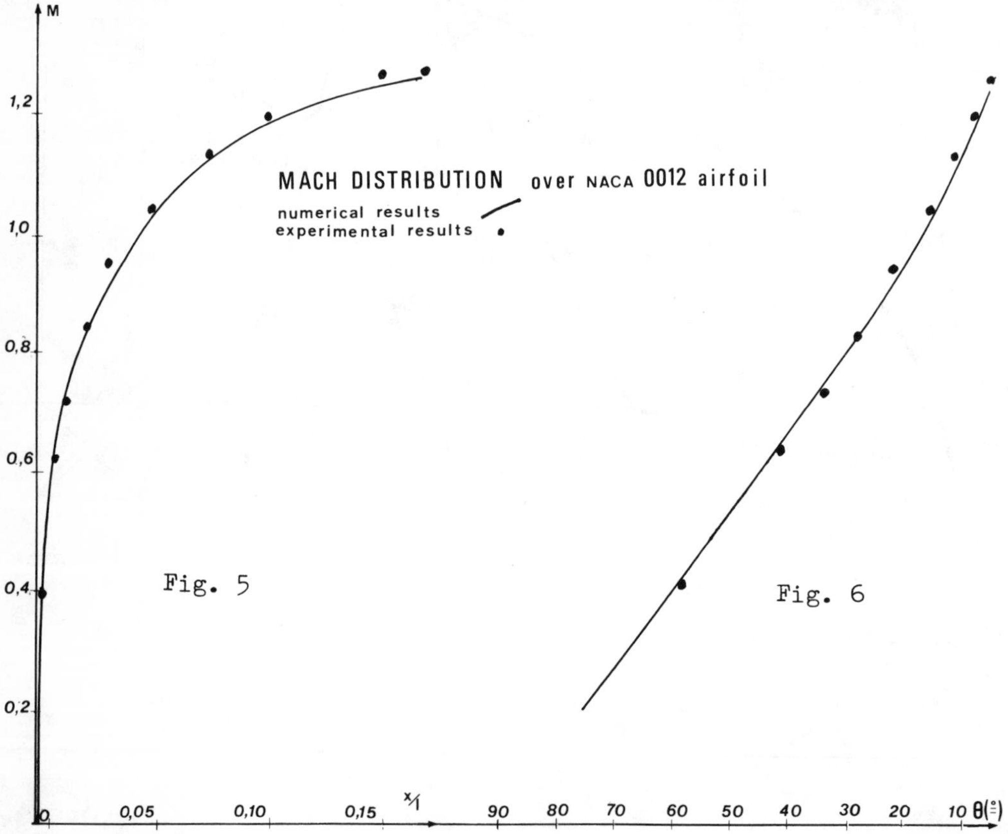

Fig. 5

Fig. 6

THE STRUCTURE OF A REFLECTING OBLIQUE SHOCK WAVE

Ernst Heinrich Hirschel

DFVLR-Institut für Angewandte Gasdynamik, Porz-Wahn, W. Germany

INTRODUCTION

In 1962 Liepmann, Narasimha and Chahine published a paper on the structure of a plane shock layer. It was shown that the Navier-Stokes solution gives reasonably good results up to approximately the Mach number $M = 2$ compared with the solution obtained with the Bhatnagar-Gross-Krook model of the Boltzmann equation. For higher Mach numbers only the high pressure region of the shock wave is properly prescribed. The low pressure region extends not far enough into the upstream region. Thus the Navier-Stokes solution is a usefull tool for the calculation of the structure of shock waves, as long as this shock waves are weak.

In investigations of the low density flat plate problem Rudman and Rubin in 1968, and Cheng, Chen, Mobley and Huber in 1969 presented independently one-layer approaches to that problem by using finite difference marching techniques in order to integrate the governing equations. The shock formation region (merged layer) in which the shock wave is build up is calculated beginning at the (sharp) leading edge of the plate with the wall as inner boundary and the undisturbed flowfield as outer boundary. At the end of the shock-formation region, the shock wave emerges with "jump conditions" approximately equal to the Rankine-Hugoniot conditions corresponding to the calculated local inclination angle of the shock wave (Rudman and Rubin (1968), Hirschel (1970)). The inclination angle of the shock wave is small and although the freestream Mach number is high, the shock shape can be considered as a good approximation, accordingly to Liepmann et alii, since the normal shock Mach number is small.

In order to study this emerging shock wave in more detail and to test a new possibility of calculating the structure of weak shock waves in general, the following numerical experiment has been set up:

Consider an oblique shock wave which is reflected at a wall without boundary layer. (This is the classical case of shock reflection studied in inviscid supersonic gas dynamics). Upstream of the reflection point the initial conditions are given in form of the shock wave profiles and the portion of the undisturbed flow between the shock wave and the wall as seen in Fig. 1 (u and v are the velocity components in x- and y-direction, respectively, ρ is the density, T the temperature, M the Mach number and β the shock wave inclination angle).

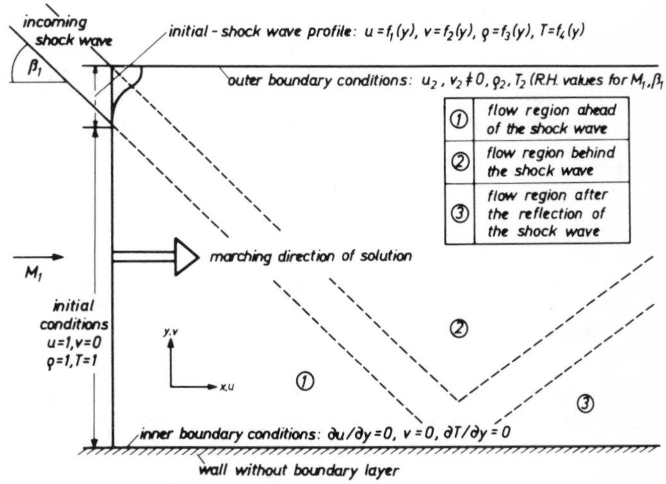

Fig. 1. Schematic of flow field

As discussed in the next section, the governing equations are of such a character that a downstream proceeding marching solution can be employed.

The solution is expected to show the shock wave extending down to the wall and its reflection at the wall. The calculated "jump conditions" after the reflection can be compared against the values corresponding to the reflection of a Rankine-Hugoniot shock wave with the freestream Mach number M_1 and the inclination angle β_1 at a wall without boundary layer. The resulting shock wave structures (incoming and reflected wave) must be onedimensional except for the region where the reflection takes place.

Two conditions are to be observed:
1) the inclination angles β of the shock waves must be small in order to allow the omission of the second order derivative terms in x-direction in the Navier-Stokes equations and the energy equation (see next section) and
2) the Mach number in direction normal to the shock wave $M_N = \sin \beta\, M$ must be small ($M_N < 2$) in each case, otherwise the Navier-Stokes approach is not justified.

GOVERNING EQUATIONS

Consider the Navier-Stokes equations and the energy equation for a thermally perfect monatomic gas in two dimensions for the stationary case (Liepmann et alii made their investigation for Argon; the same is done here in order to have the possibility of comparisons, and to avoid all the complications due to the inner degrees of freedom of polyatomic gases).

If the angle of inclination β of the shock waves is very small by an order of magnitude analysis one arrives at the equations derived first by Rudman and Rubin for the flate plate problem. Large gradients of all dependent variables are present only in y-direction (see Fig. 1), and in the y-momentum equation and in the energy equation all second order derivatives in x-direction are small. The same is true for all viscous terms except the $\partial(\mu\, \partial u/\partial y)/\partial y$-term, and even the pressure gradient term in the x-momentum equation.

Dealing with the present problem, only the second order derivatives in x-direction are neglected in the Navier-Stokes equations and the energy equation. Consequently, larger angles of inclination β can be included in the investigation and the two-point boundary value properties in x-direction are omitted.

Together with the continuity equation and the equation of state, a set of governing equations is obtained in this way, which is in all probability essentially of hyperbolic character if the flow is supersonic, and elliptic if the flow is subsonic. Although this has not been proved as yet, certain clues point to this fact:

1) In the incompressible case, the Navier-Stokes equations and the continuity equation form a system, the type of which can be determined by the technique described for instance by Courant and Hilbert. Omitting the second order derivatives in x-direction ($\mu\, \partial^2 u/\partial x^2$ and $\mu\, \partial^2 v/\partial x^2$) of the Navier-Stokes equations, the system remains elliptic. Only after omitting the pressure gradient term $\partial p/\partial x$ of the x-momentum equation, the system becomes parabolic in x-direction.
2) If the pressure gradient term $\partial p/\partial x$ of the x-momentum equation is not neglected, the governing equations for the flat plate problems are not solvable in the subsonic part of the flow. In the supersonic part no difficulties arise, when both pressure gradients are retained.

This problem has been noted or investigated by several authors, for instance: Baum and Denison (1967), Ferri and Dash (1969), Cheng et alii (1970), Rubin and Lin (1971) and Hirschel (1971a) and is not solved.

For the present investigation no problems arose in the solution (see next chapter), probably because the whole flow field is always supersonic.

The transport properties for Argon used in the calculations are approximations of the Chapman-

Enskog data for the Lennard-Jones (6-12) potential. The viscosity is given by

$$\mu = 0.1444 \cdot 10^{-4} (11.9886\, T^{0.75} - 53.96549\, T^{0.7} + 81.31478\, T^{0.65} - \tag{1}$$
$$- 40.38637\, T^{0.6})\ [\mathrm{g\ cm^{-1}\ s^{-1}}]$$

and the heat conductivity by

$$K = 0.78053 \cdot 10^{7} \mu\ [\mathrm{g\ s^{-3}\ K^{-1}}] \tag{2}$$

The initial conditions are prescribed at a point $x = 0$ upstream of the reflection region (see Fig. 1). In the part of the flow field between shock wave and wall the initial conditions are the free stream conditions given, in dimensionless form, for $0 \leq y \leq y_s$:

$$u = 1, \quad v = 0, \quad \rho = 1, \quad T = 1, \tag{3}$$

and in the shock wave, for $y > y_s$:

$$u = f_1(y), \quad v = f_2(y), \quad \rho = f_3(y), \quad T = f_4(y). \tag{4}$$

The functions $f_1 - f_4$ are found by transforming the profiles u and T calculated by Liepmann et alii for the plane normal shock layer into profiles for the oblique shock wave with the angle of inclination β_1.

The inner boundary is the wall (without a boundary layer), which intercepts the shock wave, and since the flow field in this case is identical to either half of the flow field generated by the intersection of two oblique shock waves of equal strength but of opposite families, the proper boundary conditions are:

$$x > 0, \quad y = 0: \quad \partial u/\partial y = 0, \quad v = 0, \quad \partial T/\partial y = 0. \tag{5}$$

The outer boundary lies at the station y where the shock wave profiles have assumed the Rankine-Hugoniot jump conditions after the shock according to the free stream shock Mach number M_1 and the angle of inclination β_1 (Fig. 1: R. H. values for M_1, β_1). The calculation ends when the reflected shock wave intersects the outer boundary.

NUMERICAL SOLUTION

The set of governing equation is integrated by an implicite finite difference solution, having the features of a typical marching technique calculation. The Crank-Nicholson scheme and the Richtmeyer algorithmen are employed with the solution vector of the form

$$W = \begin{bmatrix} u \\ v \\ T \end{bmatrix} . \tag{6}$$

The density is determined from the continuity equation by using the iteration method of Krause, (1967), (originally Krause calculates the normal component of velocity v in boundary layer problems by means of the continuity equation). The method of solution for two momentum equations, although for slip flow boundary conditions (flate-plate problem), is described in a paper by Hirschel (1971b).

RESULTS

The flow conditions in the following example are $M_1 = 10$, $\beta_1 = 8.6269^\circ$, so that the normal shock Mach number is $M_N = 1.5$, which is the Mach number of the first case considered in the paper of Liepmann et alii. The stagnation condition $T_0 = 5807$ K yields the same free stream temperature; only the free stream pressure is lower, since it was intended to get a shock wave thickness of the same order of magnitude as observed in a typical hypersonic low density flat plate problem. Nevertheless a comparison with the data of Liepmann et alii is possible by simply non-dimensionalizing the coordinates with the mean free path calculated from the free stream conditions. Only the transport properties are different: Liepmann et alii employ Sutherlands law and a Prandtl number Pr = 1 because of their comparison with the B-G-K-solution.

Fig. 2 and 3 show the comparison of pressure and velocity profiles of the oblique shock wave in y-direction (see Fig. 1).

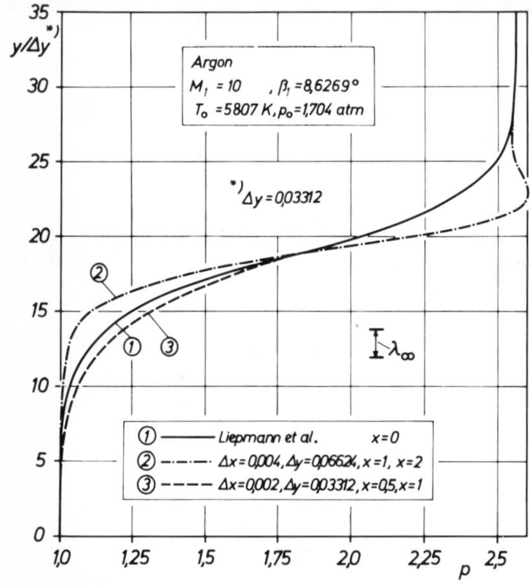

Fig. 2. Pressure p as function of y/Δy

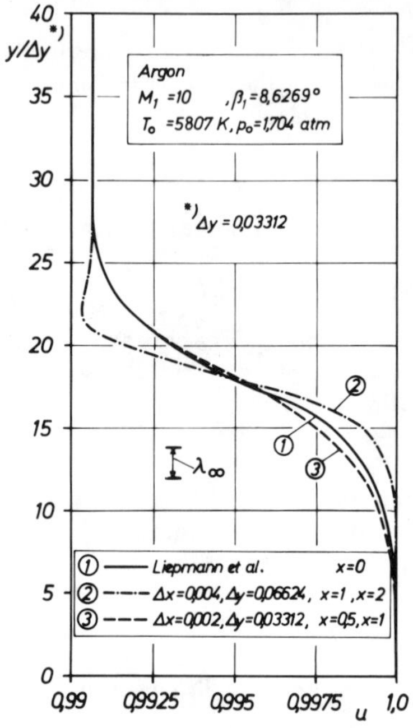

Fig. 3. Velocity component u as function of y/Δy

The pressure p and the velocity u are non-dimensionalized with the free stream pressure p_1 and the free stream velocity u_1, respectively. The shock thickness is approximately equal to eighteen free stream mean free pathes λ_∞ = 0.0134 cm (Δy = 0.03312 ≈ 0.00843 cm).

Curve 1 in both figures represents the initial conditions obtained from the first case (M_1 = 1.5) in the paper of Liepmann et alii. Covering the shock layer with 14 points in y-direction one obtains the curves 2, which appear in this form after about 100 steps in x-direction and remain constant for any station downstream. The stepsize Δy is obviously too large for this case. Reducing Δy and Δx to half of the values used first yields the curves 3. Further reducing does not alter the shape of the curves, showing that 25 to 30 steps throughout the shock layer are sufficient.

The difference between the curves 1 and 3 is due only to the different transport properties. This has been verified by applying the Sutherland viscosity and the Prandtl number Pr = 1 in the present approach.

Furthermore it has been found that any arbitrary initial profiles as long as they exhibit the general features of shock wave profiles (monotonic behaviour, the right asymptotic values upstream and downstream) converge after, say, one hundred steps in x-direction to the accurate profiles (in this case the profiles 3 in Figs. 2 and 3). It can be concluded that shock wave structures can be calculated with the present method with rough initial guesses without solving a two-point boundary value problem if the conditions outlined in the first chapter are satisfied.

Fig. 4 shows the pressure profiles at different stations x. Both the x- and the y-coordinate are expressed in the number of steps and in physical coordinates. All dependent variables are non-dimensionalized with the corresponding values in the flow region ahead of the incoming shock wave (region 1 in Fig. 1).

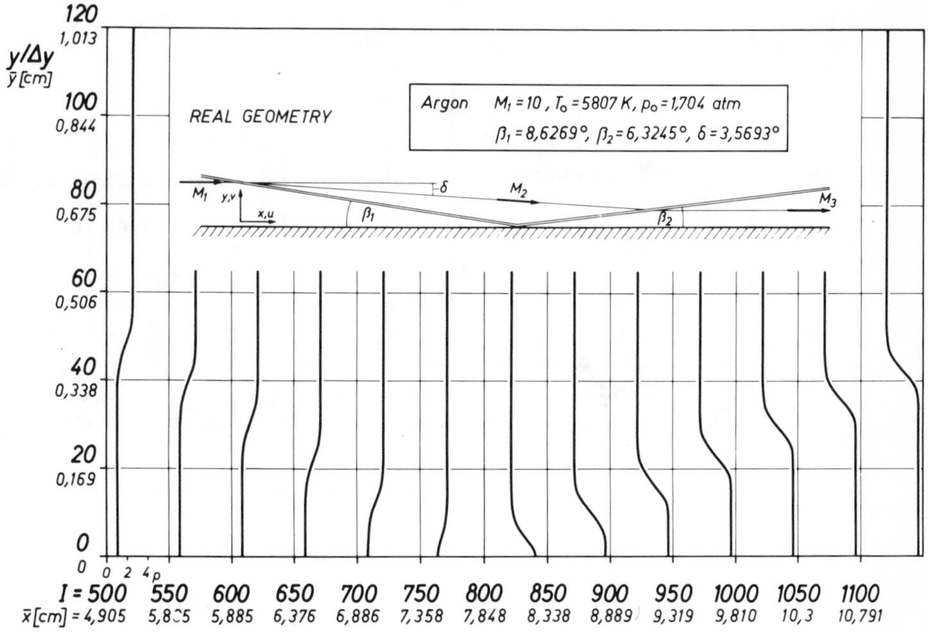

Fig. 4. Pressure profiles at different stations x

The extent of the reflection region in x-direction is approximately seventy free stream mean free pathes, compare Figs. 5 and 6. The extension in y-direction is approximately equal to the shock thickness. The schematic of the flow is sketched in the upper part of the figures.

Figs. 7 to 10 show the dependent variables as function of x at different distances from the wall.

It is worthwhile to point out that the Rankine-Hugoniot values after the reflection are computed with a very small error. Table 1 gives the relative error

$$\varepsilon = \frac{f - f_{RH}}{f_{RH}} , \tag{7}$$

where f stands for the dimensionless dependent variables obtained by the present method, each at the point of its largest deviation from the Rankine-Hugoniot value, and f_{RH} for the corresponding Rankine-Hugoniot values of the reflection process (M = Mach number).

Although in this example the inclination angles are very small, the method is applicable to larger angles, and also larger normal shock wave Mach numbers, yielding still satisfactory accurate results. Since no body length or boundary-layer thickness is introduced no lower limit is given concerning the rarefaction of the considered gas. A paper dealing with the range of applicability and the resulting errors together with other examples, as for instance the intersection of two shock waves with unequal strength is in preparation.

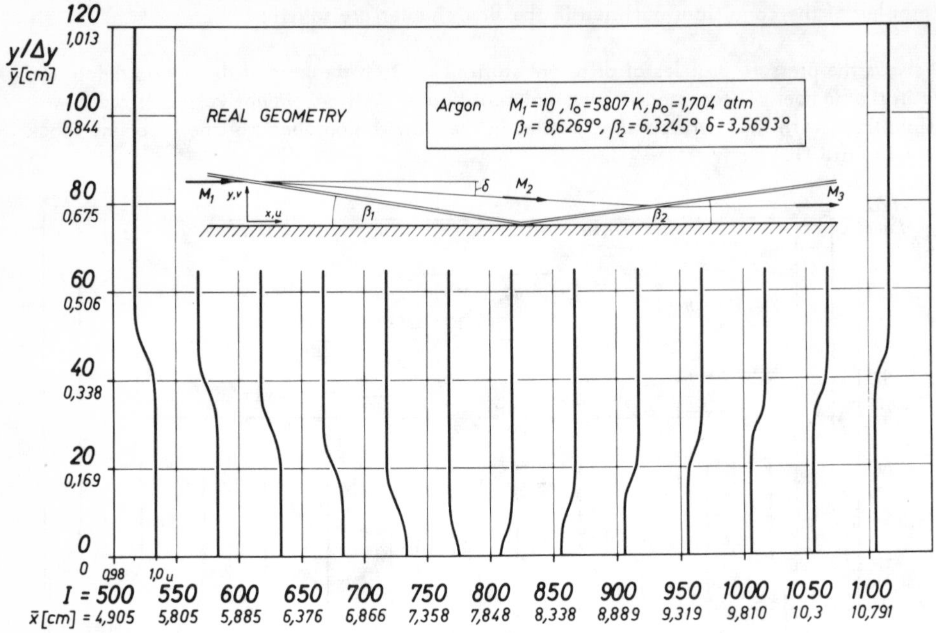

Fig. 5. Profiles of the u-component of the velocity at different stations x

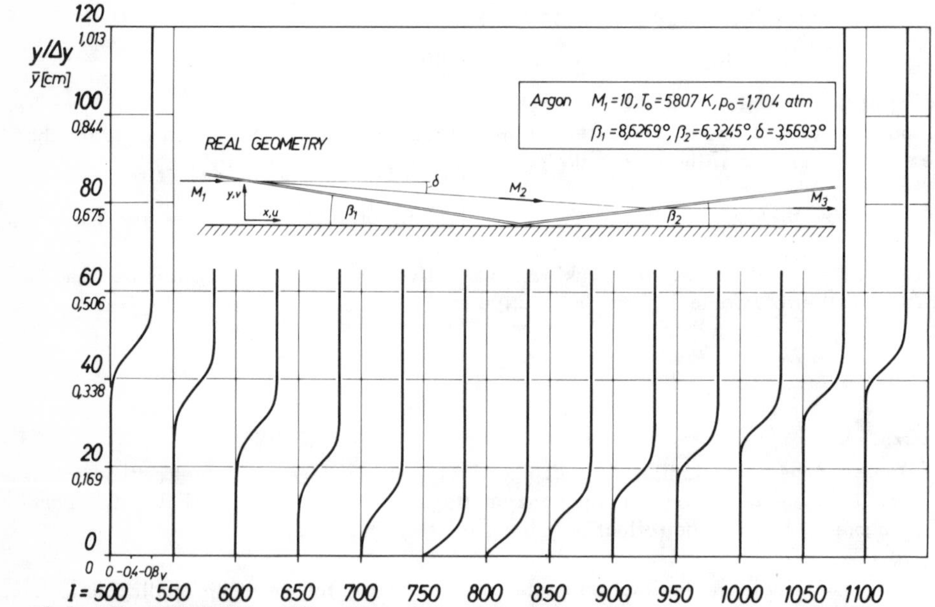

Fig. 6. Profiles of the v-component of the velocity at different stations x

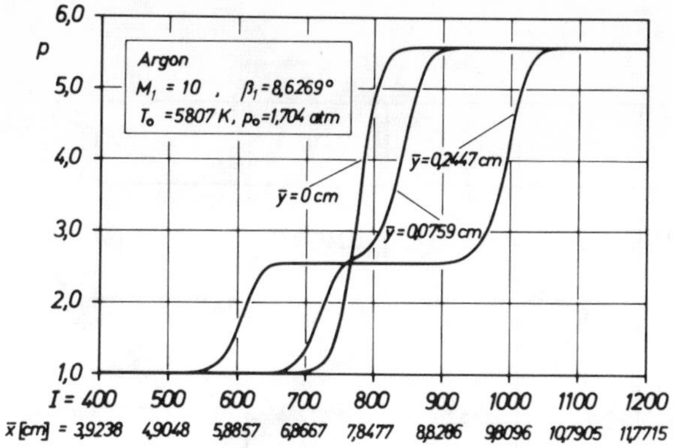

Fig. 7. Pressure profiles at different stations y

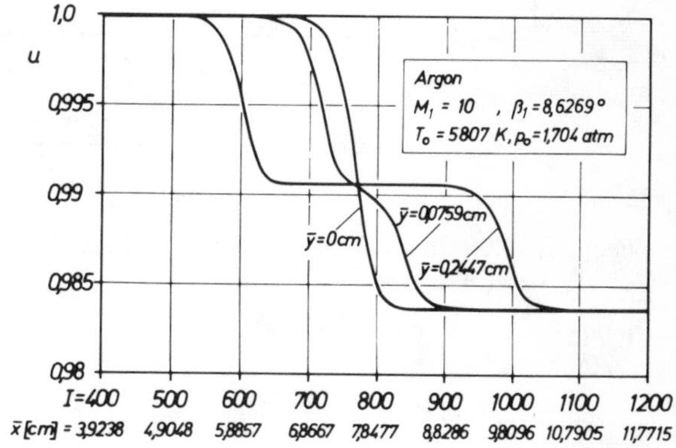

Fig. 8. Profiles of the u-component of the velocity at different stations y

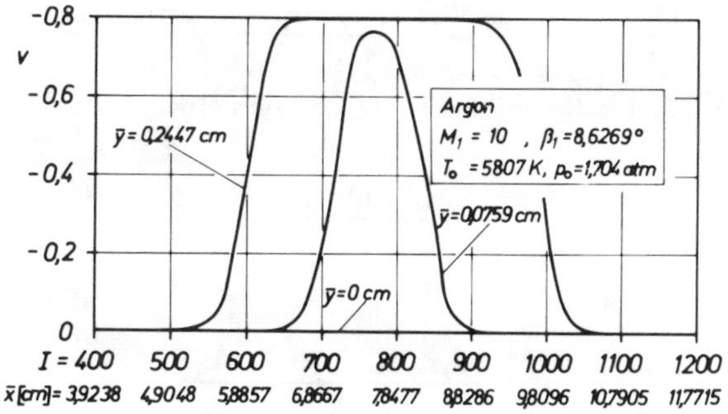

Fig. 9. Profiles of the v-component of the velocity at different stations y

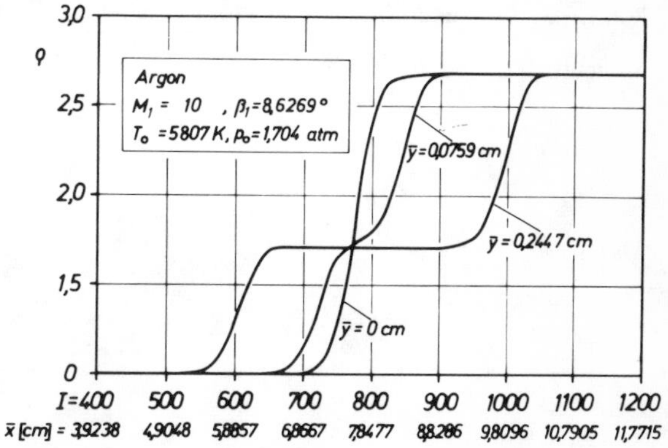

Fig. 10. Density profiles at different stations y

TABLE 1

ε of	u	v	ρ	T	p	M
Region 2 in Fig. 1	0.00004	0.00023	0.002	0.0023	0.00035	0.0011
Region 3 in Fig. 1	0.0019	absolute error: 0.0002 instead of 0.0	0.002	0.0018	0.00017	0.00094

REFERENCES

Baum, E., and Denison, M. R.: AIAA-J. 5, 1224-1230 (1967)

Cheng, H. K., Chen, S. Y., Mobley, R., and Huber, C. R.: Rarefied Gasdynamics, Suppl. 5, Vol. 1, Academic Press, New York (1969)

Cheng, H. K., Chen, S. Y., Mobley, R., and Huber, C. R.: Rand Corporation RM-6193-PR (1970)

Courant, R., and Hilbert, D.: Methoden der Mathematischen Physik, Bd. 2, Springer Verlag, Berlin (1937)

Ferri, A., and Dash, S.: ARL-Proc. of the 1969-Symp. on Visc. Interact. Phenom. in Supersonic and Hypersonic Flow, 271-317 (1970)

Hirschel, E. H.: Astronautical Research 1970, North-Holland Publ. Co. Amsterdam, 158-171 (1971)

Hirschel, E. H.: a) Laminare und turbulente Grenzschichten, DLR Mitt. 71-13, 51-73 (1971)

Hirschel, E. H.: b) DLR FB 71-97 (1971)

Krause, E.: AIAA-J. 5, 1231-1237 (1967)

Liepmann, H. W., Narasimha, R., and Chahine, M. T.: Phys. of Fluids, Vol. 5, 1313-1324 (1962)

Rubin, S. G., and Lin, T. C.: Pibal Rept. No. 71-8 (1971)

Rudman, S., and Rubin, S. G.: AIAA-J. 6, 1883-1890 (1968)

CALCULATION OF SEPARATED FLOWS AT SUBSONIC AND TRANSONIC SPEEDS

John M. Klineberg and Joseph L. Steger

Ames Research Center, NASA

Moffett Field, California, U.S.A. 94035

I. INTRODUCTION

One of the most important limitations of recently developed finite-difference methods for calculating transonic flows is that viscous effects are neglected. Although it is possible to correctly predict transonic drag rise, inviscid theory does not produce acceptable results for low Reynolds number flows or for airfoil sections at high lift coefficients. Experimental investigations of airfoils and wings at transonic speeds have indicated that the occurrence of significant scale effects is often related to the onset of trailing-edge separation. (Pearcy et al., 1968). The Reynolds number dependence of mixed subsonic-supersonic flows, in particular, is associated with the sensitivity of the shock-wave location to the pressures over the rear portion of the airfoil. The pressure distribution near the trailing edge, however, depends on the development of the boundary layer along the airfoil surface and on the extent of the region of separated flow.

This type of viscous-inviscid interaction cannot be analyzed by assuming that the viscous stresses result in perturbations on an inviscid flow field. There is a strong coupling between the inner boundary layer and the outer inviscid fluid, and the solution must be developed simultaneously in both regions. For supersonic flow, related strong-interaction problems have been studied using integral methods to describe the viscous layer and the Prandtl-Meyer relation for the external flow (e.g., Lees and Reeves, 1964). In the present investigation, a boundary-layer integral approach is combined with a finite-difference relaxation method to calculate viscous interactions at subsonic and transonic speeds. Results are obtained for separated laminar flows on circular-arc airfoils at zero angle of attack and are compared to the recent data of Collins (1972).

II. THE INVISCID FLOW

The flow field outside the viscous region is considered irrotational and the transonic small-disturbance approximation and thin-airfoil boundary conditions are assumed. The governing differential equations for the inviscid flow can be written

Continuity

$$\frac{\partial}{\partial x}\left[(1-M_\infty{}^2)\tilde{u}+\frac{\gamma+1}{2}M_\infty{}^2\tilde{u}^2\right]+\frac{\partial\tilde{v}}{\partial y}=0 \qquad (1)$$

Vorticity

$$\frac{\partial\tilde{u}}{\partial y}-\frac{\partial\tilde{v}}{\partial x}=0 \qquad (2)$$

where $\tilde{u}$ and $\tilde{v}$ are the normalized perturbation velocities with respect to the uniform stream. The finite-difference analogs of these equations are solved for the potential function and for the velocity components using the appropriate relaxation schemes developed by Steger and Lomax (1971). Because of the nature of the equations used for the viscous layer, mixed boundary conditions corresponding to $\tilde{v}$ in one region and $\tilde{u}$ in the other are specified for the inviscid flow (see Fig. 1). The selection of compatible differencing schemes for the direct and inverse calculations and the details of the relaxation method are described in Steger and Klineberg (1972).

III. THE VISCOUS FLOW

Differential Equations

For the inner flow, the governing system of equations consists of the conservation equations for mass and momentum and the equation for the rate of change of mechanical energy, obtained by multiplying the momentum equation by the velocity u. The boundary-layer approximations are assumed valid and the equations are integrated across the viscous layer and transformed into an equivalent incompressible form. The resulting system of three first-order nonlinear ordinary differential equation is as follows:

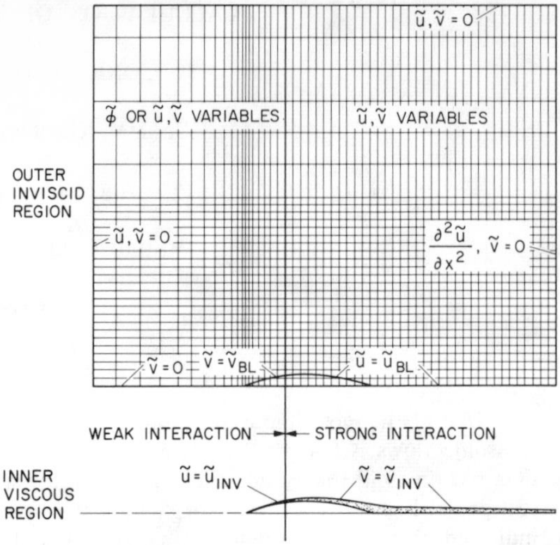

Figure 1. Map of flow regions showing boundary conditions

Continuity

$$f\delta_i^*(1+m_\infty)\frac{d\tilde{u}}{dx} + F\frac{d\delta_i^*}{dx} + \delta_i^*\frac{dH}{dx} = \frac{\tilde{v}-\alpha_B}{m_\infty} \tag{3}$$

Momentum

$$(2H+1)\,\delta_i^*(1+m_\infty)\frac{d\tilde{u}}{dx} + H\frac{d\delta_i^*}{dx} + \delta_i^*\frac{dH}{dx} = \frac{P}{Re_{\delta_i^*}} \tag{4}$$

Moment of momentum (mechanical energy)

$$3J\delta_i^*(1+m_\infty)\frac{d\tilde{u}}{dx} + J\frac{d\delta_i^*}{dx} + \frac{dJ}{dH}\,\delta_i^*\frac{dH}{dx} = \frac{R}{Re_{\delta_i^*}} \tag{5}$$

These equations have been written in terms of the velocity perturbations to be consistent with the transonic small-disturbance approximation used in the outer flow. The boundary-layer integral quantities H,J,P and R are functions of the velocity profile shape while f and F depend also on the free-stream Mach number, M_∞. The interdependence of the different profile quantities (e.g., $J = J(H)$) is approximated by the equivalent relation in similar flow. The boundary-layer and wake-like similarity solutions are used to express the various profile quantities as polynomial functions of a profile-shape parameter related to the inverse form factor, H (see Klineberg and Lees, 1969). With this approximation, the system of equations, Eqs. (3) - (5), contains four unknowns: the two perturbation velocity components, $\tilde{u}$ and $\tilde{v}$; the incompressible displacement thickness, δ_i^*; and the shape of the velocity profile, i.e., H. The system is closed by obtaining either $\tilde{u}$ or $\tilde{v}$ from the inviscid calculation. All physical variables are determined by transforming back to the compressible plane, e.g.:

$$\delta^* = \left[1 + m_\infty(1 + H)\right] \delta_i^*$$

$$C_f = \frac{2P}{Re_{\delta_i^*}}$$

where

$$m_\infty = \frac{\gamma - 1}{2} M_\infty^2, \quad Re_{\delta_i^*} = \frac{u_\infty c}{\nu_\infty} \frac{\delta_i^*}{c} (x)$$

and

$$H = H(x), \quad P = P[H(x)]$$

Method of Solution

The flow over the airfoil is divided into weak and strong interaction regions, as shown in Fig. 1. In the weak interaction region, near the forward part of the profile, the boundary layer is computed by an essentially standard procedure. The pressure distribution is obtained from the relaxation calculation of the outer flow field and the reduced system of equations, Eqs. (4) and (5), are solved by Runge-Kutta integration for $\delta_i^*(x)$ and $H(x)$. The continuity equation, Eq. (3), is uncoupled from the remaining equations and is solved for $\tilde{\nu}$, the inclination of the inviscid streamline at the edge of the viscous layer. Initial conditions near the leading edge are obtained from the approximation that the flow is locally similar ($\mathrm{d}H/\mathrm{d}x = 0$). This requirement provides relations for the initial profile quantities and displacement thickness:

$$PJ - HR = \frac{2C_o}{1+5C_o}(1 - H)R$$

$$Re_c\left(\frac{\delta_i^*}{c}\right)_0^2 = \frac{2}{1+5C_o}\frac{R}{J}\left(\frac{x}{c}\right)_0$$

where

$$C_o = (1 + m_\infty)x_o\left(\frac{\mathrm{d}\tilde{u}}{\mathrm{d}x}\right)_0$$

Note that for constant-pressure flows, ($\mathrm{d}\tilde{u}/\mathrm{d}x = 0$), the continuity equation, Eq. (3) can be written:

$$\tilde{\nu} = \alpha_B + \frac{\mathrm{d}\delta^*}{\mathrm{d}x}$$

and the present weak-interaction formulation is equivalent to a method that corrects surface slopes for the growth of the displacement thickness. The reduced system of equations has a saddle-type singularity where $J = H\,\mathrm{d}J/\mathrm{d}H$ that is a function of the profile shape. This point corresponds to the location of separation (P = 0) and is related to the separation-point singularity of classical boundary-layer theory.

The strong interaction region begins upstream of the profile maximum thickness and includes the separated boundary-layer flow along the surface and the wake downstream of the airfoil (see Fig. 1). In this region, the streamline angles are given by the outer inviscid computation and the full set of equations, Eqs. (3) - (5), are integrated to obtain $\tilde{u}$, δ_i^* and H. As opposed to the usual Prandtl formulation of boundary-layer theory, the pressure distribution is determined directly from the equations describing the viscous flow and is not specified in advance. The coupling between the two flow fields is determined by the variation in $\tilde{v}$ and is related to the mass entrained by the viscous layer. This formulation removes the singularities associated with both the location of boundary-layer separation and the rear stagnation point in the wake.

The strong-interaction equations are joined to the solution of the reduced system and the three equations are integrated smoothly through separation and into an almost constant-pressure plateau region. The value of the body slope, α_B, is assumed to remain constant downstream of the point where $d\tilde{u}/dx = 0$ to approximate the decreased dependence of the highly-separated boundary layer to the curvature of the body. This assumption is made because the slope of the dividing streamline, which is not specified in advance, is needed as a reference angle to determine the boundary-layer entrainment. The approximation is valid for thin airfoils with moderately separated zones, and the present method is not expected to produce accurate results for very extensive regions of reversed flow.

The integration is continued with constant α_B from the plateau region to the trailing edge, where the boundary layer is joined to the reversed-flow wake region. Three joining conditions are needed to determine the initial values for the wake computation, and in the present method, the pressure, displacement thickness and momentum thickness are taken to be continuous. For the dependent variables, these joining conditions require continuity of $\tilde{u}$, δ_i^*, and H at the airfoil trailing edge. The profile functions are changed to the relations for wake-like flow and the integration is reinitialized with $\alpha_B = 0$ and continued in the downstream direction. The correct solution passes through the wake rear-stagnation point and asymptotically approaches the far-field condition $\tilde{u} = 0$.

For many of the calculations presented in this paper, the separated laminar boundary layer at the airfoil trailing edge is joined to a turbulence-modeled wake to approximate the effects of transition that are present in low-speed flows. A constant eddy viscosity is assumed and the system of equations, Eqs. (3) - (5), is retained with the dissipation term on the right-hand side of Eq. (5) replaced by 0.03 HR. This approximation is described in Klineberg et al. (1972).

IV. THE COMPLETE INTERACTION

A complete interaction is calculated by alternately iterating the solutions to the viscous and inviscid equations. The initialization of a typical iteration sequence is illustrated in Fig. 2. Starting with the given airfoil slopes (curve 1), Eqs. (1) and (2) are solved by relaxation to determine the corresponding inviscid flow field. The computed surface pressure distribution (curve 2) is then specified for the viscous calculation. The weak-interaction equations, Eqs. (4) and (5), are integrated and the continuity equation, Eq. (3), is solved for $\tilde{v}$ (curve 3). These equations have a singularity near the separation point, and the values of $\tilde{v}$ diverge downstream of the airfoil maximum thickness. A joining point is selected upstream of this location and, to continue the iteration sequence, the distribution of $\tilde{v}$ is arbitrarily specified over the remainder of the flow field. For the present calculations on biconvex airfoils, a joining point at the 40 percent chord location and a cubic extrapolation extending one chord length downstream were used (curve 4).

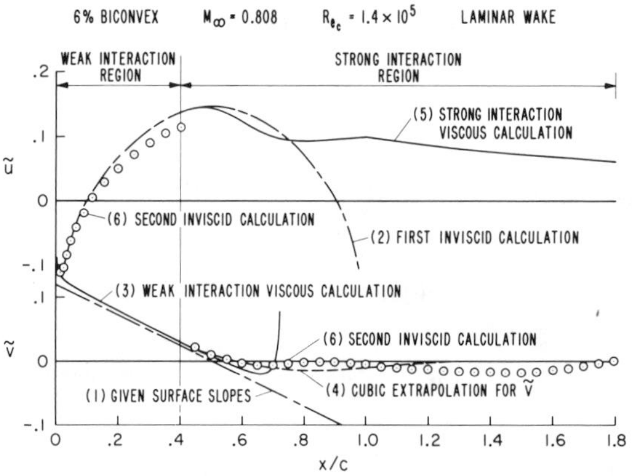

Figure 2. Iteration initialization

The assumed value of $\tilde{v}$ are specified for the integration of the strong-interaction equations, Eqs. (3) - (5), and the corresponding $\tilde{u}$ distribution is computed (curve 5). At this point in the iteration sequence, the solution for the viscous layer usually does not pass through the wake rear stagnation point and the flow does not return to free-stream conditions. This solution is sufficient to continue the iteration sequence, however, and a new relaxation calculation of the inviscid flow field is performed with mixed boundary conditions corresponding to $\tilde{v}$ in the weak-interaction region (curve 3) and $\tilde{u}$ in the strong interaction region (curve 5). The relaxation solution for the inviscid flow provides corrected values for $\tilde{u}$ and $\tilde{v}$ in the weak and strong interaction regions, respectively (shown by the symbols, curve 6). These distributions are then specified for the boundary-layer calculation, and the alternate iteration of the inner and outer solutions is continued until convergence is achieved.

The coupling of the two flow fields is particularly complicated because the inviscid equations are solved as a boundary-value problem and the viscous equations as an initial-value problem. Until a converged solution is obtained, the boundary-layer integration often diverges downstream of the airfoil and predicts unreasonable values of $\tilde{u}$ in the wake. At the same time, the inverse calculation for the outer flow can give discontinuous streamline slopes if the pressures specified by the viscous solution are imcompatible with the requirement for zero $\tilde{v}$ in the far field. The calculations are simplified by allowing a finite pressure at the downstream boundary, and the condition $\tilde{u}_{xx} = 0$ is imposed for the inviscid flow (see Fig. 1).

After the initialization of the iteration procedure, numerical instability is controlled by combining two techniques. First, the downstream boundary is taken relatively close to the airfoil trailing edge during the initial phase of the iteration and is subsequently moved downstream as the solutions become more stable. In this manner the divergent portion of the viscous calculation is not specified as a boundary condition for the inviscid flow. The second procedure is to under-relax the iteration sequence when the boundary plane is located downstream of the wake rear stagnation point. With this technique, a fraction of each new viscous solution (sometimes only 10 percent) is used to update the boundary conditions of the inviscid calculation. Convergence is difficult to obtain without this procedure because the solutions for the viscous flow often oscillate slightly about free-stream conditions in the far wake.

V. RESULTS AND DISCUSSION

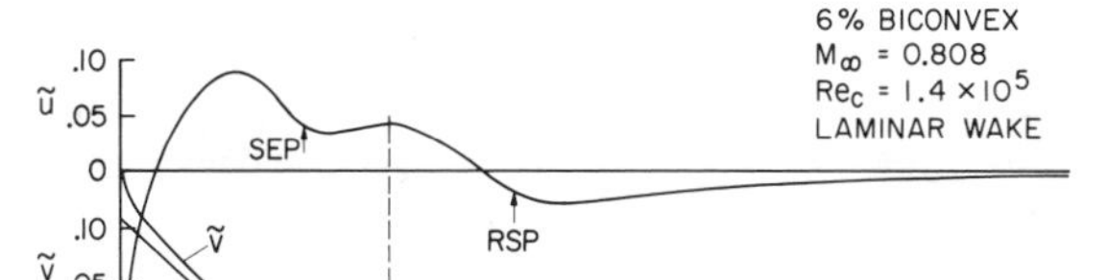

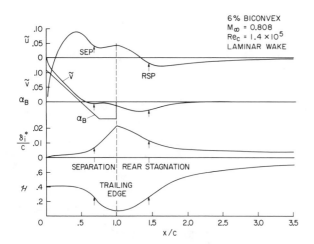

Figure 3. Trajectories for four dependent variables

The calculations presented in this paper were performed dynamically, with the aid of an interactive computer graphics system. A typical solution is shown in Fig. 3, illustrating the distributions of the four dependent variables on the airfoil and in the wake. For this particular calculation, convergence was obtained after 12 complete iterations although 24 were actually performed, requiring less than 30 minutes of IBM 360/67 CPU time. The total number of mesh points was increased from approximately 2300 to 3800 during the calculations, with the final grid boundaries located 1-1/2 chords in front, 3 chords above and 2-1/2 chords behind the airfoil. The converged solution does not depend on the original estimate for $\tilde{v}$ (shown in Fig. 2), and is insensitive to the location of the joining point provided it is chosen upstream of separation.

Near the leading edge, the rapid initial growth of the displacement surfaces causes a relatively large change in the effective body shape. The parameter H remains almost constant in this region, indicating that the transformed velocity profiles do not vary over the forward portion of the airfoil. The viscous layer goes smoothly through separation and accelerates downstream of the trailing edge, passing through the wake rear-stagnation point. The values of α_B are taken to be constant in the plateau region and zero downstream of the airfoil as previously described. In the far wake, the flow asymptotically approaches free-stream conditions and the product $H\delta_i{}^*$ (the momentum thickness θ_i) becomes constant.

The results of the present theory and the recent, unpublished data of Collins (1972) are compared in Fig. 4. Two viscous calculations are shown, for a laminar and a turbulent wake, illustrating the upstream influence that exists in subsonic flows. The agreement between the theory and measurements is quite good, with the experimentally observed separation and rear stagnation points coinciding with the locations predicted by the turbulent-wake approximation. The airfoil and displacement thickness are also shown, and the calculated skin-friction coefficients are compared to the equivalent laminar flat-plate distribution.

The predicted Reynolds number variation at $M_\infty = 0.8$ for completely laminar interactions is shown in Fig. 5. The most important effect of increasing the Reynolds number is to move the separation point upstream along the airfoil surface, lengthening the region of reversed flow. The corresponding pressure distributions exhibit a strong dependence on the location of separation. Although the displacement surface becomes thinner with increasing Reynolds number, the flow over the forward portion of the airfoil remains unchanged and the peak pressure coefficients decrease as the separation point moves upstream.

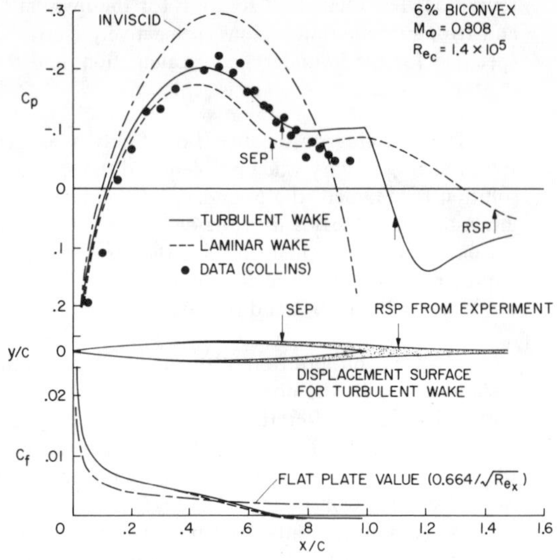

Figure 4. Companion with experiment

Figure 6 illustrates the effects of Mach number for completely laminar, subcritical interactions. As the Mach number is increased, the inviscid flow does not expand to much higher velocities because of the growth of the separated zone, and there is only a small change in the pressure distributions on the airfoil surface. At higher, supercritical Mach numbers, the turbulent approximation for the wake was used because of the extent of the reversed-flow region. As indicated in Fig. 7, the outer stream remained shock-free for these calculations, even at $M_\infty = 0.94$ where a large portion of the flow field was supersonic. For a method incorporating turbulence-modeled boundary-layer flow, calculations should show that the separation point does not move far upstream of the trailing edge and that shock waves form in the field at considerably lower free-stream Mach numbers.

A second comparison between the theory and the data of Collins (1972) is shown for a 12 percent biconvex airfoil in Fig. 8. For this configuration, boundary-layer separation occurred near the airfoil mid-chord, and the rear stagnation point was located well downstream of the trailing edge. Although the present relatively simple method is not designed to provide accurate

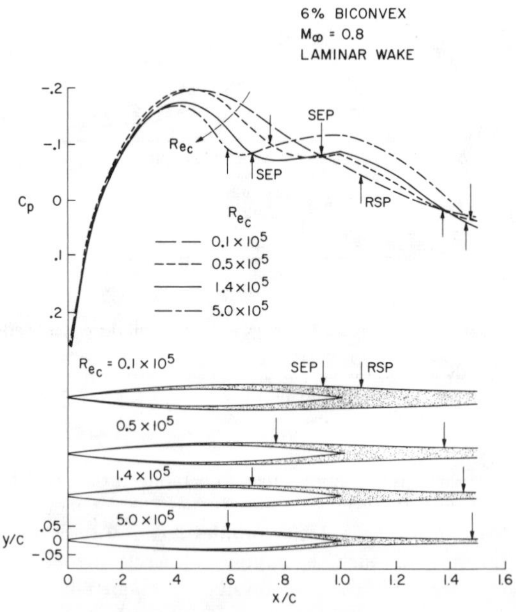

Figure 5. Effect of Reynolds number

calculations of such highly separated flow fields, the theory and measurements are in reasonably good agreement.

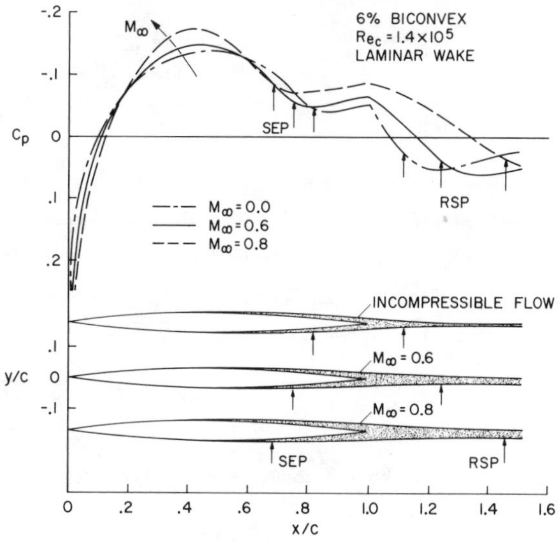

Figure 6. Effect of mach number for subcritical flow

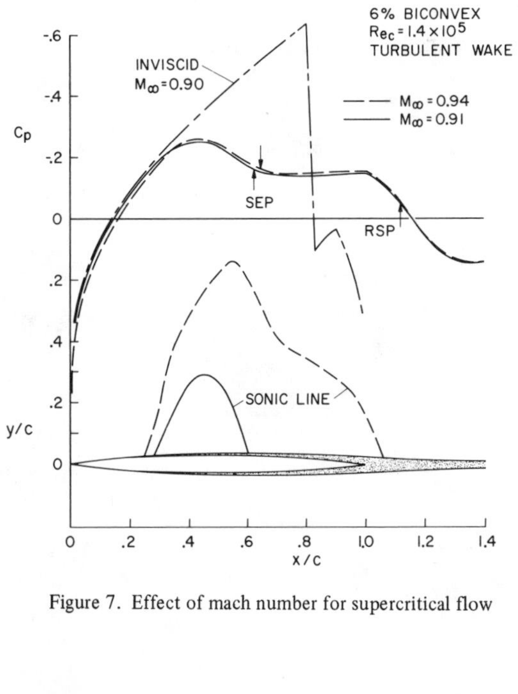

Figure 7. Effect of mach number for supercritical flow

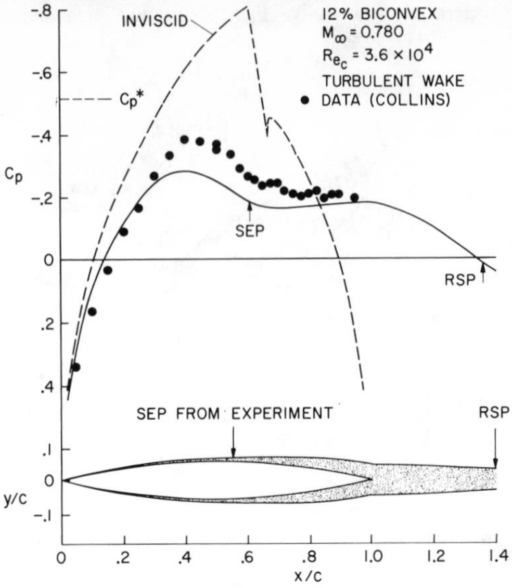

Figure 8. Comparison with experiment

REFERENCES

Collins, D. J., private communication, April 1972, NASA Ames Research Center, Moffett Field, California.

Klineberg, J. M., Kubota, T., and Lees, L., AIAA J. 10, pp. 581-588 (1972).

Klineberg, J. M., and Lees, L., AIAA J. 7, pp. 2211-2221 (1969).

Lees, L., and Reeves, B.L., AIAA J. 2, pp. 1907-1920 (1964).

Pearcy, H. H., Osborne, J., and Haines, A. B., NPL Aero Note 1071, A.R.C. 30477 (1968).

Steger, J. L., and Klineberg, J. M., AIAA paper 72-679 (1972).

Steger, J. L., and Lomax, H., *Lecture Notes in Physics,* Vol. 8, pp. 193-197, Springer-Verlag, (1971).

NUMERICAL SOLUTIONS OF BLAST WAVE PROPAGATION PROBLEMS

V. P. Korobeinikov, V. V. Markov, P. I. Chushkin,
L. V. Shurshalov

Academy of Sciences of the U.S.S.R., Moscow

The development and applications of numerical methods for the solutions of a number of gasdynamic problems arising at explosions in different media are considered. The idealized formulations of these problems taking into account the principal features of the phenomenon are used. The paper is a survey of some results obtained by the authors during recent years.

1. Point Explosion in a Perfect Gas with Counterpressure

Korobeinikov and Chushkin (1964; 1967) developed the method of integral relations for the numerical solution of the problem of point explosion with plane, cylindrical and spherical symmetries ($\nu = 1,2,3$, respectively) in a perfect gas with counterpressure. The system of equations for unsteady one-dimensional gas flows is written in the variables $\xi = (r/r_s)^\nu$, $g = a_\infty^2/D^2$, where r is the space coordinate, a_∞ is the sound velocity in an ambient medium, r_s and D are the coordinate and the velocity of the shock wave, respectively. Near the explosion center an interval is introduced which is bounded by the line $\xi = \xi_0(g)$ corresponding to a fixed Lagrangian coordinate. An asymptotic solution is used within this central interval. The region between the shock wave $\xi = 1$ and the line $\xi = \xi_0(g)$ is subdivided into strips and calculated by the method of integral relations with the approximation by polynomials in ξ across all the strips. The approximating system of equations with respect to g is integrated from a small value of $g = g_*$ (at which the linearized blast solution is used) to a value of g close to one.

The numerical solution with a large number of strips (eight strips) is obtained for the plane, cylindrical and spherical explosion in a gas for different values of adiabatic exponent γ. The distributions of basic gasdynamic functions and the time history of the shock wave motion are published in tables (Korobeinikov, Chushkin, Sharovatova, 1969). Earlier the corresponding tables for the initial stage of a point explosion (Korobeinikov, Chushkin, Sharovatova, 1963) were computed on the basis of the linearized solution (Korobeinikov, Chushkin, 1963).

For purposes of illustration, the variation of relative pressure at the explosion center versus the dimensionless time τ (referred to the dynamic time) is shown in Fig. 1 for a monoatomic gas and different ν.

The method was extended to the case of a blast in an electrically conducting gas with a magnetic field (Korobeinikov, 1965) and to the case of propagation of a detonation wave from a point blast in a detonating gas (Korobeinikov, 1969).

2. Point Explosion in a Combustible Gas Mixture with Account of Kinetics

A numerical analysis of flows of gas mixtures is carried out in which exothermal chemical reactions can occur behind a shock wave arising as a result of instantaneous energy release at a point, along an axis or on a plane (Korobeinikov, Levin, Markov, Chernyi, 1972). Formerly the splitting process of a strong detonation wave into shock wave and ignition front was studied for various initiation energies and activation energies (Bishimov, Korobeinikov, Levin, 1970).

To investigate the dynamics of the propagation of shock wave and ignition front, the finite-difference method (Godunov, Prokopov, 1961) is used, in which the relations for decompositions of discontinuities on cell boundaries are applied. Only the case of small explosion energies is studied in detail. To describe the chemical kinetics, a model of total direct and inverse reaction is assumed, which is initiated in a gas particle after an induction period. The numerical solution of the problem is obtained up to the stage of strong decay of shock wave for a cylindrical blast in an oxygen-hydrogen mixture when the specific explosion energy is equal to $3 \cdot 10^6$ erg/cm and the initial pressure is the normal atmospheric pressure p_o. It is found that for small explosion energies and uniform initial conditions the ignition zone continues to come off the shock wave even at late times.

The space distribution of mixture concentration β and relative pressure p/p_s, density ρ/ρ_s and velocity u/D are presented in Fig. 2 for the time instant $t/t_* = 279$. Here $t_* = 10^{-7}$ sec, Q is the heat of combustion, the subscript s denotes the shock front value of the functions. In this case the shock front pressure $p_s = 1.34\ p_o$, the shock wave radius $r_s = 113\ t_* \sqrt{Q}$ and the combustion wave radius $r_c = 28\ t_* \sqrt{Q}$.

3. Underwater Explosion

The one-dimensional problem of an underwater explosion is considered with spherical, cylindrical and plane symmetries in unbounded space. The effect of gravity is neglected. It is supposed that in the water there is an envelope of corresponding shape, containing highly heated gaseous products of explosion. The distribution of gasdynamic functions inside the envelope corresponds to that produced by a detonation initiated at the center of symmetry of the charge. Alternatively, for simplicity, the distribution is supposed to be uniform (i.e., the detonation velocity is infinite). The envelope is assumed to vanish instantaneously at the initial time instant.

The thermodynamic properties of the water and the gas are described by the following equation of state

$$p = A(v) + B(v)\,E \qquad ,$$

where p, v, E denote the pressure, specific volume and specific internal energy, respectively. It is assumed that for the gaseous explosion products $A = 0$, $B = (\gamma-1)/v$ with the adiabatic exponent γ depending on v. Treating the experimental data obtained for the isothermal and shock compressions, Shurshalov (1967; 1971b) derived the following expressions for water

$$A = -\left(\frac{dk}{dv} + \frac{k}{C_v}\frac{dS}{dv}\right)\rho_o \quad , \quad B = \frac{\rho_o}{C_v}\frac{dS}{dv} \qquad ,$$

where S is the entropy, k is a prescribed function of v, C_v is the specific heat at constant volume, $\rho_o = 1\text{g/cm}^3$. The present equation of state describes rather accurately the properties of water at pressures higher than 1000 atm. At lower pressures the well-known Tait equation of state is used, and a linear combination of these two equations is used in some intermediate zone.

The solution of the underwater explosion problem (Shurshalov, 1971b) was carried out by the finite-difference method with artificial viscosity (Wilkins, 1966).

The underwater explosion of a trotyl spherical charge with the radius r_o is calculated. The initial distribution of variables inside the sphere is taken from the computation of the detonation phase (Fonarev, Chernjavskii, 1968). In the early stages of the explosion, Fig. 3 shows the relative pressure $P = p/p_o$ versus the Lagrangian coordinate $R_1 = r/r_o$ for several values of dimensionless time $\tau = t/t_o$, where $t_o = r_o \rho_o^{\frac{1}{2}}\, p_o^{-\frac{1}{2}}$ and $p_o = 1\ \text{kg/cm}^2$. The feature of flow in this case is the formation of secondary shock waves inside the gas bubble, which, however, have no important effect on the water motion outside the bubble.

A study is also made of a water flow due to the expansion of spherical, cylindrical and plane volumes of compressed gaseous detonation products, assuming a uniform initial distribution of gasdynamic parameters. In these cases the gas energy per unit volume is taken to be much larger ($P = 5 \cdot 10^5$) than in the preceding example. The variations of some gasdynamic functions at the shock versus time are given for different τ in Fig. 4, where the pressure, P, the velocity, $U = ut_0/r_0$ and the density, $G = \rho/\rho_0$, are represented by solid, dashed and dash-dotted lines, respectively. Here new waves occur inside the gas bubble; this fact is confirmed by the sharp points on the curves presented. In the plane case a rarefaction wave appears, weakening the main shock wave, while in the cylindrical and spherical cases a secondary shock wave appears, somewhat strengthening the main shock wave.

4. Analysis of Singularity in the Problem of a Point Explosion at a Free Surface

In the self-similar problem of a point explosion at the boundary of a semi-space containing a perfect gas, the point of intersection of the shock wave and the free surface is singular. Approaching this point along different directions, the hydrodynamic variables have different limiting values. Investigating the local properties of flow in the neighborhood of this point, it is possible to obtain some information useful for the numerical solution of the problem.

The flow in the neighborhood of this singular point is studied for the cases of regular and critical reflections of shock from the free surface (Shurshalov, 1969; 1971a). The basic functions are represented by series expansions with respect to the radial distance from the singular point. The determination of coefficients (depending on polar angle) in these series is reduced to the solution of a series of boundary-value problems for the system of ordinary differential equations with unknown boundary. The coefficients for the two first terms of the series are found. The value of shock inclination angle at the free surface remains an undetermined quantity (it is necessary to solve the full problem to fix its value). As to the critical value of shock inclination angle at the free surface ϕ_*, dividing the regions of regular and irregular reflections, here the following formula is derived

$$\cos \phi_* = \sqrt{\frac{\gamma+1}{2\gamma}}$$

It is proved that for any value of shock inclination angle at the free surface, the flow region cannot be entirely disposed along one side of the undisturbed free surface.

5. Explosion of Cylindrical Charge of Finite Length

An explosion of cylindrical charge of finite length and finite radius produces a perturbed gas flow with shock wave and contact discontinuity separating the explosion products from the ambient medium. The initial blast stage may be simulated by the instantaneous expansion of the corresponding volume of a compressed gas.

The solution of the axisymmetrical unsteady problem of expansion of a cylindrical compressed gas volume of finite length in a medium at rest was carried out by a numerical method (Godunov et al., 1961). A scheme was developed in which the main shock wave and the contact interface are treated as the boundary lines of a network with the exact satisfaction of all the necessary conditions on them. The first results concerning the solution of the problem under consideration were published by Shurshalov (1971c).

Let us give as an example some numerical results in the case when the radius/length ratio for the cylinder is equal to 0.05, the ratio of pressure and density in the compressed gas to the corresponding ambient values are equal to 10^4 and 10^2, respectively, and $\gamma = 1.4$ for both media. The shock wave (solid line) and the interface (dashed line) are plotted in Fig. 5 for several dimensionless moments of time τ (Figs. 1-6 correspond to $\tau = 0.02, 0.06, 0.13, 0.23, 0.34$ and 0.45).

The region initially occupied by the compressed gas is shaded.

It is interesting to notice a feature of the interface motion due to the effect of the secondary shock wave. The secondary shock formed near the interface propagates into the inner region, gradually increasing its strength. It drags the divergent gas flow and in our case the secondary shock wave is so strong that the flow is reversed by it. This feature is well illustrated in Fig 5, which shows the velocity field for $\tau = 0.23$. A bend of the interface is observed at the position where the secondary shock occurs for the first time. This bend arises as the contact surface, moving together with the gas particles, slows down owing to the action of the secondary shock wave. Actually it is observed that some portion of the contact discontinuity comes to rest. Then after the reflection of the secondary shock at the origin, the gas inside the contact surface begins to expand again and the interface continues its divergent motion.

6. Explosion of Flying Meteorite

As a particular case of the preceding problem we consider the problem of blast of a semi-infinite cylindrical charge with variable specific energy along its axis. The solution obtained will be applied to simulate the blast phenomenon of a large meteorite flying in the atmosphere, that is accompanied by interaction of ballistic and explosion waves. The inclination and the front point location of the charge are taken according to the meteorite path before the blast. Using the hypersonic-unsteady analogy, the energy in the tail part of the charge is estimated through the drag force of the leading meteorite surface. The energy in the head part of the charge can be estimated with the aid of some recorded total effects due to the explosion (seismograms, barograms, destructions on the ground). Generally speaking, the specific energy in the head part of the charge must be essentially larger than the specific energy in the tail part.

To solve this unsteady three-dimensional problem (in the case of the stratified atmosphere) a numerical approach together with an analytical approximation is used (Korobeinikov, Chushkin, Shurshalov, 1971; 1972). At first the non-homogeneity of the Earth's atmosphere is ignored and the flow is computed by a numerical method (Godunov et al., 1961) during the period when the shock wave is still sufficiently strong. After the weakening of the shock, the explosion is calculated using the sector approximation and the fact that the shock shape approaches a hemisphere-cylinder combination. The behavior of the blast wave in a uniform atmosphere is illustrated in Fig. 7.

The non-homogeneity of the Earth's atmosphere is taken into account on the basis of the quasi-one-dimensional ray theory, assuming a plane sections hypothesis. We use the fact that the pressure and density, referred to local values at a point reached by the shock wave, depend weakly on the atmospheric non-homogeneity (Korobeinikov, Karlikov, 1963; Korobeinikov, 1971). Then the initial stage of shock reflection from the Earth's surface is calculated. Here we use in the case of irregular reflection the data by Brode (1968) giving the overpressure behind the reflected wave as a function of the inclination angle and the overpressure of the incident wave. The field of dynamic pressures found in such a way defines the size and character of destruction on the ground.

The method developed was applied to the case of the blast of the Tunguska meteorite. The parameters which are not known in advance (the meteorite path inclination α, the explosion height H, the energies of ballistic E_b and explosion E_e waves) are determined so that the calculated picture of ground destructions for the model blast corresponds to the picture observed. The calculated picture of flattened forest at the Tunguska impact is presented in Fig. 8 for the case of $\alpha = 30°$, $H = 5$ km, $E_b = 2{\cdot}10^{16}$ erg/cm, $E_e = 4{\cdot}10^{22}$ erg. Here is shown the isodynams $\rho u^2 = 0.008$ kg/cm^2 (solid line), the isochrones (dashed lines) and the directions of flow on the Earth (arrows) immediately after the shock reflection that define the directions of felled trees. These results are in good agreement with the actual measured data available.

References

Bishimov, E., Korobeinikov, V. P., Levin, V. A. Astr. Acta 15, 267-273 (1970).

Brode, H. L. Ann. Rev. Nucl. Sci. 18, 153-202 (1968).

Fonarev, A. S., Chernyavskii, S. Yu. Izv. AN SSSR, Mekh. zhidk. gaza No. 5, 169-174 (1968).

Godunov, S. K., Zabrodin, A. V., Prokopov, G. P. Zh. vych. mat. mat. fiz. 1, 1020-1050 (1961).

Korobeinikov, V. P. Dokl. AN SSSR 165, 1019-1022 (1965).

Korobeinikov, V. P. Astr. Acta 14, 411-419 (1969).

Korobeinikov, V. P. Ann. Rev. Fluid Mech. 3, 317-346 (1971).

Korobeinikov, V. P., Chushkin, P. I. Zh. prikl. mekh. tekh. fiz. No. 4, 48-57 (1963).

Korobeinikov, V. P., Chushkin, P. I. Dokl. AN SSSR 154, 549-552 (1964).

Korobeinikov, V. P., Chushkin, P. I. Trudy Matem. inst. AN SSSR 87, 4-34 (1967).

Korobeinikov, V. P., Chushkin, P. I., Sharovatova, K. V. "Tables of gasdynamic functions for initial stage of point explosion," Vych. Tsentr AN SSSR (1963).

Korobeinikov, V. P., Chushkin, P. I., Sharovatova, K. V. "Gasdynamic functions of point explosion," Vych. Tsentr AN SSSR (1969).

Korobeinikov, V. P., Chushkin, P. I., Shurshalov, L. V. Fluid Dynamics Transactions 6, pt. 2, 351-359, PWN, Warsawa (1971).

Korobeinikov, V. P., Chushkin, P. I., Shurshalov, L. V. Astr. Acta 17 (1972).

Korobeinikov, V. P., Karlikov, V. P. Dokl. AN SSSR 167, 59-62 (1966).

Korobeinikov, V. P., Levin, V. A., Markov, V. V., Chernyi, G. G. Astr. Acta 17 (1972).

Shurshalov, L. V. Izv. AN SSSR, Mekh. zhidk. gaza No. 4, 184-186 (1967).

Shurshalov, L. V. Prikl. mat. mekh. 33, 358-363 (1969).

Shurshalov, L. V. Izv. AN SSSR, Mekh. zhidk. gaza No. 3, 3-7 (1971a).

Shurshalov, L. V. Izv. AN SSSR, Mekh. zhidk. gaza No. 5, 36-40 (1971b).

Shurshalov, L. V. Dokl. AN SSSR, 199, 1262-1264 (1971c).

Wilkins, M. L. Fundamental Methods in Hydrodynamics, 211-263, Academic Press, New York - London (1966).

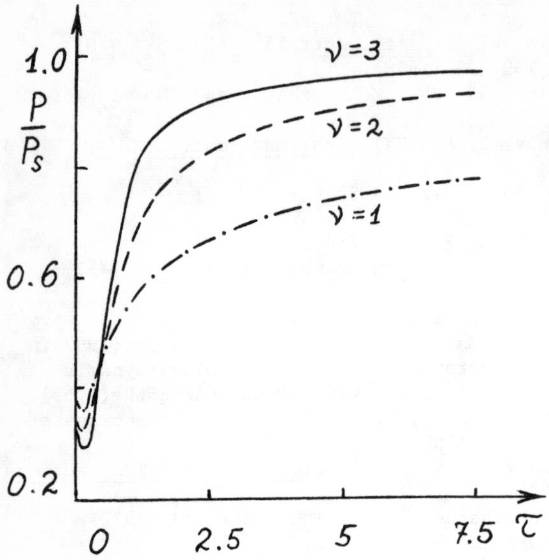

Fig. 1

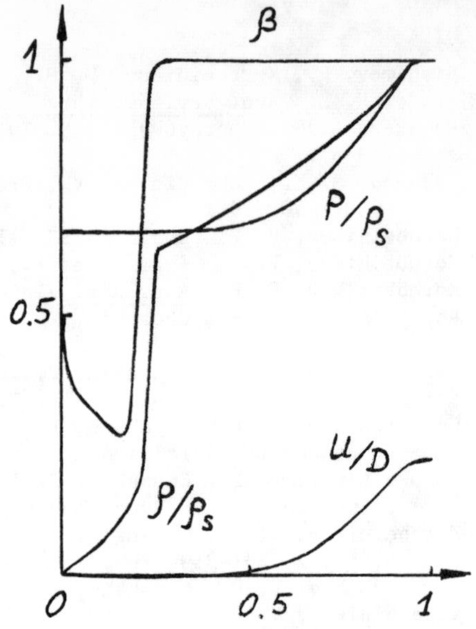

Fig. 2

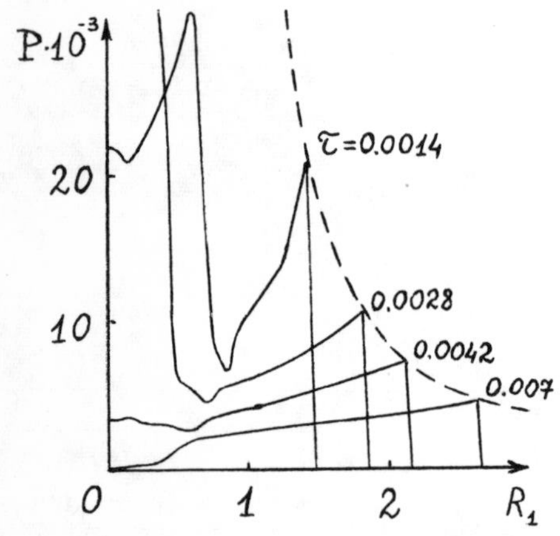

Fig. 3

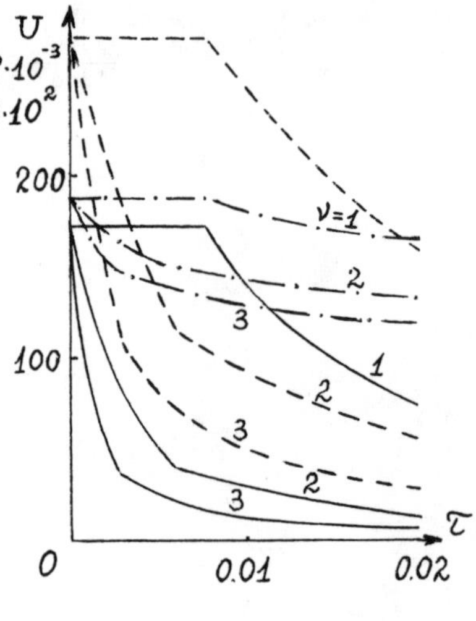

Fig. 4

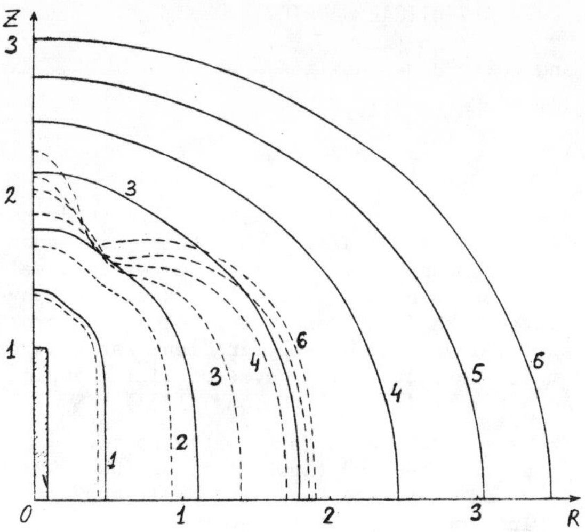

Fig. 5

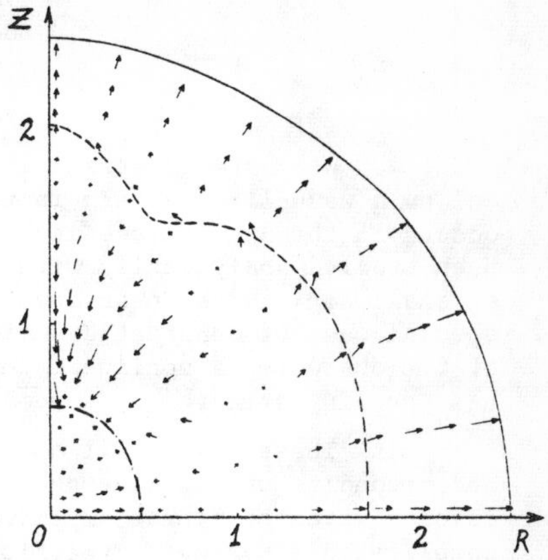

Fig. 6

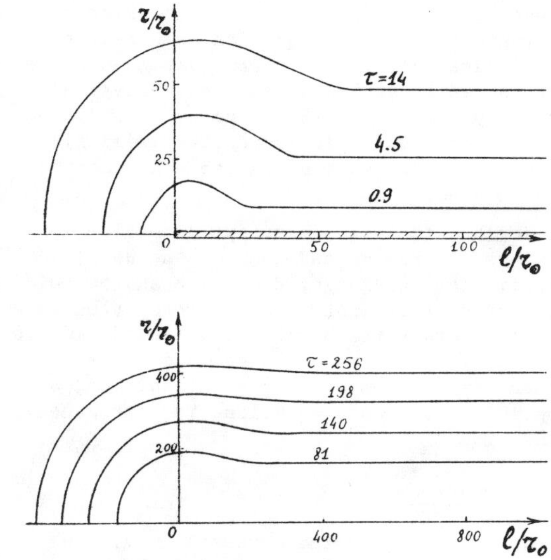

Fig. 7

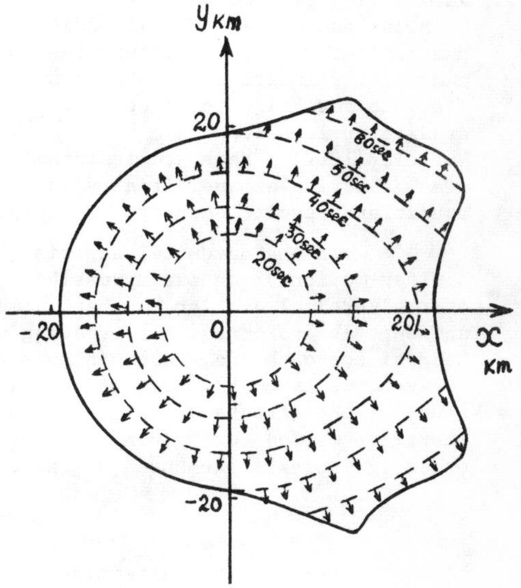

Fig. 8

STRATIFIED FLOW OVER A VERTICAL BARRIER

Jung-Tai Lin and Colin James Apelt*
Colorado State University

INTRODUCTION

When stably stratified fluid passes over an obstacle, the flow tends to follow a wave-like pattern in the lee side of the obstacle. When this phenomenon occurs in the atmosphere, it is described as mountain lee-waves. Lee-waves have been studied analytically with the help of perturbation theory for steady lee-waves of small amplitudes in an inviscid fluid by Lyra (1943) and later authors. For the special case of constant dynamic pressure and uniform density gradient far upstream of the obstacle, a nonlinear inviscid model for steady lee-waves of finite amplitude has been investigated by a number of authors, beginning with Long (1953).

These analytical approaches have contributed significantly to the understanding of lee-wave phenomena. However, lee-wave flows in nature usually involve waves of finite amplitude and, in addition, the boundary conditions are complicated. Moreover, the flow may be in a transient state. For the investigation of lee-wave phenomena in these more complex cases, laboratory simulation and numerical modelling with a computer are the only approaches currently available apart from prototype study in the field. Laboratory simulations have been conducted by Long (1959) and Davis (1969) with a liquid model, and by Lin and Binder (1967) in a wind tunnel. In laboratory simulation, one usually encounters limitations in the variability of flow parameters such as velocity and stratification. In particular, Reynolds numbers are usually quite low when high Froude numbers (or low Richardson numbers) are required. Since the governing Navier-Stokes equation is non-linear, numerical simulation is a far from trivial task but, if successful, it enjoys the distinct advantage that the flow parameters can be varied with ease. Numerical studies of lee-waves have been carried out by Hovermale (1965) and by Foldvik and Wurtele (1967). Both studies treated transient flow of inviscid stratified fluid over a barrier as an initial value problem. For the lack of suitable boundary conditions, cyclic boundary conditions were used.

In the numerical experiment described in this paper, two-dimensional laminar motion of an incompressible, viscous, diffusive and nonhomogeneous fluid over a vertical barrier has been simulated. The studies by Hovermale and by Foldvik and Wurtele indicated that treating a transient flow instead of a steady flow is a helpful approach and, thus, in this numerical study transient flows are considered. However, we have abandoned the cyclic boundary condition and have made a detailed study of the effects of boundary conditions on the computation. In addition, a new iterative method for solving sets of implicit difference equations is introduced in order to increase the speed of the computation.

GOVERNING EQUATIONS

Consider two-dimensional laminar motion of an incompressible, viscous, diffusive and nonhomogeneous fluid in a uniform gravitational field. The fluid properties other than the density, such as the viscosity μ_o and thermal conductivity k_o, are assumed to be uniform in the flow field. After use of the Boussinesq approximation, the equations of motion, the continuity equation and the equation of energy take the form,

$$\rho_o(\partial\tilde{u}/\partial\tilde{t} + \tilde{u}\,\partial\tilde{u}/\partial\tilde{x} + \tilde{w}\,\partial\tilde{u}/\partial\tilde{z}) = -\,\partial\tilde{p}/\partial\tilde{x} + \mu_o\tilde{\nabla}^2\tilde{u}\ , \tag{1}$$

$$\rho_o(\partial\tilde{w}/\partial\tilde{t} + \tilde{u}\,\partial\tilde{w}/\partial\tilde{x} + \tilde{w}\,\partial\tilde{w}/\partial\tilde{z}) = -\,\partial\tilde{p}/\partial\tilde{z} - \tilde{\rho}g + \mu_o\tilde{\nabla}^2\tilde{w}\ , \tag{2}$$

$$\partial\tilde{u}/\partial\tilde{x} + \partial\tilde{w}/\partial\tilde{z} = 0, \tag{3}$$

$$\rho_o C_p(\partial\tilde{T}/\partial\tilde{t} + \tilde{u}\,\partial\tilde{T}/\partial\tilde{x} + \tilde{w}\,\partial\tilde{T}/\partial\tilde{z}) = k_o\,\tilde{\nabla}^2\,\tilde{T}\ , \tag{4}$$

where $\tilde{u}$ and $\tilde{w}$ are the velocity components in the Cartesian coordinates $\tilde{x}$ and $\tilde{z}$, $\tilde{\rho}$

*Present address: University of Queensland.

is the density, ρ_o is the reference density, $\tilde{p}$ is the pressure, C_p is the specific heat capacity at constant pressure, $\tilde{T}$ is the temperature, $\tilde{t}$ is the time, and $\tilde{\nabla}^2 = \partial^2/\partial\tilde{x}^2 + \partial^2/\partial\tilde{z}^2$ is the Laplacian operator. In equation (4) the heat dissipation due to molecular viscosity is neglected.

For numerical integration, the primitive equations have been replaced by the single vorticity transport equation and the equations have been made dimensionless by the substitutions,

$$x = \tilde{x}/L, \quad z = \tilde{z}/L, \quad u = \tilde{u}/U, \quad w = \tilde{w}/U, \quad t = \tilde{t}/(L/U), \quad T = (\tilde{T} - T_o)/\Delta T,$$

$$\zeta = \tilde{\zeta}(L/U), \quad \psi = \psi/UL, \quad R_e = UL/\nu, \quad P_r = \nu/\kappa, \quad F_r = U/(gL\Delta T/T_o)^{\frac{1}{2}} = R_i^{-\frac{1}{2}} .$$

L, U and ΔT are the characteristic length, velocity and temperature difference respectively; $\nu = \mu_o/\rho_o$ is the kinematic viscosity; $\kappa = k_o/\rho C_p$ is the thermal diffusivity; R_e, P_r, F_r and R_i are the Reynolds, Prandtl, Froude and Richardson numbers respectively. The set of dimensionless equations which has been integrated is

$$\partial\zeta/\partial t + u\ \partial\zeta/\partial x + w\ \partial\zeta/\partial z = R_i\ \partial T/\partial x + \nabla^2\zeta/R_e\ , \tag{5}$$

$$\partial T/\partial t + u\ \partial T/\partial x + w\ \partial T/\partial z = \nabla^2 T/(P_r R_e), \tag{6}$$

$$\zeta = \partial w/\partial x - \partial u/\partial z = -\nabla^2\psi\ . \tag{7}$$

The stream function ψ is defined by $u = \partial\psi/\partial z$, $w = -\partial\psi/\partial x$.

Discretisation of Governing Equations

Since it was desired to calculate asymptotic steady states as well as transient phenomena associated with lee-waves, one objective was to develop a numerical integration process for the governing equations in which reasonably large time intervals could be used. The phenomena themselves do not require very small time intervals to resolve them since transient lee-waves propagate slowly and have a characteristic length comparable to, or greater than, the characteristic dimension of the obstacle which initiates them. For these reasons an implicit scheme in which the time derivatives were replaced by central difference approximations was used for discretisation of the equations (5) and (6) and, after some experimentation, the Crank-Nicolson implicit scheme was adopted. Even when an implicit scheme is used the non-linearity of the governing equations can cause difficulties when R_e becomes large. If the convective terms of the equations (5) and (6) are discretised with central space differences the off-diagonal terms of the amplification matrix of the difference equations can become large in amplitude as R_e increases and the eigenvalues of the matrix may be greater than unity in amplitude. This problem has been dealt with by approximating $\partial\zeta/\partial x$ (or $\partial\zeta/\partial z$) with the backward or forward difference according as u (or w) is positive or negative, a device which has been used by a number of authors. The resulting difference equation which approximates equation (5) is,

$$\begin{aligned}
&(\zeta^{n+1}_{i,j} - \zeta^{n}_{i,j})/\Delta t + \tfrac{1}{2}\, u^{n}_{i,j}[A_2(\zeta^{n+1}_{i,j} - \zeta^{n+1}_{i-1,j} + \zeta^{n}_{i,j} - \zeta^{n}_{i-1,j})/\Delta x \\
&\qquad + A_3(\zeta^{n+1}_{i+1,j} - \zeta^{n+1}_{i,j} + \zeta^{n}_{i+1,j} - \zeta^{n}_{i,j})/\Delta x] \\
&+ \tfrac{1}{2}\, w^{n}_{i,j}[A_4(\zeta^{n+1}_{i,j} - \zeta^{n+1}_{i,j-1} + \zeta^{n}_{i,j} - \zeta^{n}_{i,j-1})/\Delta z \\
&\qquad + A_5(\zeta^{n+1}_{i,j+1} - \zeta^{n+1}_{i,j} + \zeta^{n}_{i,j+1} - \zeta^{n}_{i,j})/\Delta z] \\
&= [(\zeta^{n+1}_{i+1,j} - 2\zeta^{n+1}_{i,j} + \zeta^{n+1}_{i-1,j} + \zeta^{n}_{i+1,j} - 2\zeta^{n}_{i,j} + \zeta^{n}_{i-1,j})/\Delta x^2 \\
&+ (\zeta^{n+1}_{i,j+1} - 2\zeta^{n+1}_{i,j} + \zeta^{n+1}_{i,j-1} + \zeta^{n}_{i,j+1} - 2\zeta^{n}_{i,j} + \zeta^{n}_{i,j-1})/\Delta z^2]/2R_e \\
&+ \tfrac{1}{4}\, R_i[(T^{n+1}_{i+1,j} - T^{n+1}_{i-1,j} + T^{n}_{i+1,j} - T^{n}_{i-1,j})/\Delta x]\ ,
\end{aligned} \tag{8}$$

in which $A_2 = 1$, $A_3 = 0$ for $u^n_{i,j} \geq 0$; $A_2 = 0$, $A_3 = 1$ for $u^n_{i,j} < 0$;

$$A_4 = 1, A_5 = 0 \text{ for } w^n_{i,j} \geq 0; A_4 = 0, A_5 = 1 \text{ for } w^n_{i,j} < 0 . \quad (9)$$

The difference equation which approximates equation (6) is essentially similar to equation (8) and it is not reproduced here. Complete details are given in Lin and Apelt (1970). The equation (7) is approximated by the central difference expression,

$$(\psi_{i+1,j} - 2\psi_{i,j} + \psi_{i-1,j})/\Delta x^2 + (\psi_{i,j+1} - 2\psi_{i,j} + \psi_{i,j-1})/\Delta z^2 = - \zeta_{i,j} . \quad (10)$$

In the difference equations the convention is used that superscripts refer to the time variable and subscripts to the space variables such that, for example,

$$\zeta^n_{i,j} \equiv \zeta(i\Delta x, j\Delta z, n\Delta t).$$

SOLUTION OF IMPLICIT DIFFERENCE EQUATIONS

The penalty associated with the choice of the Crank-Nicolson method for integrating the equations (5) and (6) is the necessity to solve a large set of implicit difference equations in producing the ζ and T fields at each time step. The difference equations for ψ also must be solved as a simultaneous set. A number of iterative methods has been used for the solution of such large sets of equations, the majority of authors using either the Gauss-Seidel process, the successive over relaxation (SOR) process or the alternating direction iterative (ADI) process. However, Stone (1968) demonstrated that a strongly implicit iterative method based on matrix resolution enjoyed a significantly faster convergence rate than all of these three iterative methods when applied to some quite substantial test problems. Consequently, the method described by Stone was used in the solution of each set of simultaneous difference equations arising from the discretisation of equations (5), (6) and (7).

At each time step of the integration process it is necessary to solve three sets of difference equations. Each set has been linearized, as illustrated in the equation (8), but the three sets are coupled together in a non-linear way both in themselves and through the boundary conditions. Because of this coupling it is necessary to calculate the solutions to the three sets of equations in an iterative manner and this outer iteration is discussed in a later section. At this stage, the description of Stone's method is given for the solution of a single set of equations. If the equations are represented in matrix form as

$$[M] \vec{T} = \vec{q} , \quad (11)$$

the matrix of coefficients for each set of difference equations corresponding to the equations (5), (6) and (7) can be arranged to have the same sparse structure. Each difference equation can be written in the form,

$$B_{i,j} T_{i,j-1} + D_{i,j} T_{i-1,j} + E_{i,j} T_{i,j} + F_{i,j} T_{i+1,j} + H_{i,j} T_{i,j+1} = q_{i,j} , \quad (12)$$

and the matrix [M] has non-zero elements along only five diagonal lines, as illustrated in Fig.1(a). In general, when such a matrix is resolved into a lower triangular matrix [L] and an upper triangular matrix [U] these do not preserve the sparse structure of [M]. However, if a matrix [M+N], which approximates [M], can be found, such that it is the product of a lower and upper triangular matrix, each of which has non-zero elements on only the three diagonal lines corresponding to the non-zero diagonal lines in [M], as illustrated in Fig.1(b), then it is possible to solve the set of equations by an iterative method combined with triangular resolution and to preserve the benefits arising from the sparse structure of [M]. The iterative process is derived by augmenting the equation (11) to give,

$$[M + N] \vec{T} = [M + N] \vec{T} + (\vec{q} - [M] \vec{T}) . \quad (13)$$

[M + N] can easily be resolved into lower and upper triangular matrices of sparse structure and, so, the left hand side of equation (13) can be solved for $\vec{T}$ very economically by the direct method often described as the Choleski method if the right hand side of the equation is known. The iterative procedure is therefore established in the form,

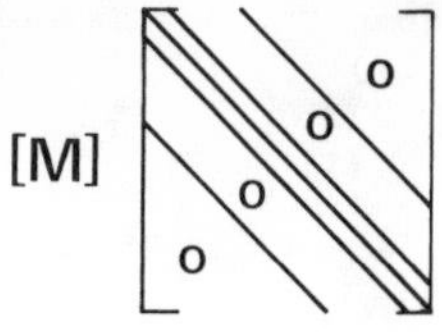

Fig.1a

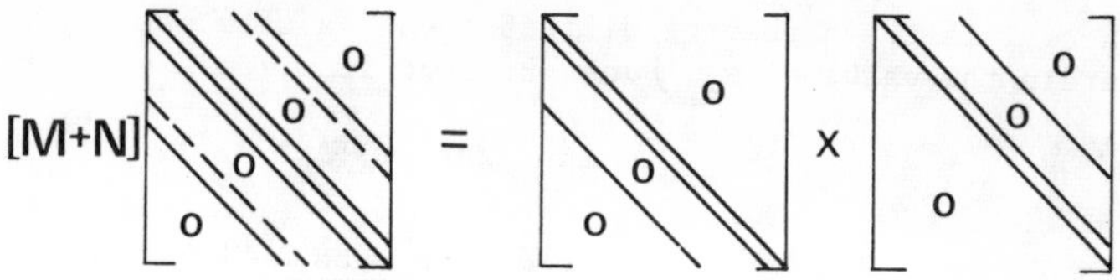

Fig.1b

$$[M + N]\vec{T}^{(m+1)} = [M + N]\vec{T}^{(m)} + (\vec{q} - [M]\vec{T}^{(m)}) , \tag{14}$$

in which the superscripts refer to the cycle of iteration. The most significant contribution of Stone (1968) was to devise a simple and very effective means for setting up the matrix [M + N]. This is achieved by modifying the left hand side of equation (12) to become,

$$\begin{aligned} &B_{i,j}\, T_{i,j-1} + D_{i,j}\, T_{i-1,j} + E_{i,j}\, T_{i,j} + F_{i,j}\, T_{i+1,j} + H_{i,j}\, T_{i,j+1} \\ &-B_{i,j}\, e_{i,j-1}\, [T_{i+1,j-1} - \alpha(-T_{i,j} + T_{i+1,j} + T_{i,j-1})] \\ &-D_{i,j}\, f_{i-1,j}\, [T_{i-1,j+1} - \alpha(-T_{i,j} + T_{i,j+1} + T_{i-1,j})] \; . \end{aligned} \tag{15}$$

For the case, $\alpha = 1$, the sum of each set of four values of the variable which are bracketed together is of order $(\Delta x.\Delta z)$ and it is this fact which makes (15) such a good approximation to the left hand side of the equation (12).

The approximation (15) in conjunction with the iterative process defined by (14) leads to the following algorithm for solution of the equations (11),

$$\begin{aligned} \gamma_{i,j} = E_{i,j} &+ D_{i,j}\, (e_{i-1,j} - \alpha\, f_{i-1,j})/(1 - \alpha\, f_{i-1,j}) \\ &+ B_{i,j}\, (f_{i,j-1} - \alpha\, e_{i,j-1})/(1 - \alpha\, e_{i,j-1}) \; , \end{aligned} \tag{16}$$

$$e_{i,j} = -[F_{i,j} + \alpha\, B_{i,j}\, e_{i,j-1}/(1 - \alpha\, e_{i,j-1})]/\gamma_{i,j} \; , \tag{17}$$

$$f_{i,j} = -[H_{i,j} + \alpha\, D_{i,j}\, f_{i-1,j}/(1 - \alpha\, f_{i-1,j})]/\gamma_{i,j} \; , \tag{18}$$

$$d_{i,j} = [q_{i,j} - D_{i,j}\, d_{i-1,j}/(1 - \alpha\, f_{i-1,j}) - B_{i,j}\, d_{i,j-1}/(1 - \alpha\, e_{i,j-1})]/\gamma_{i,j} \; , \tag{19}$$

$$T_{i,j} = f_{i,j}\, T_{i,j+1} + e_{i,j}\, T_{i+1,j} + d_{i,j} \; . \tag{20}$$

The computational procedure is as follows:-

(1) All the capitalised elements and $q_{i,j}$ are known quantities at all grid points and the equations (16), (17) and (18) allow $e_{i,j}$, $f_{i,j}$ and $d_{i,j}$ to be computed in sequence (i = 1(1)imax, j = 1(1)jmax), i.e. by sweeping the field of grid points from the lower left corner to the upper right corner.

(2) When all $e_{i,j}$, $f_{i,j}$ and $d_{i,j}$ have been computed, $T_{i,j}$ can be computed directly from the equation (20) in the reverse order.

In the double sweep described, the boundary conditions are introduced at boundary points in the appropriate manner. Each double sweep constitutes one cycle of iteration of the process established by equation (14) and the process is repeated until successive iterates of $T_{i,j}$ coincide within the desired tolerance.

Convergence of iterative process

In order that the iteration be convergent it is necessary that $0 < \alpha < 1$. The details of the convergence analysis are given by Stone (1968) and by Lin and Apelt (1970). In fact, it is found advantageous to use a range of values of α rather than a fixed value and Stone's suggestion was adopted in that the individual values of α were geometrically spaced as follows,

$$1 - \alpha_\ell = (1 - \alpha_{max})^{\ell/(L-1)} , \quad \ell = 0,1,\ldots,L-1$$

where L is the number of different values of α used. In the computations described herein the value of α_{max} was obtained from,

$$1-\alpha_{max} = \min\,[2\Delta x^2/(1+\Delta x^2/\Delta z^2),\ 2\Delta z^2/(1+\Delta z^2/\Delta x^2)]\ .$$

TRANSIENT FLOW OVER A VERTICAL BARRIER

The flow which has been modelled is that which occurs when stratified fluid, initially at rest, is impulsively accelerated in the horizontal direction past a vertical barrier of infinitesimal thickness, so that the velocity far from the barrier has magnitude U which, thereafter, is maintained constant. Initially, the fluid is uniformly stratified in the vertical direction. The height of the vertical barrier is $2\tilde{L}$ and the characteristic temperature difference is defined by $\Delta T = \tilde{T}_L - T_o$, where $\tilde{T}_L$ is the temperature far upstream at height L, measured from the centre of the vertical barrier. The boundary conditions imposed on the surface of the barrier are a uniform temperature, i.e. $\tilde{T} = T_o$ or $T = 0$, and the no-slip boundary conditions for the viscous fluid. The steep local gradients in ζ at the sharp edge of the barrier were smeared out by computing ζ at the edge of the plate from the expression,

$$\zeta^{n+1}_{i,j} = \frac{\Delta t}{R_e \Delta x^2}\,(-2\zeta^n_{i,j+1} + \zeta^n_{i,j+2} + \zeta^n_{i+1,j} + \zeta^n_{i-1,j} - \zeta^n_{i,j}) + \zeta^n_{i,j} + O(\Delta t^2) + O(\Delta x \Delta z). \qquad (21)$$

The derivation of the formulae which approximate the several boundary conditions is given in detail in Lin and Apelt (1970).

The fluid is nominally of unlimited extent but the region modelled was restricted to a rectangular domain of dimensions 15L vertically and 45L horizontally, as shown in Fig.2. In order further to reduce the computational task, symmetry conditions were imposed along the horizontal axis through the midpoint of the barrier. This excludes the possibility of alternate vortex shedding occurring but the model still applies to a wide range of flows. Pao et al (1968) established that vortex shedding does not occur in stratified flows past a circular cylinder if a stability parameter, σ, $(= R_i^{\frac{1}{2}}/\mathrm{Log}_{10}(R_e/40);\ R_e > 40)$ exceeds unity. Since gravity waves may propagate both upstream and downstream from the barrier, the conditions at the inflow and outflow boundaries must be specified in a way which will permit such disturbances to pass out of the computational region. At each of these boundaries ψ, ζ and T were extrapolated with a formula derived from the Milne fourth order predictor formula. As shown in Fig.2, the inflow boundary was set further from the barrier than was the outflow boundary because the former proved to be the more troublesome. At the boundaries parallel to the horizontal axis the conditions imposed were $w=0$, $\zeta=0$ and $\partial T/\partial x=0$.

The typical integration step which extends the solution from time $n\Delta t$ to $(n+1)\Delta t$ involves a nested iterative procedure as follows:

ζ^{n+1} is computed on the barrier;

The (m+1) iterates of the following quantities are computed in the order given, using the most recently available values of other quantities:

$T^{n+1\ (m+1)}$ at interior points, then on inflow and outflow boundaries,

$\zeta^{n+1\ (m+1)}$ at interior points, then on the barrier,

$\zeta^{n+1\ (m+1)}$ on inflow and outflow boundaries,

$\psi^{(m+1)}$, $u^{(m+1)}$, $w^{(m+1)}$.

Each step of this procedure involves one cycle of the iterative process defined by the equation (14) except that in each cycle of the outer iteration, the iteration for ψ is carried to completion. The outer iteration is repeated until successive iterates for T and ζ coincide within the specified tolerance.

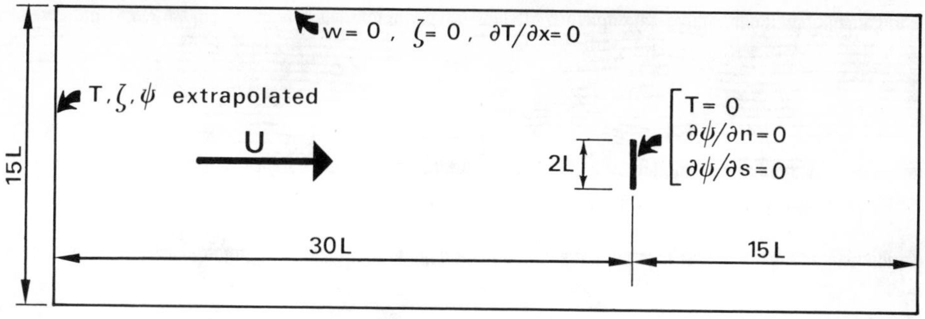

Fig. 2 EXTENT OF NUMERICAL MODEL

Two cases were computed corresponding to the following set of values of parameters: Case I. (R_e = 397, P_r = 10, R_i = 1.58) and Case II. (R_e = 5000, P_r = 1, R_i = 1.58). In each case $\Delta x = \Delta z = 0.25$ and Δt was given an initial value of 0.03 which was incremented by 0.02 at each successive step until $\Delta t > 0.2$ was reached and, thereafter, Δt = 0.20 or 0.22. The tolerances for the several iterations were set at 0.01 times the local value of the variable or 0.001, whichever was the larger for $t < 20$; for later times the tolerances were reduced tenfold. The results of the computations are presented in detail in Lin and Apelt (1970) and in a movie film produced from the computer-drawn contour plots. The most interesting aspect of the transient solutions is the development of vorticity. Initially, the flow is irrotational but the no-slip condition on the barrier generates vorticity which is convected and diffused throughout the flow. Further, the density (or temperature) inhomogeneity provides a second mechanism for creation of vorticity and, from quite early times, vorticity is found in the flow both upstream and downstream from the barrier.

The late time result for Case I is shown in Fig.3, which is a set of computer drawn contour plots. At this stage there is very little change in the magnitudes of all variables from one time step to the next and the flow appears to have approached very closely to the final steady state. The streamlines are lifted up on the upstream side of the barrier while, on the lee side, they converge to produce a strong downslope current. Two lee-waves are clearly visible and the crests and troughs tilt toward upstream, which is in agreement with observation. There is, in fact, a third wave which is too weak to be distinguished from the streamlines but which can be seen from the alternating positive and negative regions of vorticity downstream of the barrier. The vorticity contours also clearly show how vorticity exists throughout most of the flow field, extending a long way upstream of the barrier and throughout the whole of the downstream region, instead of being confined to a distinct wake as would be the case for a homogeneous fluid. There is a small stationary vortex attached to the downstream side of the barrier but the small scale of the figures prevent it from being shown clearly in Fig.3. The lines of constant temperature are very similar in shape to the streamlines, as would be expected. The parameters of Case I were chosen to enable comparison to be made with an experimental result of Pao et al (1968) for flow of stratified salt solution past a circular cylinder, and this comparison shows that the flow patterns are very similar. The parameters of Case II were chosen to test the computation method at large R_e. The late time result is shown in Fig.4 and the flow pattern has approached very close to the steady state. The pattern of streamlines and the vorticity distribution show some differences from the results of Case I but the two sets of results are quite similar, indicating that viscous effects are secondary in such flows. Of course, it must be pointed out that the details of the boundary layer and of the separating shear layer are only crudely modelled at the spatial resolution used. A much finer resolution in the vicinity of the barrier would be required if details of these regions were required.

The treatment of the inflow and outflow boundaries appeared to be quite satisfactory for Case I. However, in Case II at much later times than that

STREAM FUNCTION Contour from 0 to 7.2, interval 0.4

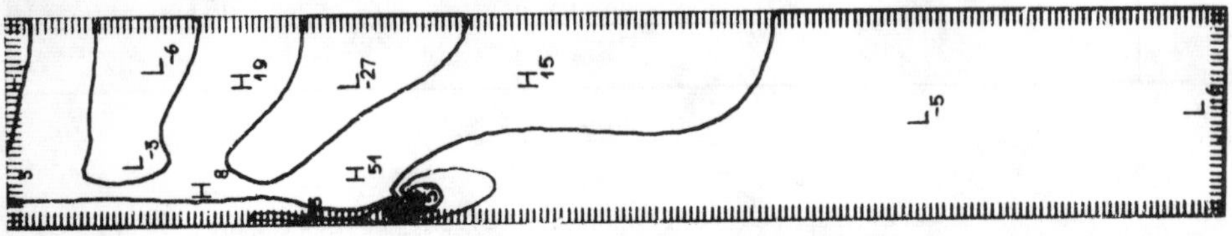

VORTICITY Contour from -4.5 to 0.3, interval 0.3

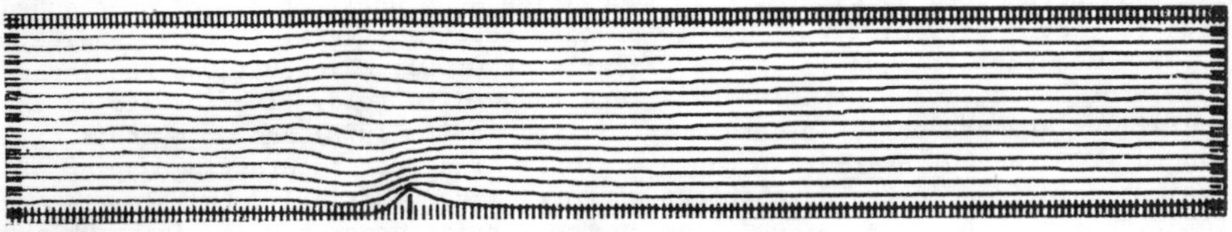

TEMPERATURE Contour from 0 to 7.2, interval 0.4

FIG. 3. FLOW PATTERNS AT $t = 15.68$, $Re = 397$, $Pr = 10$, $Ri = 1.58$, $\Delta X = 0.25$, $\Delta Z = 0.25$, $\Delta t = 0.20$, grid points (180 x 30)

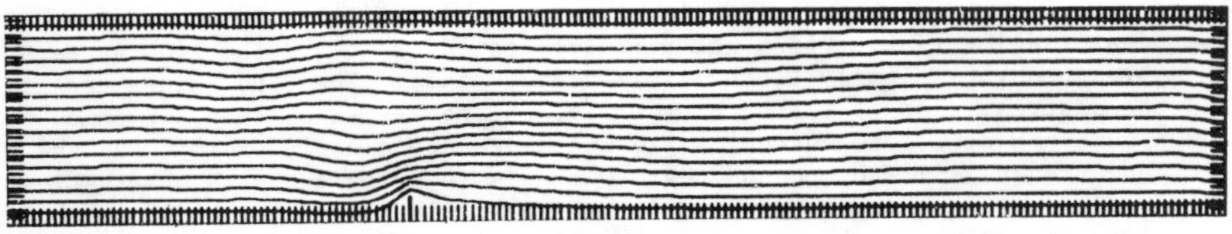

STREAM FUNCTION Contour from 0 to 7.2, interval 0.4

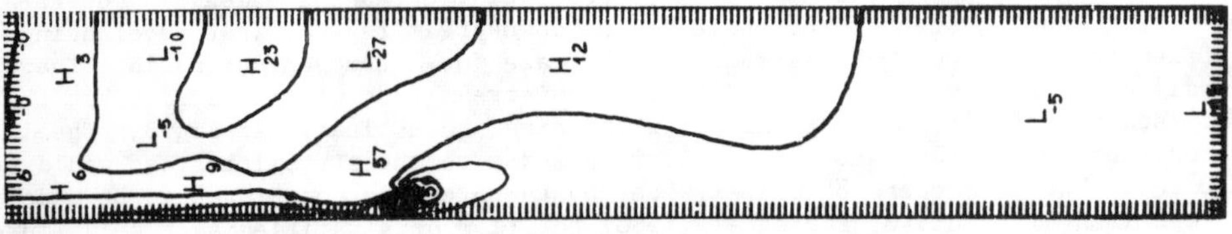

VORTICITY Contour from -4.5 to 0.3, interval 0.3

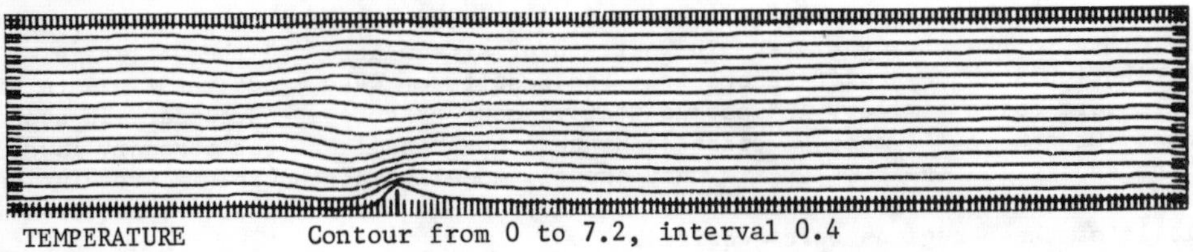

TEMPERATURE Contour from 0 to 7.2, interval 0.4

FIG. 4. FLOW PATTERNS AT $t = 18.42$, $Re = 5000$, $Pr = 1$, $Ri = 1.58$, $\Delta X = 0.25$, $\Delta Z = 0.25$, $\Delta t = 0.22$, grid points (180 x 30) corresponding to Fig.4, i.e. $t > 21$ approximately, disturbances began to appear at the inflow boundary and, if the computation is continued, the disturbance gradually propagates downstream. No such phenomenon was detected at the outflow boundary. Numerical experiments which are described in Lin and Apelt (1970) indicated clearly that these disturbances were associated with the treatment of the inflow boundary condition, not with the method of computing the solution, and that they occur in stratified flow but not in homogeneous flows with the same boundary treatment. It was because of these difficulties that the inflow boundary was set twice as far from the obstacle as was the outflow boundary. A refinement which would remove the inflow boundary even further from the barrier without increasing the computational effort, i.e. with the same number of mesh points, would be the use of transformed coordinates such as elliptical coordinates. The solutions obtained prior to the appearance of the disturbance at the inflow boundary seem to be satisfactory and, in the case under discussion, the steady state solution was achieved within quite small tolerances before the disturbance appeared. Nevertheless, if it is desired for some reason to extend the computations for even longer times it will be necessary to develop a more satisfactory formulation of the inflow boundary condition.

The experience with the method of integration itself was very satisfactory. The maximum time step used, $\Delta t = 0.22$, is approximately four times the maximum value which could be used if an explicit scheme were used in conjunction with directional differencing of the convective terms of the equations (5) and (6) and no stability problem was encountered. In each case computed, the number of iterations required for T and ζ was 5 at the first time step and for ψ it was 14 and, quite quickly, the number of iterations required decreased to 2 for all three variables. The steady state solution shown in Fig.3 was produced with a total of 213 iterations for T and ζ and 272 for ψ. The corresponding iteration numbers for the solution shown in Fig. 4 are 227 and 270 respectively. This situation indicates clearly that the strongly implicit iterative process is indeed a good method for solving the implicit difference equations which arise in this case. The calculations were done on a CDC 6600 computer and the computer time required per time step, averaged over all time steps, was approximately 25 seconds.

ACKNOWLEDGEMENT

Acknowledgement is made to the National Center for Atmospheric Research, which is sponsored by the National Science Foundation (U.S.A.), for the use of its Control Data 6600 computer; and to the financial support by THEMIS Project under grant N00014-68-A-0493-0001. One of us (CJA) was supported by a National Science Foundation Senior Foreign Scientist Fellowship award at Colorado State University.

REFERENCES

Davis, R.E. J. Fluid Mech. 36, 127-143 (1969).

Foldvik, A., and Wurtele, M.G. Geophys. J. Roy. Astr. Soc. 13, 161-185 (1967).

Hovermale, J.B. Dept Meteor., Pennsylvania State University (1965).

Lin, J.T., and Apelt, C.J. Fluid Dynamics and Diffusion Laboratory, Colorado State University THEMIS Tech. Rep. 7, 1-78 (1970).

Lin, J.T., and Binder, G.J. Fluid Dynamics and Diffusion Laboratory, Colorado State University (1967).

Long, R.R. Tellus 5, 42-48 (1953).

Long, R.R. The Rossby Memorial Volume, 372-380 (1959).

Lyra, G. Zeit. Angnew. Math. Mech. 23, 1-28 (1943).

Pao, Y.H., Callahan, M.E., and Timm, G.K. Boeing Sci. Res. Lab. Document D1-82-0736, 1-30 (1968).

Stone, H.L. J. Num. Anal. SIAM 5, 530-558 (1968).

Note: In Figs 3 and 4 the direction of flow is from right to left.

DEVELOPEMENT DE LA METHODE DES SINGULARITES A REPARTITION DISCRETISEE POUR L'ETUDE DES ECOULEMENTS INCOMPRESSIBLES ET COMPRESSIBLES

par LUU Thoai-Sum et COULMY Geneviève

Maîtres de Recherche au L.I.M.S.I., C.N.R.S.

1. INTRODUCTION

A la suite des travaux de Hess, Smith (5) et Giesing (4), qui ont introduit la méthode des singularités à répartition discrétisée pour le calcul des écoulements à potentiel, nous avons pu élaborer, durant ces dernières années, divers développements permettant d'étendre la portée de cette technique, et qui font l'objet de la présente communication.

Les problèmes du champ rencontrés dans le domaine des écoulements à potentiel sont très diversifiés en fonction du type de conditions aux limites imposées. En raison de la non-unicité de la répartition de singularité pour créer le potentiel, nous montrons qu'il est possible, par un choix approprié de la nature des singularités à répartir sur les frontières, d'aborder une grande variété de problèmes d'hydro ou d'aérodynamique. Nous présentons également une extension de la méthode pour traiter les problèmes du champ poissonnien en donnant, comme application, le calcul de l'écoulement compressible autour d'un profil pour les régimes subsonique et transsonique sans choc.

2. NON UNICITE DE LA REPARTITION DE SINGULARITE DANS LA CREATION D'UN POTENTIEL

D'après la formule de Green, on peut montrer qu'un potentiel harmonique, régulier et uniforme dans le domaine D^+ extérieur à une enveloppe fermée C , s'écrit :

$$\varphi_M = \int_C \left[(\varphi'^{+}_{n} - \varphi'^{-}_{n})_P \, G(P,M) + (\varphi^+ - \varphi^-)_P \frac{\partial G(P,M)}{\partial n_P} \right] dS \qquad (2.1)$$

où $\vec{n}$ est la direction de la normale extérieure à C . Si $r = |\vec{PM}|$, la fonction $G(P,M)$ est égale à $-1/4\pi r$ ou $\log r/2\pi$ suivant que l'espace considéré est à 3 ou 2 dimensions. L'expression (2.1) montre que le potentiel est produit par une répartition de source et une répartition de doublet, et que la densité de ces répartitions est définie par la discontinuité de φ'_n et de φ à travers le contour. Or, si l'on suppose que φ , défini dans D^+ , correspond à une solution d'un problème aux limites bien déterminé, rien n'empêche de définir dans le domaine D^- intérieur à C , le champ harmonique φ^- de façon arbitraire. En particulier, on peut définir φ^- à partir de φ^+ sur la frontière, soit par une condition de Neumann avec $\varphi_n^{-'} = \varphi_n^{+'}$, ce qui conduit à éliminer la répartition de source dans (2.1) à condition que $\int_C \varphi_n^{+'} dS = 0$, soit par une condition de Dirichlet avec $\varphi^- = \varphi^+$, ce qui élimine la répartition de doublet. Comme le champ φ^- dans D^- est arbitraire, il existe donc par combinaison des deux cas précédents, une infinité de manières pour représenter la fonction φ^+ dans D^+ bien que celle-ci doive répondre à des conditions aux limites bien définies par le problème posé.

De plus, une intégration par partie de la répartition sur C suivant un sens tangentiel particulier permet de transformer l'une en l'autre, d'une part la répartition de doublet à axe normal et une répartition de tourbillon, d'autre part la répartition de source et une répartition de doublet à axe tangentiel. En effet, dans l'espace à deux dimensions par exemple, on a la relation :

$$\int_A^B \frac{(q+i\gamma)_P}{2\pi} \log(z-z_P) e^{-i\beta_P} dz_P = \frac{Q_B+i\Gamma_B}{2\pi} \log(z-z_B) - \int_A^B \frac{Q_P+i\Gamma_P}{2\pi} \frac{1}{z-z_P} dz_P \qquad (2.2)$$

le premier membre représentant une répartition de source et de tourbillon sur la courbe AB, d'intensités respectives q et γ ; au second membre, le deuxième terme représente une répartition de doublet à axe tangentiel et de doublet à axe normal, d'intensités respectives Q_P et Γ_P telles que :

$$Q_P + i\Gamma_P = \int_A^P (q+i\gamma)_P \, e^{-i\beta_P} dz_P$$

et le premier terme une source et un tourbillon d'intensités Q_B et Γ_B placés en B.

L'intérêt de la non-unicité de la répartition pour créer le potentiel est de permettre, dans une certaine mesure, le choix de la nature de la singularité à répartir sur les frontières de façon à réaliser au mieux le type de conditions aux limites imposées.

3. METHODE DES SINGULARITES A REPARTITION DISCRETISEE

En supposant que la nature de la singularité est appropriée au problème aux limites posé, concernant le champ harmonique, on peut formuler ce problème par une équation intégrale servant à déterminer sa densité inconnue X_P :

$$\int_C X_P \, K(P,P') \, dS_P = f(P') \qquad \text{où } P \text{ et } P' \in C \qquad (3.1)$$

Le noyau $K(P,P')$, qui dépend des données aux limites et de la nature de la singularité répartie sur C, n'est pas toujours régulier lorsque $P \to P'$. La méthode proposée consiste à découper C en petits éléments de contour numérotés, sur chacun desquels on définit X_P par sa valeur moyenne X_j considérée comme constante sur toute l'étendue de l'élément, son numéro d'ordre étant j. On traduit la condition

(3.1) en écrivant que cette relation est satisfaite aux points de contrôle fixés au centre de chacun des éléments, ce qui conduit à un système d'équations linéaires définissant les X_j :

$$[A_{ij}][X_j] = [F_i] \quad \text{où} \quad F_i = f(P_i) \quad \text{et} \quad A_{ij} = \int_{\Delta C_j} K(P, P_i)\, dS_P$$

Même si $K(P, P_i)$ est singulier au voisinage de P_i le passage aux limites permet en général de définir A_{ij} en ce point. Le choix de la nature de la singularité sera réalisé, en fonction des données aux limites à traiter, de manière à rendre la diagonale principale de la matrice $[A_{ij}]$ prépondérante, ce qui permet l'utilisation de méthodes itératives, rapides et précises comme celle de Gauss Siedel, pour résoudre le système.

C'est ainsi qu'on utilisera une répartition de source ou de doublet à axe normal pour des conditions de Neumann, Dirichlet ou Fourier, et une répartition de tourbillon ou de doublet à axe tangentiel lorsque la valeur de la dérivée tangentielle du potentiel est imposée sur la frontière.

Par ailleurs, comme certains problèmes du champ impliquent une symétrie paire ou impaire du potentiel par rapport à l'élément porteur, on notera que sources et doublets à axe tangentiel respectent la première symétrie tandis que la seconde est réalisée par les tourbillons et les doublets à axe normal. Pour créer un champ avec circulation, on pourra utiliser soit une répartition de tourbillon, soit une répartition de doublet à axe normal sur une surface s'étendant jusqu'à l'infini.

4. EXEMPLES D'APPLICATION DE LA METHODE A DIVERS ECOULEMENTS INCOMPRESSIBLES A POTENTIEL

Il est bien connu que la difficulté rencontrée dans les problèmes du champ laplacien réside essentiellement dans la réalisation des conditions aux limites. Nous allons donner un aperçu des divers problèmes aux limites que notre méthode a permis d'aborder dans les domaines d'aéro et d'hydrodynamique.

4.1. Problème de Neumann, écoulement sans circulation

Le mouvement du fluide provoqué par le déplacement de corps solides sans circulation peut être défini par son potentiel des vitesses φ. Si $\overrightarrow{V_{ent}}$ désigne la vitesse d'entraînement d'un des corps, la vitesse relative du fluide par rapport à ce corps s'écrit $\overrightarrow{V_r} = \text{grad}\,\varphi - \overrightarrow{V_{ent}}$. La condition de glissement du fluide sur la superficie du corps implique que la composante normale de $\overrightarrow{V_r}$ soit nulle, ce qui conduit à une donnée aux limites de type Neumann pour le potentiel φ : $\varphi'_n = \overrightarrow{V_{ent}} \cdot \vec{n}$. Le fluide étant au repos à l'infini on a $(\text{grad}\,\varphi)_\infty \to 0$. Le potentiel peut être généré par une répartition de source sur la superficie des corps, ce qui est un procédé classique. Mais il peut être généré également par une répartition de doublet, quoique la représentation soit moins fine dans ce cas; on obtient cependant des résultats satisfaisants, si l'on prend soin de déterminer la vitesse tangentielle par différences finies sur φ pris au centre de chaque élément.

On notera que la densité de doublet $\Delta\varphi$ est définie à une constante additive près. Cette indétermination est mise en évidence par le fait que les équations du système (3.2) traduisant la condition de Neumann ne sont pas indépendantes. En effet, si à chaque équation est affecté un coefficient égal à la surface de l'élément correspondant, la somme de chacune des colonnes du système représente le flux de chaque élément de doublet à travers la surface fermée de l'obstacle, et, pour le second membre, celui de l'écoulement de vitesse $\vec{V}_{ent}$ à travers ce même obstacle; ces flux étant conservatifs, ces différentes colonnes sont nulles. Pour lever l'indétermination, il suffit de supprimer une des équations et d'imposer une valeur (par exemple zéro) au $\Delta\varphi$ correspondant.

La méthode a permis de traiter avec succès, dans les domaines bi et tridimensionnels, différents problèmes d'écoulement stationnaire ou instationnaire mettant en présence plusieurs corps se déplaçant à des vitesses différentes (2). Dans ce dernier cas, le phénomène doit être découpé dans le temps en un certain nombre de séquences pour chacune desquelles le champ est résolu; par différences finies, on peut alors évaluer le terme instationnaire $(\partial\varphi/\partial t)$ de la formule de Bernoulli conduisant à la détermination des répartitions de pression.

4.2. Problème de Neumann, écoulement avec circulation

Le calcul de l'aile épaisse d'envergure finie (12) constitue un exemple typique de ce genre de problème. Comme dans le cas précédent, le potentiel est régi par une donnée aux limites de type Neumann traduisant la condition de glissement. La circulation le long de l'envergure $\Gamma(y)$ qui s'établit autour de chaque section de l'aile, doit être déterminée dès qu'est imposée la condition de Kutta-Joukowsky. Cette condition exige que le bord de fuite soit le lieu des points d'arrêt, c'est-à-dire que la vitesse soit tangentielle au plan bissecteur du bord de fuite dans le voisinage immédiat de ce dernier. Le potentiel est alors représenté par des répartitions, de source et de doublet sur la superficie de l'aile d'une part, de doublet sur la nappe de tourbillons libres d'autre part, la densité de cette dernière étant égale à la loi de circulation $\Gamma(y)$. Sur la superficie de l'aile, la densité de doublet $\Delta\varphi$ se déduit de $\Gamma(y)$ par une loi d'affinité respectant la continuité au bord de fuite. On discrétise les répartitions en divisant la surface de l'aile et du sillage en N bandes parallèles à l'axe des x , sur chacune desquelles on considère $\Delta\varphi_j$ constante. Sur l'aile elle-même, chaque bande est subdivisée en petits éléments de surface sur chacun desquels on applique une répartition de source de densité uniforme q_j et une répartition de doublet dont l'intensité se déduit de $\Delta\varphi_j$ par une loi d'affinité. N_1 désignant le nombre total d'éléments sur l'aile, les $N_1 + N$ densités inconnues q_j et $\Delta\varphi_j$ sont déterminées lorsqu'on écrit la condition de glissement aux points de contrôle situés au centre de chacun des éléments et la condition de Kutta-Joukowsky en des points situés sur chaque bande au voisinage immédiat du bord de fuite.

Par un procédé analogue, nous avons pu de même aborder le calcul du champ dû au fonctionnement d'une hélice marine, effectué dans le cadre de la théorie linéaire de la surface portante (6). Dans le domaine bidimensionnel, nous avons montré également qu'il était possible d'adapter la méthode à l'étude des écoulements instationnaires autour de profils ou d'aubes passantes avec ou sans lâchage de tourbillons libres (9), (10).

4.3. Problèmes de Dirichlet, Fourier, ou mixte (7), (8)

Dans le cas bidimensionnel, le problème de Dirichlet concerne en particulier les transformations conformes et les écoulements autour de profils, traités par la fonction de courant. On résout ce type de problème en utilisant une répartition de source, partie réelle de $\log(z - z_p)$.

La théorie de la ligne portante appliquée à l'aile de grande envergure ou à l'hélice aérienne conduit à des problèmes aux limites de type Fourier; sur la nappe de tourbillons libres à l'infini aval, le potentiel des vitesses est régi par la relation de Prandlt que l'on peut interprêter comme une donnée aux limites de type Fourier, reliant la valeur du potentiel à sa dérivée normale. Ce genre de problème a été résolu par la technique des singularités, en utilisant une répartition discrétisée de doublet.

Indiquons en outre que la théorie linéarisée de l'aile avec soufflage au bord de fuite conduit à des problèmes ayant une condition de Neumann sur la trace de l'aile et une condition de type Fourier sur la trace de la nappe du jet. La théorie linéaire des écoulements subcavitants donne lieu à des problèmes ayant une condition de Neumann sur la trace du profil et une condition fixant la dérivée tangentielle sur la trace de la cavité. Avec un choix approprié de la nature des singularités pour représenter le potentiel, nous avons également réussi à adapter la méthode à l'étude de ces problèmes.

5. CALCUL DE L'ECOULEMENT COMPRESSIBLE

Nous avons également développé la méthode des singularités pour traiter les écoulements à potentiel permanents et compressibles (11). Le mouvement peut être défini par son potentiel des vitesses de perturbation φ . Si $\vec{V}_0$ désigne la vitesse uniforme à l'infini, la vitesse du fluide est donnée par $\vec{V} = \vec{V}_0 + \mathrm{grad}\,\varphi$, et φ est régi par une équation aux dérivées partielles que l'on peut mettre sous forme poissonnienne :

$$\nabla^2 \varphi = M^2 \frac{\partial V}{\partial s} = Q \qquad (5.1)$$

où M représente le nombre de Mach local et $\partial V / \partial s$ la dérivée du module de la vitesse suivant le sens tangentiel $\vec{s} = \vec{V}/|\vec{V}|$. D'après la théorie de Green, on peut représenter φ , d'une part par une répartition de source de densité Q dans l'espace, d'autre part par une répartition de source et de tourbillon sur la superficie de l'obstacle. Dans le cas de l'écoulement autour d'un profil bidimensionnel, on peut écrire :

$$\varphi(z) = \iint_D \frac{Q}{2\pi} \mathcal{R}_e \left[\log(z - z_p) \right] dS_p + \int_C \mathcal{R}_e \left[\frac{q + i\gamma}{2\pi} \log(z - z_p) \right] ds_p \qquad (5.2)$$

où $q(s)$ et γ désignent respectivement les densités de source et de tourbillon, γ pouvant être supposé à priori constant sur tout le contour du profil C . La condition de glissement du fluide sur le profil exige que $\varphi'_n = \vec{V} \cdot \vec{n}$. La condition de Kutta-Joukowsky équivaut à écrire que la somme algébrique des vitesses tangentielles, aux points d'extrados z_{ex} et d'intrados z_{in} , est nulle lorsque ces deux points se rapprochent du bord de fuite z_f tels que $|z_{in} - z_f| = |z_{ex} - z_f| \to 0$. Ces deux conditions sont traduites par les équations intégrales suivantes :

$$\int_C - \mathcal{I}_m \left[\frac{q + i\gamma}{2\pi} \frac{e^{i(\beta - \beta_p)}}{z - z_p} dz_p \right] = - \vec{V}_o \cdot \vec{n} + \iint_D \frac{Q}{2\pi} \mathcal{I}_m \left[\frac{e^{i\beta}}{z - z_p} \right] dS_p \qquad z \in C^+ \qquad (5.3)$$

$$\int_C \mathcal{R}_e \left[\frac{q + i\gamma}{2\pi} \left(\frac{e^{i\beta_{ex}}}{z_{ex} - z_p} - \frac{e^{i\beta_{in}}}{z_{in} - z_p} \right) e^{i\beta_p} dz_p \right] = - \vec{V}_o (\vec{s}_{ex} + \vec{s}_{in}) - \iint_D \frac{Q}{2\pi} \mathcal{R}_e \left[\frac{e^{i\beta_{ex}}}{z_{ex} - z_p} - \frac{e^{i\beta_{in}}}{z_{in} - z_p} \right] dS_p \quad (5.4)$$

où β, β_{ex}, β_{in} et β_p sont respectivement les angles que font avec l'axe réel les tangentes à C^+ aux points z, z_{ex}, z_{in} et z_p et où $\vec{s}_{ex}$ et $\vec{s}_{in}$ désignent les vecteurs tangentiels aux points z_{ex} et z_{in} .

Dans le calcul numérique, on divise le domaine D extérieur au profil en petits pavés sur la superficie de chacun desquels Q est considéré comme constant. La vitesse induite en un point quelconque z , par un de ces pavés est donné par (1) :

$$\varphi'_x - i\varphi'_y = \frac{Q}{2\pi} \iint_\Delta \frac{1}{z - z_p} dS_p = \frac{Q}{\pi} \sum_{m=1}^{4} \frac{S_{mn}}{z_n - z_m} \log \frac{z - z_m}{z - z_n}$$

z_m et z_n désignent les affixes des sommets du pavé, avec $n = 2, 3, 4, 1$ pour $m = 1, 2, 3, 4$ respectivement. S_{mn} représente la surface du triangle z_m, z_n, z , positive si ces 3 points sont placés suivant le sens trigonométrique, négative dans le cas contraire.

Après discrétisation, les intégrales des seconds membres de (5.3) et (5.4) sont remplacées par des sommations que l'on peut évaluer dès que la valeur de Q est connue sur chaque pavé.

Le contour C du profil est approché par un polygone de N segments sur chacun desquels la densité de source q_j est considérée constante. Au centre des segments, on fixe un point de contrôle en lequel on écrit la condition (5.3) qui se trouve alors remplacée par un ensemble de N équations auxquelles s'ajoute (5.4) pour former un système de $N+1$ équations linéaires permettant de déterminer les inconnues $q_1, \dots, q_N$ et γ.

La résolution du problème est effectuée suivant un procédé itératif. On commence, à la première itération, par traiter l'écoulement comme incompressible, c'est-à-dire en admettant Q nul, ce qui permet d'évaluer la vitesse et le nombre de Mach au centre de chaque pavé de D ; Q est déterminé à partir de la dérivée $\partial V/\partial s$, évaluée par différence finie pour chacun des pavés en fonction des valeurs de $\vec{V}$ sur les pavés adjacents. On répète le calcul itératif jusqu'à ce que la convergence soit atteinte.

Nous avons utilisé cette méthode pour traiter les écoulements subsoniques autour de profils d'ailes jusqu'à des régimes transsoniques supercritiques sans choc. Les résultats obtenus concordent bien avec ceux déduits des théories de Nieuwland (13), de Garabedian et Korn (3). Une mise au point est actuellement en cours pour tenir compte de la formation de l'onde de choc.

REFERENCES

(1) Coulmy, G., C.R.Ac.Sc., 1972.

(2) Coulmy, G., Bellevaux, C., Luu, T.S., Marty, P., Ta Phuoc Loc, Sagnard, J., Martinod, R., AFITAE, Nov. 1971.

(3) Garabedian, P.R., et Korn, D.G., Numerical Solution of Partial Differential Equation, II Academic Press, 1971, pp 253-271.

(4) Giesing, J.P., Trans. ASME, Series D, vol. 90, n° 3, Sept. 1968.

(5) Hess, J.L., Smith A.M.O. Progress in Aeronautical Sciences, vol. 8, 1967.

(6) Luu, T.S., ATMA, 1970.

(7) Luu, T.S., Colloque du CNRS, Marseille, Nov. 1971.

(8) Luu, T.S., Coulmy, G., et Corniglion, J., ATMA, 1969.

(9) Luu, T.S., Coulmy, G., et Corniglion, J., ATMA, 1971.

(10) Luu, T.S., et Corniglion, J., ATMA, 1972.

(11) Luu, T.S., Coulmy, G., et Dulieu, A., ATMA, 1972.

(12) Luu, T.S., Coulmy; G., et Sagnard, J., ATMA, 1971.

(13) Nieuwland, G.Y., NLR-TR T. 172, 1967.

THE NUMERICAL SOLUTION OF CONVECTIVE HEAT TRANSFER IN THE SPACE SHUTTLE BASE REGION BY TELENIN'S METHOD

James C. S. Meng*

NASA-George C. Marshall Space Flight Center
Marshall Space Flight Center, Alabama 35812

I. INTRODUCTION

With the advent of the concept of the reusable space shuttle vehicle, emphasis has been put upon its minimum weight and reusability. Problems related to the accurate prediction of the base thermal environment wherein engine nozzles are exposed to trapped recirculating hot air at hypersonic reentry conditions have come into prominence. The study aims at eliminating excessive conservatism in the design of base heat shields and engine insulation.

Atmospheric reentry involves a total temperature and Mach number condition that cannot be effectively simulated by known experimental devices for an extended period. Numerical schemes which can yield accurate solutions without requiring large storage capacity and long execution time for computers are desirable. One such scheme, established by Telenin, exploits the obvious numerical advantages of working with Cauchy-type problems for the present elliptic system of equations. It was first proposed for axisymmetric blunt body problems, and later adopted for conical flow problems by Holt and Ndefo (1970). It is well known that Cauchy's problems are, in general, improperly posed for an elliptic system of equations. However, for an *a priori* restricted class of solutions (such as the class of bounded analytic functions), Cauchy's problems become correctly posed for the elliptic systems.

The base region, in the form of a cavity, is composed of the base wall and two protruding shrouds (Fig. 1). The cavity walls, the free mixing layer, and the near wake region define the bounded domain wherein Telenin's scheme applies. The equations are transformed so that the region of interest becomes a rectangle which is subsequently divided into strips along the shrouds; Lagrange interpolation polynomials of degree four and seven are applied in the cavity and the near wake region across the strips. Augmented first-order ordinary differential equations are obtained. The problem is then reduced to a two-point boundary value problem.

It is known that errors committed in the arbitrary trial data increase exponentially with the number of trial variables and the physical dimension of the integration domain, so that Telenin's scheme is not immune to instability. Especially in the present case, a singular layer exists right at the initial surface. As a matter of fact, instability was encountered as the major difficulty during integration across the boundary layer on the base wall. A useful tool, the multiple shooting method which was originally developed for trajectory optimization, is found effective in suppressing this instability. In essence, it subdivides the domain into subintervals, and the number and their location are determined by optimizing storage and stability of integration. Each subinterval was treated by the simple shooting method, with an overall satisfaction of boundary conditions on both ends and elimination of jump across the subinterval realized by solving a linear system.

---

* NAS-NRC Post Doctoral Associate.

There are several important advantages of the present scheme over the finite-difference type computation. First, it occupies one or two orders of magnitude smaller storage space; secondly, it consumes at least two orders of magnitude less computer time; thirdly, the analyticity of the solution is guaranteed; and finally, the equations are satisfied exactly on the strips. A simple estimate is given to suppprt these assertions. The storage required for the present scheme is only that for storing variables at the intersection points of the strips and the subintervals; it is one or two orders of magnitude smaller than the number of grid points for the finite-difference scheme. The computation time for the present scheme is needed for the following three types of operations:

(a) Integration of N (number of variables) · S (number of strips) · M (number of subintervals) equation.

(b) Integration of $M \cdot (N \cdot S)^2$ variational equations.

(c) Inversion of M N · S by N · S matrices.

The number of operations for one iteration is then approximately $M^3 N^3 S^3 + M(NS + N^2 S^2) T$ (number of integration step), or $\cong 10^8$ with M = 12, N = 7, S = 7, T = 60. The total computation time per iteration is about 10^2 seconds, while a finite-difference scheme would have to invert a NTS by NTS matrix, or about $N^3 T^3 S^3$ operations, or 10^4 seconds per iteration.

The full Navier-Stokes equations and energy equations are solved for flows at zero angle of attack in a free stream of Mach number 11, Reynolds number 10^5, total temperature 10^4 °R, and the walls are cooled. It was assumed that heat conductivity and viscosity are linearly proportional to temperature, the specific heat is constant, and Prandtl number is unity. The detailed flow field and thermal environment in the base region are presented in the form of temperature contours and velocity vectors.

II. FORMULATION OF THE PROBLEM

A physical sketch is given in Fig. 1, the origin is set at the bottom corner of the base wall, and the region of interest is surrounded by the base wall, the boundary layers on both shrouds, and the far wake region. By transforming this region from physical plane to ξ, η-plane defined by

$$\xi = \frac{x}{H}$$

$$\eta = \frac{y - \Psi_b(x)}{\Psi_t(x) - \Psi_b(x)} , \qquad (1)$$

where $\Psi_t(x)$, $\Psi_b(x)$ are top and bottom boundary layer edges, the region of interest becomes a rectangle bounded by $\eta = 0$, $\eta = 1$, $\xi = 0$ and the far-wake region (Fig. 2).

<u>a. Governing Equations</u>. By nondimensionalizing all flow variables by their corresponding free stream values, pressure is made dimensionless in terms of $\rho_\infty U_\infty^2$. Replace the first-order derivatives by new variables:

$$\begin{pmatrix}\epsilon\\ \sigma\\ \beta\end{pmatrix} = \frac{\partial}{\partial x}\begin{pmatrix}u\\ v\\ T\end{pmatrix} . \tag{2}$$

Substituting these into Navier-Stokes equations and energy equations and rearranging, we obtain:

$$\zeta u \rho_\xi = (\dot{\tau} + \eta\dot{\zeta}) u \rho_\eta - \rho\epsilon\zeta - \rho_\eta v - \rho v_\eta \tag{3}$$

$$\frac{4}{3}\frac{T\zeta^2}{Re}\epsilon_\xi = -(\dot{\tau} + \eta\dot{\zeta})\left(\frac{\zeta T}{\gamma M_\infty^2}\rho_\eta - \frac{4}{3}\frac{\zeta T}{Re}\epsilon_\eta\right) + \frac{\zeta^2 T}{\gamma M_\infty^2}\rho_\xi$$

$$+ \rho u \epsilon \zeta^2 + \rho v u_\eta \zeta + \frac{1}{\gamma M_\infty^2}\rho\beta\zeta^2$$

$$- \frac{1}{Re}\left[\frac{2}{3}\zeta\beta(2\epsilon\zeta - v_\eta) + T_\eta(u_\eta + \sigma\zeta) + T(u_{\eta\eta} + \frac{1}{3}\sigma_\eta\zeta)\right] \tag{4}$$

$$\frac{T\zeta^2}{Re}\sigma_\xi = (\dot{\tau} + \eta\dot{\zeta})\frac{\zeta T}{Re}\sigma_\eta + \rho u\sigma\zeta^2 + \rho v v_\eta\zeta + \frac{\zeta}{\gamma M_\infty^2}(\rho_\eta T + \rho T_\eta)$$

$$- \frac{2}{3Re}[T_\eta(2v_\eta - \epsilon\zeta) + T(2v_{\eta\eta} - \zeta\epsilon_\eta)]$$

$$- \frac{1}{Re}[\beta(u_\eta\zeta + \sigma\zeta^2) + T\epsilon_\eta\zeta] \tag{5}$$

$$\frac{T\zeta^2}{Re\,Pr}\beta_\xi = (\dot{\tau} + \eta\dot{\zeta})\left(\frac{\gamma - 1}{\gamma}uT\zeta\rho_\eta + \frac{1}{Re\,Pr}\zeta T\beta_\eta\right)$$

$$- \frac{\gamma - 1}{\gamma}uT\zeta^2\rho_\xi + \rho u\beta\zeta^2 + \rho v\zeta T_\eta$$

$$- \frac{\gamma - 1}{\gamma}(\rho u\beta\zeta^2 + vT\zeta\rho_\eta + \rho v\zeta T_\eta)$$

$$- \frac{1}{Re\,Pr}\beta^2\zeta^2 - \frac{1}{Re\,Pr}(TT_\eta)_\eta$$

$$+ \frac{(\gamma - 1)M_\infty^2}{Re}T\left[\frac{4}{3}(\epsilon^2\zeta^2 + v_\eta^2 - \epsilon\zeta v_\eta) + (\sigma\zeta + u_\eta)^2\right] \tag{6}$$

$$\zeta u_\xi = (\dot{\tau} + \eta\dot{\zeta})u_\eta + \epsilon\zeta \tag{7}$$

$$\zeta v_\xi = (\dot{\tau} + \eta\dot{\zeta})v_\eta + \sigma\zeta \tag{8}$$

$$\zeta T_\xi = (\dot{\tau} + \eta\dot{\zeta})T_\eta + \beta\zeta \tag{9}$$

$$\dot{\tau} = \frac{d}{d\xi}\,\frac{\Psi_b(x)}{H} = \tan\,[\nu(M_b) - \nu(M_{bo})] \tag{10}$$

$$\dot{\zeta} = \frac{d}{d\xi}\,\frac{\Psi_t(x) - \Psi_b(x)}{H} = \tan\,[\nu(M_{to}) - \nu(M_t)] - \dot{\tau} \tag{11}$$

where subscripts ξ and η denote $d/d\xi$ and $\partial/\partial\eta$. $\nu(M)$ is the Prandtl-Meyer function. The subscripts b and bo indicate bottom boundary layer edge conditions at arbitrary ξ and at $\xi = 0$; similar conditions on the top boundary are denoted by t and to. We shall divide the domain of interest into N - 1 strips, as shown in Fig. 2, and approximate the flow variables in terms of Lagrange interpolation polynomials across the strips; i.e.,

$$\begin{pmatrix} u(\xi,\eta) \\ v(\xi,\eta) \\ \rho(\xi,\eta) \\ T(\xi,\eta) \\ \epsilon(\xi,\eta) \\ \sigma(\xi,\eta) \\ \beta(\xi,\eta) \end{pmatrix} \cong \sum_{i=1}^{N} \begin{pmatrix} u_i(\xi) \\ v_i(\xi) \\ \rho_i(\xi) \\ T_i(\xi) \\ \epsilon_i(\xi) \\ \sigma_i(\xi) \\ \beta_i(\xi) \end{pmatrix} \prod_{\substack{k=1 \\ k\neq i}}^{N} \frac{(\eta - \eta_k)}{(\eta_i - \eta_k)} \quad . \tag{12}$$

These expressions are substituted into equations (3) through (9) with the requirement that the resulting equations be satisfied identically on each line η_i. An approximating system of 7 N first-order ordinary differential equations is then obtained for the approximate values $u_i, v_i, \rho_i, T_i, \epsilon_i, \sigma_i, \beta_i$ of the dependent variables of the N lines; η_i = constant.

b. Initial Thermal Boundary Layers. To integrate the equations toward downstream, initial conditions have to be specified. External to the shrouds, these conditions correspond to the solutions of forebody thermal boundary layers. Assuming no separation is ahead of the shroud-trailing edge, the solution of the compressible boundary layer past a flat plate is applied. With Prandtl number unity, we have

$$\frac{T}{T_e} = 1 + \frac{\gamma - 1}{2} M_e^2 \left[1 - \left(\frac{u}{u_e}\right)^2\right] + \frac{T_w - T_{ad}}{T_e}\left(1 - \frac{u}{u_e}\right) \quad . \tag{13}$$

c. Far-Wake Solution. To form a boundary-value problem, conditions on both boundaries are required. The far-wake solution provides part of the downstream boundary conditions. A linearized Oseen-type solution of compressible laminar wake was given by Kubota (1962). Defining $\bar{x}$, $\bar{y}$ as the coordinates after Stewartson-Illingworth transformation, with origin set at the neck, we have

$$\frac{u}{u_e} = 1 - \frac{A}{\sqrt{\bar{x}}}\, e^{-\frac{\bar{y}^2}{4\bar{x}}}$$

$$\frac{h}{h_e} = 1 + \frac{B}{\sqrt{\bar{x}}}\, e^{-\frac{Pr\bar{y}^2}{4\bar{x}}} , \qquad (14)$$

where

$$A = \frac{1}{2}\sqrt{\frac{Re}{\pi}} \left(\frac{\rho_e u_e \theta}{\rho_\infty u_\infty H} M_e^2 \right)_{\text{at neck}} ,$$

$$B = \frac{1}{2}\sqrt{\frac{Re\,Pr}{\pi}} \left[St + \left(\frac{\rho_e u_e^3 \theta}{\rho_\infty u_\infty h_0 H} \right)_{\text{at neck}} \right] ,$$

and θ is the momentum thickness. Since the external flow is represented by the Prandtl-Meyer solution in the present study, the neck condition corresponds to that when the flow is parallel to the centerline. For the present problem, with $M_\infty = 11$, the $M_{e\,\text{at neck}} = 9.586$. The total heat loss of the flow past the vehicle is estimated by neglecting the base heat transfer and assuming the vehicle length is 20 times the base height, so that the Stanton number is taken to be -0.07339.

III. ERROR GROWTH IN THE BOUNDARY LAYER

The major difficulty encountered in carrying out the present computation was the instabilities. Gilinskiy et al. (1965) showed that error caused by the approximation of flow variables by Lagrange interpolation polynomial across the strips may oscillate along the strips in the linear case. In present nonlinear systems, the error not only oscillates but grows rapidly. To cope with this, the author relied upon two fundamental tools; the boundary layer equations and the multiple shooting method.

The entire flow domain of interest was first conceived to be governed by the Navier-Stokes equations so that the problem can be treated through a unified point of view. However, this experienced tremendous problems of instability due to the fact that the uniform validity of the Navier-Stokes equations practically breaks down when dealing with a problem of extremely nonuniform grids. For high Reynolds number flows with large separation bubble, the boundary layer equations are more feasible for the high gradient areas. Although this will limit the accuracy of the solution to less than 1/Re there, the instability problem can be avoided partially.

The problem of error propagation in the base wall boundary layer is of special importance because almost all the physical processes in determining the base flow heat transfer properties occur here and in the free shear layers. The governing equations for flow can also be regarded as the error propagation equations, since without knowing the solution a priori the guessed initial values may contain an error of its own magnitude. We shall focus our attention upon the error growth of the heating rate across the base wall boundary layer. The following equation gives the growth rate of $\beta = \partial T/\partial x$ at $\xi = 0$ along the strip:

$$\beta_{\xi} = \frac{1}{\zeta}(\dot{\tau} + \eta\dot{\zeta})\beta_{\eta} - \beta^2 - (\gamma - 1)M_{\infty}^2 \ \mathrm{Pr}\ T\ \sigma^2$$

$$+ \mathrm{Re}\ \mathrm{Pr}\ F(\rho, T, \epsilon, \sigma, \beta) \quad . \tag{15}$$

The last term on the right side can be neglected if boundary layer equations are used; however, if it is retained on the Navier-Stokes equations, rapid error amplification caused by this term will occur since the initial values cannot always be chosen so as to guarantee the last term's smallness. The second and third terms are dominant then and remain to be negative; this will therefore reduce the danger of divergence. The approximate solution of the above equation can be represented by the following relation:

$$\frac{2}{\beta} - \frac{1}{\sqrt{(\gamma - 1)M_{\infty}^2 \ \mathrm{Pr}\ T\ \sigma^2}} \tan^{-1} \frac{\beta}{\sqrt{(\gamma - 1)M_{\infty}^2 \ \mathrm{Pr}\ T\ \sigma^2}} \cong \xi \quad , \tag{16}$$

so that β will decrease when ξ increases; in other words, the integration is stable. Similar analyses can also establish the fact that in shear layers the error is amplified slower by boundary layer equations than by Navier-Stokes equations. Based upon this result, we shall use boundary layer equations on base wall, shrouds, and in free shear layers and Navier-Stokes equations in the remaining regions. This mathematical model is depicted in Fig. 3.

IV. MULTIPLE SHOOTING METHOD – THE CONTINUATION METHOD

A serious shortcoming of the shooting method becomes apparent when the differential equations amplify the errors so rapidly that divergence occurs before the initial value problem can be completely integrated. This may happen even in the face of accurate guesses for the initial values. The multiple shooting method can frequently circumvent the difficulty, or else a finite difference scheme can be employed. The method is essentially a combination of difference scheme and initial value problems. It is designed to suppress the growth of the trial integral curves by dividing the domain of integration into a number of subintervals, integrating each individual initial value problem over its own interval, and then simultaneously adjusting all the guessed initial data to satisfy the boundary conditions and continuity conditions at the junction points.

The formulation of the multiple shooting method can be found in Osborne (1969), and a comprehensive version was given by Bulirsch (1971). From the point of view of practical calculation, Bulirsch indicated that the correction technique developed by Broyden (1965) should be applied to modify the Jacobian matrix for following iterations without going through the time consuming integration of N^2 equations in every iteration. The convergence factor (or the λ-strategy) is very useful in solving nonlinear problems, and the more nonlinear the equations are, the smaller the convergence factor is. In previous studies, it was always on the order of 10^{-1}, but in the present study, it had to be as small as 10^{-4} in the first few iterations to insure stable integration.

Let $\vec{w}$ and $\vec{y}$ be the boundary conditions and the unknown vector; therefore, the convergence sphere and rate of convergence for the shooting method are $(1 - \sqrt{1 - 2h_0})\eta_0/h_0$

and $(2h_0)^{2^k-1}\eta_0/2^{k-1}$, with the Jacobian matrix J, $||J|| \leq B_0$, $\sum_{j,s=1}^{N} ||\frac{\partial^2 w_i}{\partial y_j \partial y_s}|| \leq K$ for all i's, $||J^{-1}\vec{w}|| \leq \eta_0$, $||\vec{w}|| = \max_{1\leq i\leq N} |w_i|$, $||J|| = \max_{1\leq i\leq N} \sum_{\ell=1}^{N} |J_{i\ell}|$ and k is the number of iterations counted after the initial values fall within the convergence sphere. By Kantorovich theorem, the convergence is guaranteed as long as $h_0 = B_0\eta_0K$ is smaller or equal to 1/2. For simple problems, convergence can often be obtained by simply going through many iterations. In complex problems, one has to modify the guessed values to fulfill as many of the Kantorovich sufficient conditions as possible for convergence. Ironically, the labor required to make such test is N times more than that needed for solving the problem itself. For example, the quantity K needs integration of $MN^2(N+1)/2$ equations throughout the entire domain so that the advantage of working with the Cauchy-type problem would be greatly diminished.

The subdivision of the domain for the multiple shooting method is determined by the relation $|\xi_{j+1} - \xi_j| \sim \frac{1}{L_j}$ if all the derivatives are Lipschitz continuous, and B_0 can be taken as L_j in each subinterval j. This L_j is usually quite large in nonlinear problems; for example, it is on the order of 10^3 for the present problem. Therefore, in theory, about 10^3 subintervals to insure against the instability are required; but, in practice, the advantage of the Cauchy-type problem will be offset if the number of subintervals becomes comparable to the number of the grids by the difference scheme. This dilemma can be resolved by incorporating the continuation method developed by Roberts and Shipman (1967) to the multiple shooting method. They employed the simple shooting method and stretched the domain length to the final length in each iteration to solve a problem which could not be solved by the shooting method alone. It was shown (Roberts and Shipman, 1971) that the method will be stable if $\Delta\xi_{j\,new} \leq \frac{1}{2\bar{M}KB_o^2}$; $\bar{M}$ is the uniform bound of the derivatives over $[\xi_j, \xi_{j+1}]_{new}$. However, it is found that one should not continue the segment length this way in practice because the denominator is very large, $\sim 0(10^{10})$, but should find the $\Delta\xi_{j\,new}$ by $\Delta\xi_{j\,new} = \Delta\xi_{j\,old}\,(\bar{M}KB_o^2)_{old}/(\bar{M}KB_o^2)_{new}$. This can be simplified to $\Delta\xi_{j\,old}\,(\bar{M}B_o^2)_{old}/(\bar{M}B_o^2)_{new}$ by experience, since K was found to be relatively constant through many iterations.

By applying Broyden's correction technique, the convergence factor, and the continuation method to the multiple shooting method, the present problem was solved by using 12 subintervals.

V. RESULTS

It was noted in the last section that the Kantorovich h_o cannot be obtained economically for the present problem. To illustrate how the multiple shooting method converges according to the theorem, we carried out a two-phase stagnation point flow solution. This was a smaller system of seven equations and four subintervals; the Euclidean error norm and h_o are presented in Fig. 4. One finds that as the method converges, even the first guess falls outside the convergence sphere; as soon as it hits inside the sphere, the convergence is reached.

The present problem is reduced to a system of 33 equations after applying the symmetry condition along the centerline and the interaction equation along the shear layer. The computation was conducted on UNIVAC 1108. In initial trials, six subintervals were employed, and the convergence appeared poor. Later, double precision and 12 subintervals were used; this improved the convergence. The Jacobian matrix was first computed every five iterations with Broyden's technique applied accordingly; the solution yielded obvious errors. It appeared that the method produces the best result if the Jacobian matrix is computed every three iterations. Fig. 5 shows the error norm and the convergence factor versus the number of iterations; the error first increases then decreases steadily while the convergence factor increases monotonically. The evolution of velocity vector and the temperature contour through iterations to satisfy the boundary conditions and continuity across the subintervals are shown in Figs. 6 through 8.

The initial velocity vector plot of the forebody boundary layer revealed no turning action around the edge of the shroud, because the pressure drop had not propagated upstream. It is seen that there are discontinuities across the intervals and significant interpolation errors along $y/H = .57$ and $.71$. After several iterations, this error diminished in magnitude while the flow around the shroud edge began to incline towards the wall. The recirculating flow pattern is shown clearly in the final figure. In the initial temperature contour plot, there are negative values of temperature which are indicated by the blanks, and discontinuities also exist across the intervals. The fact that the contour lines failed to be normal to the centerline indicates that errors due to the Lagrange interpolation exist. In an area near the wall, the gap between the contours is small because of the high heating rates. The cold and hot spots emerge in the final figure and the profiles show little variations along the horizontal direction throughout the near wake region.

In conclusion, if a problem contains singular layers, then Telenin's method will be effective only if reasonable initial guesses exist and the integration domain is of the same order or smaller than the basic length of the problem. On the other hand, if it contains no singular layer, then Telenin's method should be superior to many existing methods, especially in solving boundary layer separation problems.

REFERENCES

Broyden, C. G., Math. Comp. 19, pp. 577-593, 1965.

Bulirsch, R., Vortrag im Lehrgang, Flughahroptimierung der Carl-Cranz-Gesellschafte. V., 1971.

Gilinskiy, S. M., Telenin, G. F., and Tinyakov, G. P., NASA TT F-297, 1965.

Holt, M., and Ndefo, D. E., J. Comp. Phys. 5, No. 3, pp. 463-486, 1970.

Kubota, T. GALCIT Hypersonic Research Project. Internal Memorandum No. 29, 1955.

Osborne, M. R., J. of Math, Anal. Appl. 27, pp. 417-433, 1969.

Roberts, S. M., and Shipman, J. S., J. of Math. Anal. Appl. 18, pp. 45-58, 1967.

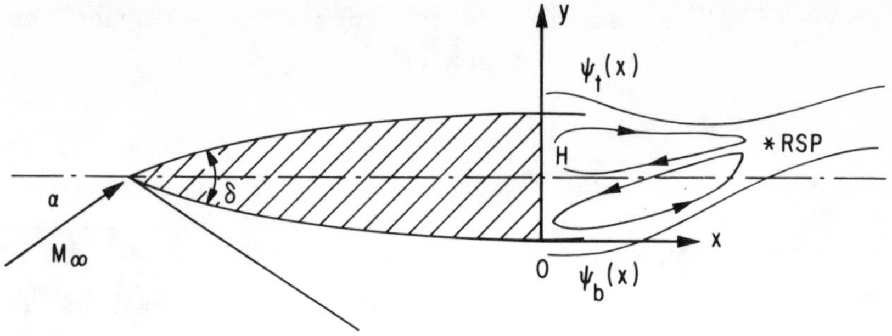

Fig. 1. Physical sketch of a two-dimensional space shuttle base region

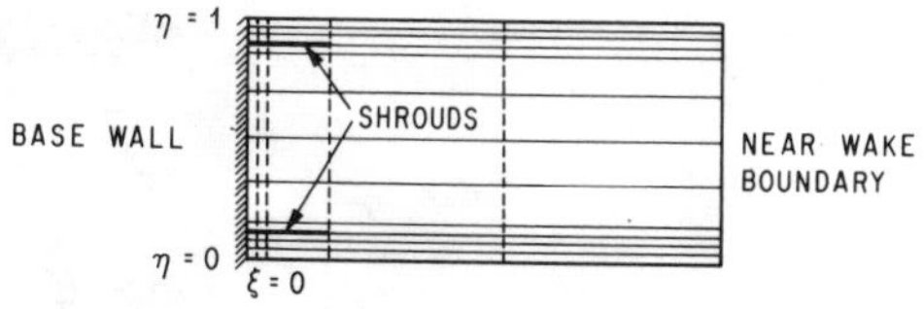

Fig. 2. Construction of strips and segments on transformed plane

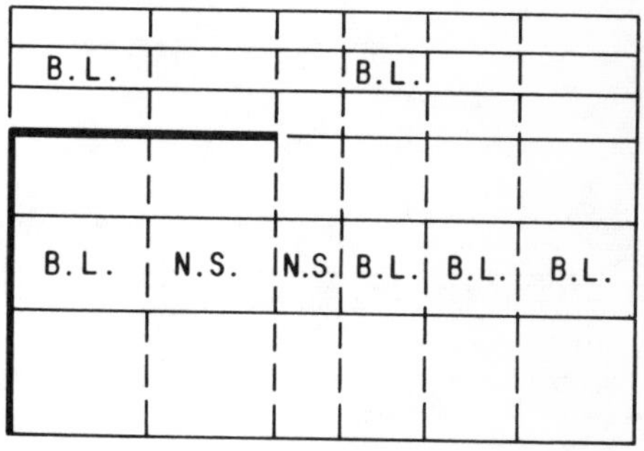

Fig. 3. Governing equations and segmentation of the base flow regions

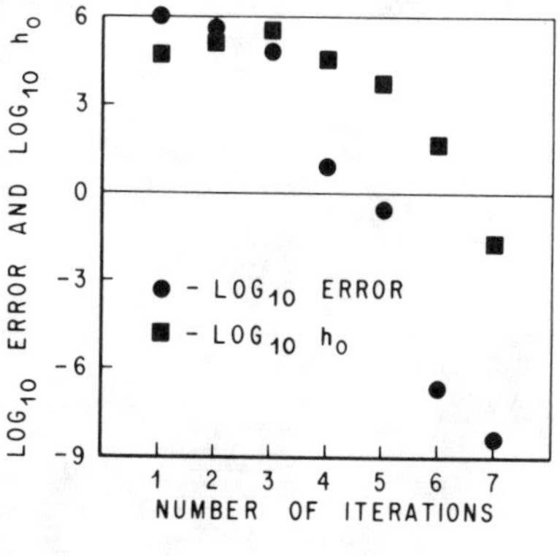

Fig. 4. Error norm and kantorovich h_o

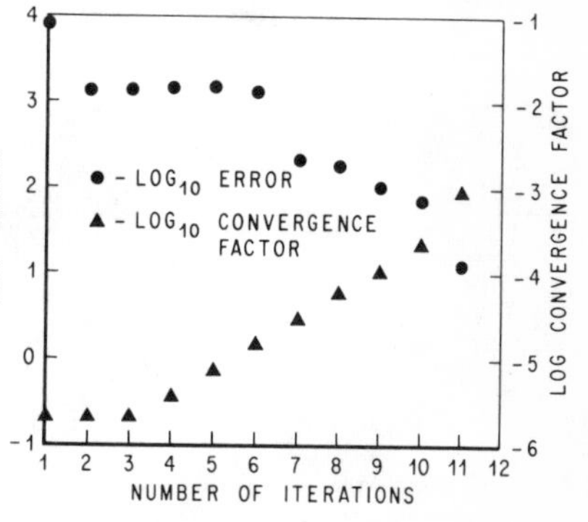

Fig. 5. Error norm and convergence factor versus iterations

Velocity vectors and temperature contours in A 2-D space shuttle base region

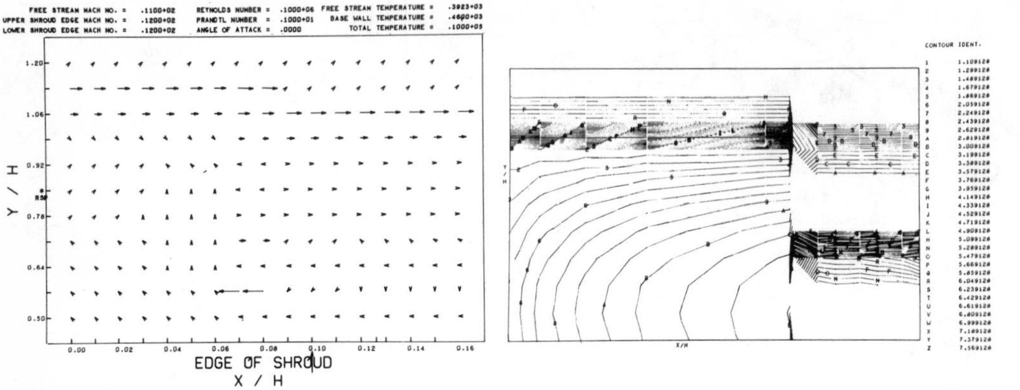

Fig. 6

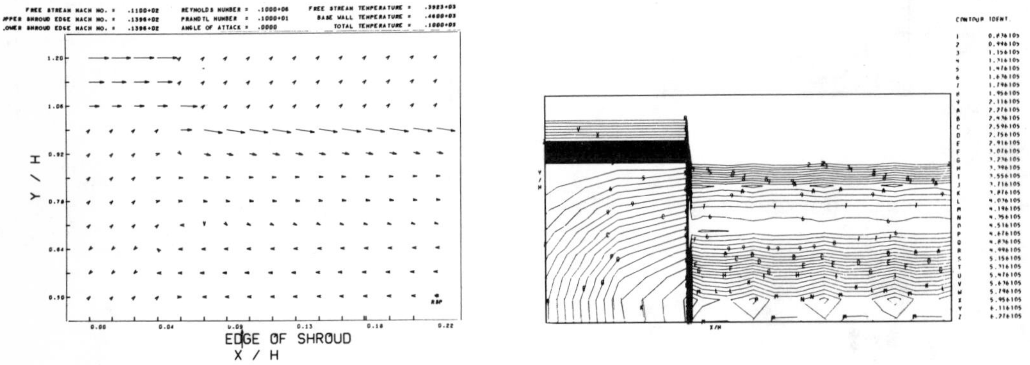

Fig. 7

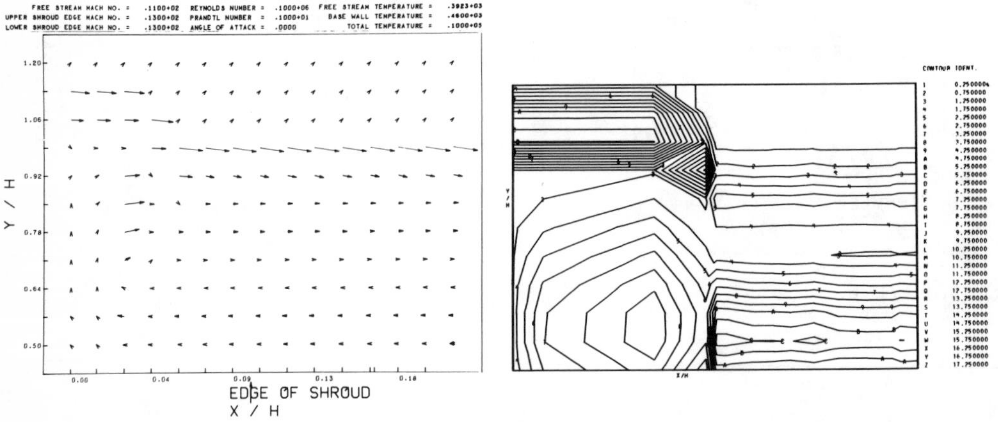

Fig. 8

A RELAXATION METHOD FOR CALCULATING TRANSONIC FLOWS WITH DETACHED BOW SHOCKS

By Earll M. Murman

Ames Research Center, NASA

Moffett Field, California, U.S.A. 94035

I. INTRODUCTION

The occurrence of detached bow shock waves in flows with supersonic free-stream Mach numbers is a familiar fluid dynamics problem. At moderate supersonic Mach numbers through hypersonic speeds, such flows are generally associated with blunt nosed shapes. For these Mach numbers, the extent of the imbedded subsonic region is of the order of the nose radius of the body. Several analytical or numerical methods have been advanced to treat this problem in either a direct or inverse fashion. However, at low supersonic Mach numbers detached bow shock waves may appear in front of both sharp and blunt nosed bodies, and often the extent of the imbedded subsonic region approaches or exceeds the overall dimensions of the body. No general analytical or numerical methods have been reported to treat this problem. One solution for such a flow past a double wedge body was presented by Vincenti and Wagoner (1). Their solution was a combined numerical and analytical one for the problem formulated in the hodograph plane and is not readily extendable to other body geometries.

In a paper at the previous conference (2) a method was presented for calculating steady, inviscid, transonic flows with imbedded shock waves directly in the physical plane. The governing mixed elliptic-hyperbolic partial differential equation is replaced by either elliptic or hyperbolic finite difference operators depending on the local velocity at each mesh point. A relaxation algorithm is used to solve the equations, and subsonic and/or supersonic regions and shock waves develop naturally during the iterative solution. The method has been used successfully to solve the boundary value problem of mixed flow past bodies with $M_\infty < 1$. Bailey (3) reported calculations using this method for flow past a parabolic arc of revolution for $M_\infty > 1$ with both attached and detached shock waves. This paper discusses the application of the method to two dimensional problems with $M_\infty > 1$ for flows past sharp and moderately blunt nosed geometries. The basic numerical method is that reported in Ref. 2 with an improved treatment of the parabolic point and the inclusion of overrelaxation of subsonic points (4).

II. THEORY

The problem to be solved is an initial value problem in the x-direction and a boundary value problem in the y direction, and is governed by a mixed elliptic-hyperbolic differential equation whose solution contains imbedded discontinuities (shock waves). Transonic small disturbance theory (5) used to model the problem leads to the governing equation

$$\left[K\phi_x - \frac{(\gamma+1)}{2}\phi_x^2\right]_x + \left(\phi_{\tilde{y}}\right)_{\tilde{y}} = 0 \tag{1a}$$

or

$$\left[K - (\gamma+1)\phi_x\right]\phi_{xx} + \phi_{\tilde{y}\tilde{y}} = 0 \tag{1b}$$

where the transonic similarity parameter is defined by[a]

[a]The definitions of these quantities differ slightly from standard definitions appearing in the literature (5), and have been found to give the best agreement between small disturbance theory and solutions for the full potential equation (4). Herein most solutions are reported in terms of the transonic parameters K, $\overline{C}_p$, $\tilde{y}$ so that other scalings may be used if desired. For results reported in terms of physical variables M_∞, C_p, y, Eqs. (2) – (4) have been used for the scaling.

$$K = \frac{1 - M_\infty^2}{M_\infty \delta^{2/3}} \tag{2}$$

The body is described by $y = \delta F(x)$, $0 \leqslant x \leqslant 1$, where δ is the usual thickness ratio and $F(x)$ is the shape function. The perturbation potential from free-stream conditions is denoted by $\delta^{2/3} M_\infty^{-3/4} \phi$ and $\tilde{y}$ is a reduced transonic coordinate [a]

$$\tilde{y} = \delta^{1/3} M_\infty^{1/2} y. \tag{3}$$

The pressure coefficient is given by [a]

$$C_p = \delta^{2/3} M_\infty^{-3/4} \bar{C}_p; \quad \bar{C}_p = -2\phi_x. \tag{4}$$

K is zero for sonic free-stream conditions and negative for $M_\infty > 1$. The gas constant γ is taken as 1.4 for all reported calculations.

Equation (1) has characteristic directions and relations given by

$$\frac{d\tilde{y}}{dx} = \frac{dv}{du} = \pm\left[(\gamma+1)u - K\right]^{-1/2} \tag{5}$$

where $u \equiv \phi_x$ and $v \equiv \phi_{\tilde{y}}$. The characteristics are real when u is greater than the sonic velocity $u^* = K/(\gamma+1)$. The weak solution of Eq. (1) corresponding to the shock jump relations is

$$\left[(\gamma+1)\frac{u_1 + u_2}{2} - K\right]\left(u_2 - u_1\right)^2 = \left(v_2 - v_1\right)^2 \tag{6a}$$

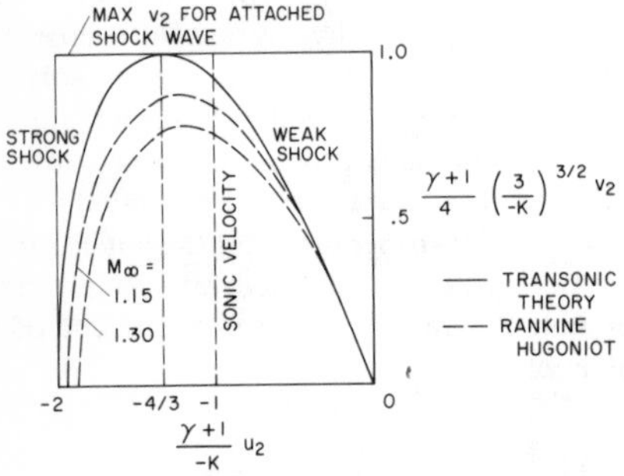

Figure 1. Shockwave polar for transonic small disturbance equation with undisturbed upstream conditions $u_1 = 0$, $v_1 = 0$

where subscripts 1 and 2 are the upstream and downstream states, respectively. The shock polar (Eq. (6a)) is plotted in normalized form in Fig. 1 for the undisturbed freestream case $u_1 = 0$, $v_1 = 0$. This curve is the transonic approximation to the familiar Rankine-Hugoniot polar. From the fact that the tangential component of momentum is preserved across a shock, if follows that

$$\phi_2 = \phi_1 . \tag{6b}$$

The boundary condition for Eq. (1) for a non-lifting symmetric body is

$$\phi_{\tilde{y}} = \frac{dF}{dx} \text{ on } \tilde{y} = 0; \ 0 \leqslant x \leqslant 1. \tag{7}$$

For a uniform supersonic free-stream

$$\phi = \phi_x = \phi_{\tilde{y}} = 0 \quad \text{at} \quad x \to -\infty \tag{8}$$

provide the initial conditions that are applied upstream at the bow shock. For numerical calculations a condition is needed on an upper boundary at some finite distance away from the body. Such a condition may be obtained by assuming that in the far field the flow is supersonic and velocity perturbations ϕ_x are small compared to the sonic velocity $K/\gamma+1$ yielding

$$\phi_{\tilde{y}} = -\sqrt{-K}\,\phi_x - \frac{\gamma+1}{4\sqrt{-K}}\phi_x^2 + \ldots \tag{9}$$

Another possible upper boundary condition may be obtained by assuming a physical boundary such as a solid or perforated wind tunnel wall or free jet. The relationship (6)

$$\phi_{\tilde{y}} = -\tilde{P}\phi_x \tag{10}$$

can be used to model all these cases. The porosity parameter $\tilde{P}$ ranges from 0 for a solid wall to ∞ for a free jet. It is interesting to note that the perforated wall boundary condition is identical to the leading term in the far field formula if $\tilde{P} = \sqrt{-K}$. No condition is needed at the downstream boundary if it is located at a sufficient distance that the outflow is entirely supersonic. For numerical applications, however, some condition is needed to treat subsonic outflow which can develop at this boundary during the relaxation process prior to convergence. The condition $\phi_x = 0$ was found to be satisfactory.

III. RESULTS

The first set of calculations to be discussed is for a parabolic arc airfoil described by $F(x) = 2x(1-x)$. The leading term of the free-air boundary condition. Eq. (9), was used on the upper boundary. Figure 2 shows results for four values of K ranging from fully supersonic flow to mixed flow with a well detached bow shock wave. For reference, a 6-percent thick body is shown and the corresponding free-stream Mach numbers are indicated. In Fig. 2(a) the flow is entirely supersonic and the shock wave is attached. The supersonic oblique leading and trailing edge shock waves are spread over four mesh points by the finite difference procedures. Also shown is the solution by Spreiter and Alksne (7):

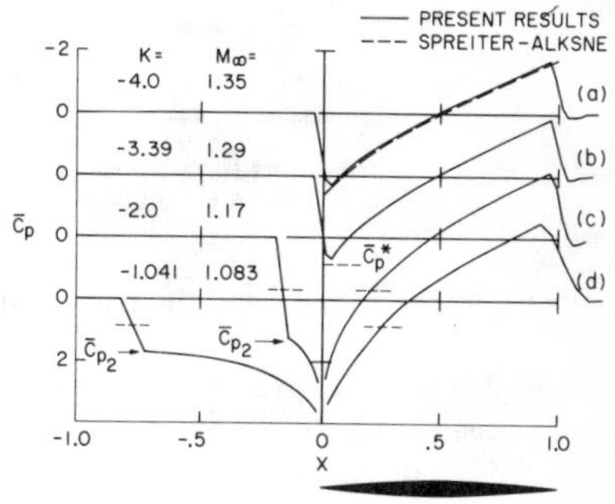

Figure 2. Pressure distributions for a parabolic arc airfoil in free air

$$\bar{C}_p = \frac{-2}{\gamma+1}\left\{K + \left[(-K)^{3/2} - \frac{3}{2}(\gamma+1)\frac{dF}{dx}\right]^{2/3}\right\}. \tag{11}$$

This equation was obtained by the local linearization technique and corresponds to the transonic approximation of simple wave theory. Equation (11) is only applicable if the flow on the body is entirely supersonic. The solution in Fig. 2(b) corresponds to the value of K for shock detachment, i.e., the maximum normalized ν_2 on the shock polar curve Fig. 1. Downstream of the leading edge the body curvature causes the flow to expand rapidly, and the imbedded subsonic zone is too small to be resolved with this mesh spacing. For the equivalent calculation past a wedge forebody the imbedded subsonic region is quite pronounced.

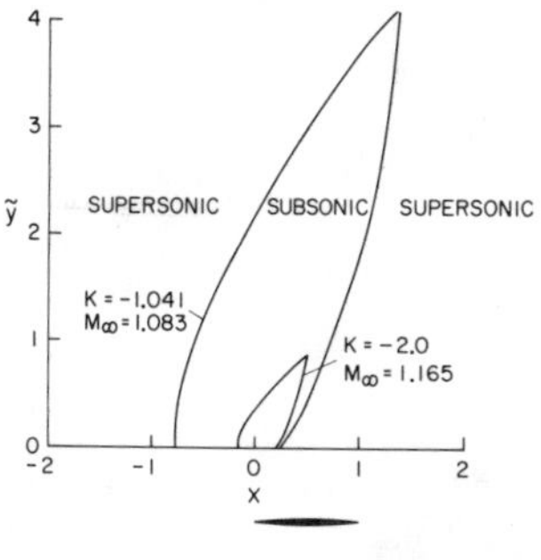

Figure 3. Sonic line locations for a parabolic arc airfoil in free air

Figures 2(c) and 2(d) are for cases where the bow shock wave is fully detached. The exact values for $\bar{C}_{p_2}$ for a normal shock wave (Eqs. (4) and (6)) are also indicated on the figure, and it is seen that the numerical method accurately predicts the pressure jump for a normal shock. In Fig. 3, the sonic line locations are shown for these values of K. The extent of the imbedded subsonic region becomes quite large with decreasing M_∞. Note that the ordinate is the transonically scaled $\tilde{y}$ variable. For the 6-percent reference body shown, the subsonic zone extends laterally about 10 chord lengths.

In Fig. 4 a second series of calculations for a parabolic arc are compared with the data of Knechtel (8). The wind tunnel in which the experiments were

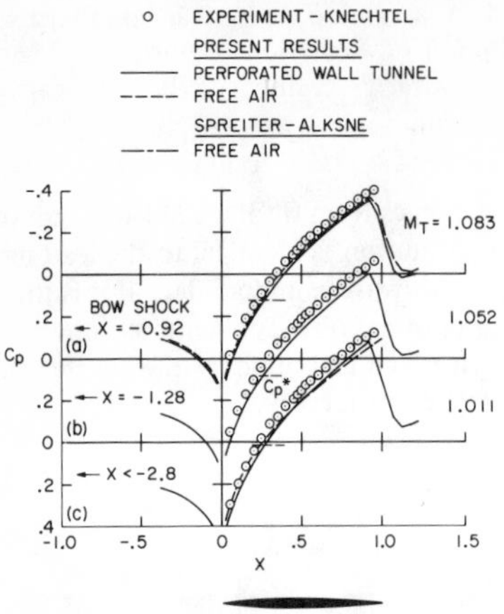

Figure 4. Pressure distributions for a parabolic arc airfoil in porous wind tunnel and comparison with free air solutions

conducted had a slightly diverging perforated wall with an open area ratio of 6 percent. In the calculations, Eq. (10) was used for the boundary condition at the tunnel half height location of $\widetilde{h} = \delta^{1/3} M_\infty^{1/2}\ h = 1.2$. The porosity parameter $\widetilde{P}$ was estimated by assuming that the perforated wall behaved as the average of a constant pressure jet $(P/P_\infty - 1 = -\gamma\delta^{2/3}\ M_\infty^{3/4}\ \phi_x = 0)$ and a solid wall $(\delta M_\infty^{-1/4}\ \phi_y = 0)$ weighted respectively, by the open area $S = 0.06$ and closed area $1\text{-}S = 0.94$ ratios. This assumption led to

$$\widetilde{P} = \frac{\gamma S}{1\text{-}S}\ \delta^{-1/3} M_\infty^{-1/2} \simeq 0.22.$$

This is undoubtedly an oversimplified model for the porosity, but it agrees in order of magnitude with experimental values and other theoretical estimates (6). The calculations for $M_T = 1.083$ were done for $\widetilde{P}$ varying from 0.05 to 0.31 and the maximum change in surface pressure was $\Delta C_p = 0.013$.

The shape of the pressure distribution predicted by the calculations is similar to the data. However, there is a near constant discrepancy in the pressure levels. For the $M_T = 1.083$ case a calculation was done halving the mesh spacing, and the change in the results was indistinguishable on the scale of Fig. 4. Several possible sources, either theoretical or experimental could be postulated to account for the discrepancy between the data and calculations. A very likely source could be the nonuniformities in the tunnel flow, which were as large as $\Delta M_\infty = 0.02$. An uncertainty in the definition of M_∞ of this order would result in a constant ΔC_p error of the order noted in Fig. 4.

The free-air calculations for $M_\infty = 1.083$ are shown in Fig. 4(a). Also shown in Fig. 4(c) is the local linearization solution of Spreiter and Alksne (7) for free air flow with $M_\infty \simeq 1$. From a comparison of the free air and wind tunnel calculations it appears that the effects of the tunnel walls on the surface pressure distributions are not large for these data. The locations of the detached bow shock waves in the tunnel calculations are noted in Fig. 4. The pressure rise across the normal shock on the tunnel axis was within a few percent of the exact value.

The application of the method for detached shocks in front of blunt nosed shapes is shown in Figs. 5 and 6. Small disturbance theory is strictly invalid at stagnation points and blunt nosed regions. However, Nonweiler (9) (see also Cole (5)) has shown that Eq. (1) poses a solution with an integrable singularity for moderately blunt nosed symmetric shapes at zero angle of attack. For a nose shape given by $y = x^n$, the pressure coefficient approaches ∞ in an integrable fashion for $0.4 < n \leqslant 1$, i.e., for shapes from slightly blunter than a parabola ($n = 0.5$) to the wedge case ($n = 1$). Based on calculations to date, this criterion appears to be applicable to numerical work also. The solutions in Figs. 5 and 6 are for an NACA four-digit airfoil described by

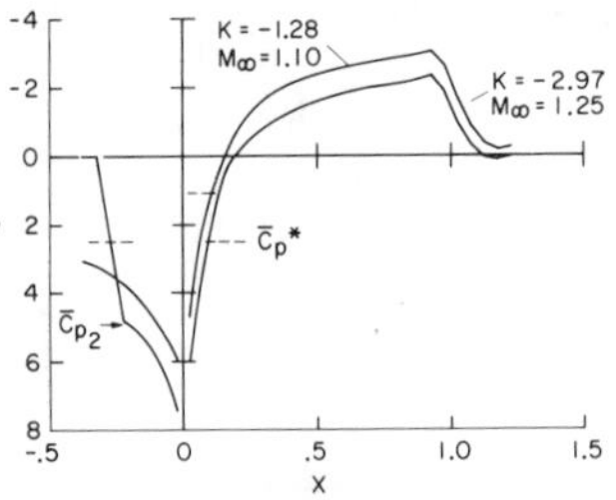

Figure 5. Pressure distributions for an NACC 4-digit series air foil in free air

$$F(x) = 2.969\sqrt{x} - 1.26x - 3.516x^2 + 2.843x^3 - 1.015x^4$$

which has a nose shape corresponding to $n = 0.5$. A 6-percent body and the corresponding values of M_∞ are indicated for reference. The calculations show the expected behavior in the surface pressure distributions and sonic line locations. Again it should be noted that the results are shown in the transonically scaled variables to facilitate application to various Mach numbers and thickness ratios. For the low Mach number case, the subsonic region extends laterally about 12 chord lengths.

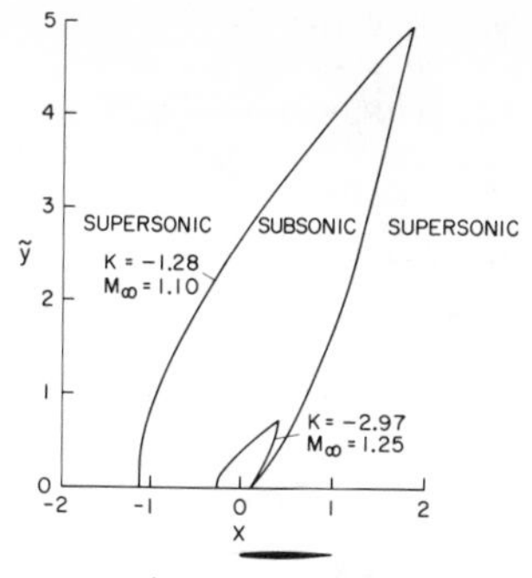

Figure 6. Sonic line locations for an NACA 4-digit series airfoil in free air

For all the calculations reported, the surface C_p values converged rapidly while the flow field structure required many iterations to converge. For example, consider the solution for $K = 1.28$ for the NACA four-digit series airfoil (Figs. 5 and 6.) The maximum change in surface C_p between iterations 200 and 1000 was less than 1 percent. However, during that time the shock detachment distance almost doubled from − 0.66 to − 1.12 and the extent of the subsonic region grew from $\tilde{y} = 3$ to $\tilde{y} = 5$. Thus, if the main interest is in calculating surface pressures, the computing times may be substantially reduced, particularly with experience and a suitable adjustment of relaxation parameters.

All the calculations discussed above were all performed using even mesh spacing with $\Delta x = 0.025$ to 0.05 and $\Delta\tilde{y} = 0.0125$ to 0.06. The total number of mesh points used in the calculations ranged from 2100 to 6700. Computing times varied from 0.8 to 68 CPU minutes on an IBM 360/67 corresponding to 30 to 1000 iterations. The longest computing times correspond to the two cases with subsonic regions which extend a large lateral distance.

IV. DISCUSSION

From the calculations of Bailey (3) for the axisymmetric case and the above results for the two-dimensional case, it can be concluded that the mixed finite difference relaxation method may be applied to transonic flow problems from completely subsonic to completely supersonic flow conditions, and that the numerical method accurately predicts the pressure rise across a normal shock. For the two-dimensional case, the extent of the imbedded subsonic region is considerably larger than for the corresponding axisymmetric body. As a result, the number of mesh points required for the calculation, and hence the computing times, are much larger. Some experimentation was done with variable mesh, but stability and accuracy problems were encountered when the mesh aspect ratio exceeded about 2.

Calculations similar to the above could probably be done more accurately and faster if a stable and convergent method was developed to fit in the bow shock wave using the shock relations Eq. (6). The finite difference procedures then need be used only downstream of the bow shock. With a suitable reformulation of the problem, the entropy increase across the shock could be taken into account in defining the potential so that cases with normal shock Mach numbers > 1.3 could be accurately calculated. The present method without shock fitting should be directly extendable to lifting airfoils at moderate angles of attack as have been the subsonic calculations (4).

REFERENCES

1. Vincenti, W. C. and Wagoner, C. B., NACA Report 1095, 1951.
2. Murman, E. M. and Krupp, J. A. Lecture Notes in Physics, Vol. 8 pp 199-206, Springer-Verlag 1971.
3. Bailey, F. R., NASA TN D-6582, 1971.
4. Krupp, J. A., Ph.D. Thesis, Univ. of Washington, 1971.
5. Cole, J. D., Boeing Scientific Research Laboratories Document DI-82-0878, 1969.
6. Gothert, B. H., *Transonic Wind Tunnel Testing*. AGARDograph No. 49, Permagon Press. 1961.
7. Spreiter, J. R. and Alksne, A. Y., NACA Report 1359, 1957.
8. Knechtel, E. D., NASA TN D-15, 1959.
9. Nonweiler, T. R. F., Journal of Fluid Mech. Vol. 4, Part 2, pp 140-148 (1958).

TRANSIENT THREE-DIMENSIONAL FLUID FLOW IN THE VICINITY OF LARGE STRUCTURES*

B. D. Nichols and C. W. Hirt**
University of California
Los Alamos Scientific Laboratory
Los Alamos, New Mexico 87544

INTRODUCTION

A numerical technique has been developed for calculating the three-dimensional, transient dynamics of an incompressible fluid with an optional free surface. This method is based on the Marker-and-Cell method developed by Harlow and Welch (1965). It uses an Eulerian finite-difference approximation to the Navier-Stokes equations for incompressible flows. Additional features include heat transport and thermal buoyancy in a Boussinesq approximation, and a particulate transport scheme is included that uses an equation of motion for each particle. Particle motions are allowed to be influenced through the effects of gravity, fluid drag, and diffusion. This paper briefly describes the basic numerical technique and each of the additional features, as well as a data display technique developed especially for the presentation of three-dimensional results. Numerous examples are presented to show the qualitative results of a variety of calculations.

THE NUMERICAL TECHNIQUE

The fluid region is divided into a stationary network of calculational cells with dimensions δx, δy, and δz. The fluid motion is described by the incompressible Navier-Stokes equations for viscous fluids and the equation of continuity. These equations are expressed in finite-difference form with respect to the Eulerian mesh of fluid cells. The primary field variables associated with each of these cells are the velocity components (u,v,w) and the pressure p. As shown in Fig. 1, each of the velocity components is specified at the center of the cell face to which it is normal and the pressure is specified at the cell center. The cells and associated field variables are identified by the indices i,j,k, which refer to the integer number of cells counted out from the mesh origin in the x, y, and z directions, respectively.

The fluid may have a free surface; therefore, each cell is flagged to denote whether it is an empty cell containing no fluid, a surface cell, which contains fluid but is next to an empty cell, or a full cell, which contains fluid and is not next to an empty cell.

---

*This work was performed under the auspices of the United States Atomic Energy Commission and was partially supported by the Office of Naval Research, Government Order NAonr-13-72.

**Now with Science Applications, Incorporated, 1250 Prospect Plaza, La Jolla, California.

The fluid motion is numerically determined by advancing the fluid configuration through a series of small time increments. During each time step the solution to the momentum equation is obtained in two phases. First, the known velocities and pressures from the previous time are used to determine the fluid velocities in each cell (initial conditions are used for the first time step). This explicit calculation does not necessarily assure incompressibility; therefore, in the second phase a solution algorithm that adjusts the tentative velocity field through changes in the pressure field is used. The algorithm that is preferred because of its relative ease in setting boundary conditions is one initially developed by Chorin (1966) and is described in detail for this technique by Hirt and Cook (1972a). The basic concept of the algorithm is that the pressure in each Eulerian cell is adjusted by an amount proportional to the negative of the velocity divergence. The physical basis of this can be understood by noting that if the velocity divergence in a cell is positive (corresponding to a net mass outflow from the cell) the pressure in the cell must be decreased to reduce the net mass loss. Conversely, with a negative velocity divergence an increase in cell pressure is required to reduce the net mass inflow. Thus pressure and velocity distributions leading to zero divergence are obtained by iteratively adjusting the pressures and velocities for each cell in the mesh.

When the fluid has a free surface the cells in which the surface is located will be only partially filled with fluid and a known pressure will be applied at the surface. A correct pressure distribution in this region of the fluid is obtained by determining the pressure in these cells by a linear interpolation or extrapolation between the pressure in the fluid cell immediately below the surface cell and the applied pressure at the surface. If the fluid depth is less than δz, so that no fluid cell exists below the surface cell the hydrostatic pressure gradient is employed for the interpolation.

The calculational mesh of fluid cells is bounded by a layer of fictitious cells that are used for setting boundary conditions. These cells are treated as walls with the following choice of boundary conditions: rigid free-slip, which requires zero normal velocity and zero shear stress at the wall; rigid no-slip, which requires zero normal and tangential velocities at the wall; input and output of fluid at a chosen velocity; and a continuative outflow boundary.

Obstacles may be created in the fluid region by designating any number or configuration of cells as obstacle cells. These may be located anywhere within the mesh and anywhere relative to the free surface. The same choice of boundary conditions is available for these obstacle cells as for the mesh boundaries.

In addition to the basic three-dimensional fluid transport, a number of auxiliary features are included that increase the technique's applicability. One such extension is the inclusion of a free surface capability. A single-valued surface is initially defined by specifying the surface height above the mesh bottom. The locations at which the surface height is specified is the center of each vertical column of cells. The change in the surface elevation is determined by the local fluid velocity, that is, by the vertical component of the fluid motion plus the horizontal convection of the surface elevation from adjacent cell columns. The complete finite-difference form of the kinematic surface equation and a discussion of the necessary stability conditions is given by Nichols and Hirt (1972).

The free surface boundary conditions require that the normal and tangential velocities immediately outside the surface be chosen to assure a zero transfer of momentum through the surface. Velocities normal to the surface are set to satisfy the incompressibility conditions in the surface cells. The tangential velocities in the cells just outside the fluid are obtained by setting them equal to the adjacent interior values. This is consistent with zero shear stress at the surface, but only approximates the complete viscous stress conditions as applied in a similar two-dimensional technique by Nichols and Hirt (1971). The conditions stated above are, however, appropriate for the nearly-inviscid free surface flows presented here.

Heat transport and thermal buoyancy are also optional features. The time dependent heat equation, with temperature convected by the flow field and with a constant diffusion coefficient, is expressed in finite-difference form and solved explicitly. The temperature is specified at the cell center and for the convection terms donor cell differencing, as described by Hirt and Cook (1972), is used to insure numerical stability. With a known temperature distribution the Boussinesq approximation is used to account for vertical accelerations in the fluid caused by slight density variations.

The heat equation described above may be used to represent concentration transport by considering the convection and diffusion of particulate concentration instead of temperature. The numerical solution of this equation contains errors that additionally diffuse concentrations, which may be undesirable. Sklarew (1970) pointed out that this can be avoided by placing particles in cells such that the number of particles in each cell is proportional to the concentration of the cell. Particles are then moved each time step according to an effective convection velocity that includes both convection and diffusion transport. This technique works well for problems in which the cell and particle resolution is good and, consequently, when concentrations do not change greatly from cell to cell. However, because of the large number of cells required in the three-dimensional calculations, the resolution often desired is not readily available.

Hotchkiss and Hirt (1972) have developed an alternative transport scheme that depends only weakly on the spatial cell resolution of the problem. The essence of the method, which is discussed in detail in the cited reference, is to replace the "diffusion velocity" of Sklarew by a random velocity. This random velocity is chosen such that the diffusional displacement of a particle each time step has a Gaussian distribution. This random velocity is applied to each particle independent of the mesh cell size and independent of the number of neighboring particles.

Inertial and gravitational effects for particle motion can be included in either the Sklarew or the Hotchkiss and Hirt transport schemes by using an equation of motion for each particle that includes accelerations caused by Stokes drag, gravity, and diffusion. A discussion of these effects is presented by Hotchkiss and Hirt (1972).

DATA DISPLAY TECHNIQUES

An effective means of displaying data from these three-dimensional calculations is to present the various data fields in perspective view plots that are computer generated. These include the velocity field, particulate distributions in the flow field, and free surface configurations. A hidden-line perspective view plot routine was especially developed for displaying these data and is fully described by Hirt and Cook (1972b).

Velocity fields are represented by velocity vectors that are drawn from the calculational cell centers. Any number of selected velocity vector plots may be displayed in perspective or plane view. Figure 2 shows the perspective view of a fluid region (with flow from the left) containing a rectangular structure with a horizontal plane of velocity vectors plotted. A plane view of the same velocity field is plotted with magnified vector magnitude in Fig. 3. The central vertical plane of the velocity field is shown in Fig. 4, and the vertical plane directly behind the rectangular structure, as viewed from downstream, is shown in Fig. 5. The essential character of the velocity field, such as the eddy formed on the leeward side of the structure, is effectively displayed in these three orthogonal plane plots.

Figure 6 shows the perspective view of the particulate distribution that results when particles are emitted from a point on the top of the structure into the flow field shown in Figs. 2-5.

The free surface configuration can be plotted in perspective by computing the elevation of the surface at each vertical mesh line. The surface is then constructed by connecting lines between pairs of elevation points. The plotted free surface is assumed to be transparent, which allows the complete surface deformation to be seen.

CALCULATIONAL EXAMPLES

Many examples have been calculated to demonstrate the capability of this technique. Quantitative comparison with theory has been made in some cases and these show good agreement, as reported by Hotchkiss and Hirt (1972) and Nichols and Hirt (1972). The free surface capability was initially proof-tested by calculating the propagation of low-amplitude waves that are generated by simulated piston motion at one end of a rectangular mesh. The motion of the piston is simulated by applying an acceleration to the x-component of the fluid velocity at each time cycle. This assumes the fluid mesh is attached to the piston and is not, therefore, in the laboratory frame of reference. A solitary wave is generated by applying an acceleration such that the "piston" position corresponds to a hyperbolic tangent function, as suggested by Stoker (1957). Figure 7 shows successive time plots of the surface configuration of a solitary wave as it moves down the "wave tank." The ratio of the wave amplitude to undisturbed fluid depth is 0.4 and the ratio of "effective wavelength as defined by Chan and Street (1970), to fluid depth is 15.58. The calculated wave characteristics compare very well with those predicted by theory.

A solitary wave interacting with a rectangular structure is shown in Fig. 8. In non-dimensional units, the wave speed is 0.93, the undisturbed fluid depth is 0.51, the wave amplitude is 0.32, and gravity is unity. The structure has 3 cells on a side in the x and y directions and extends upward to the top of the mesh. The mesh cells have edge lengths of 0.4 for δx and δy and 0.1 for δz. The net force on the structure acting in the direction of wave propagation has a maximum and positive value of 0.20 as the wave crest passes the upstream face of the structure, which is the time of plots b and c in Fig. 8. As the wave moves past the structure the net force decreases, eventually becoming negative and reaching its greatest negative value of 0.18 at a time slightly later than plot d in Fig. 8. At times corresponding to these maximum forces the torque about an axis parallel to the y-axis and through the structure center at $z = 0.0$ is 0.11 and -0.05, respectively.

Figure 9 shows two views of the free surface configuration resulting from fluid flowing from left to right past a partially submerged blunt body with a length to width ratio of 2.0, a width to draft ratio of 1.5, and a draft Froude number (based on the obstacle draft and input velocity) of 2.0. No frictional drag forces are assumed in the calculation. At the calculation time of 3.5 (35 cycles) that is shown in Fig. 9 the flow has reached steady state.

The particle distribution in a flow field downstream from a complicated structure is shown in Fig. 10. There is a continuous flow of fluid from left to right with particles being vented from a point source on top of the structure. Figure 11 shows the velocity field (in a plane near the top of the structure) by which particles are transported. The random velocity transport scheme is used for particles shown in Fig. 10 and its effects are readily seen by comparing the particle distribution in Fig. 10 with that in Fig. 6, which is a similar problem calculated with the Sklarew technique. Notice that the plume spread near the source in Fig. 6 is quite linear, a result of using the Sklarew technique with poor spatial resolutions. The random technique overcomes this and the consequences are clearly shown in Fig. 10.

ACKNOWLEDGMENTS

The authors would like to express their appreciation to J. L. Cook for writing the initial three-dimensional code and to R. S. Hotchkiss for many valuable suggestions relating to the development of the technique and for running the particle flow calculations.

REFERENCES

Chan, R. K.-C. and R. L. Street, Stanford University Technical Report No. 135 (1970).

Chorin, A. J., A.E.C. Research and Development Report, NYO-1480-61 (1966).

Harlow, F. H. and J. E. Welch, Phys. Fluids 8, 2182 (1965); J. E. Welch, F. H. Harlow, J. P. Shannon, and B. J. Daly, Los Alamos Scientific Laboratory Report, LA-3425 (1966).

Hirt, C. W. and J. L. Cook, To be published in Jour. Comp. Phys. (1972a).

Hirt, C. W. and J. L. Cook, Submitted for publication to Jour. Comp. Phys. (1972b).

Hotchkiss, R. S. and C. W. Hirt, Proceedings of the Summer Simulation Conference, San Diego, California (1972).

Nichols, B. D. and C. W. Hirt, Jour. Comp. Phys. 8, 434 (1971).

Nichols, B. D. and C. W. Hirt, Submitted for publication to Jour. Comp. Phys. (1972).

Sklarew, R. C., Proceedings of the 63rd Annual Meeting of the Air Pollution Control Association, St. Louis, Missouri (1970).

Stoker, J. J., "Water Waves," Interscience, New York (1957).

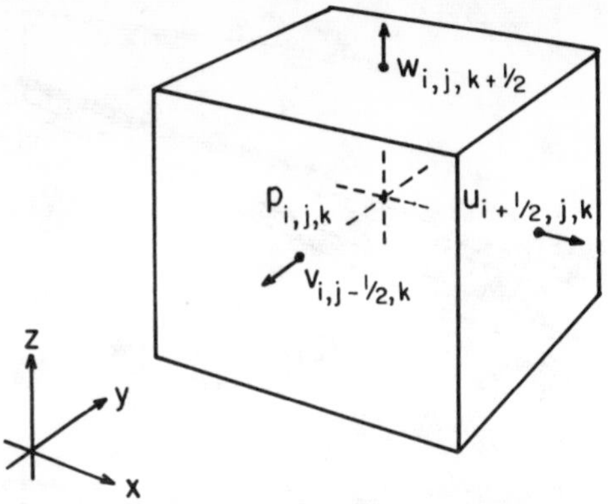

Fig. 1. Field variable locations on an Eulerian cell

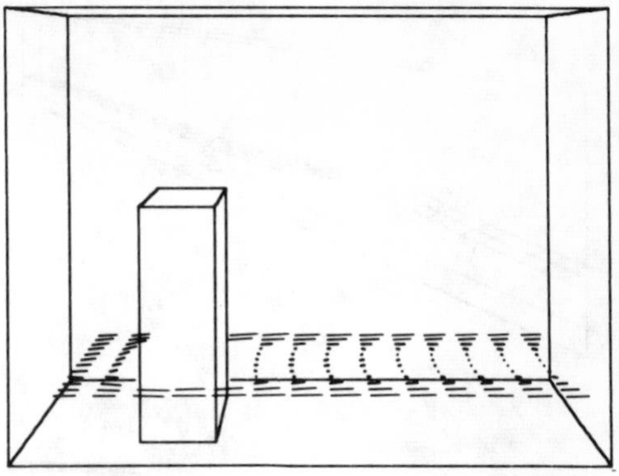

Fig. 2. Perspective view of a horizontal plane of velocity vectors around a rectangular structure

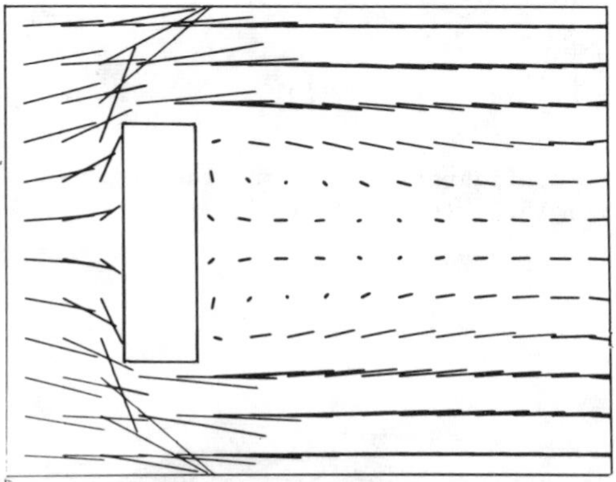

Fig. 3. Plane view of the velocity vectors shown in Fig. 2

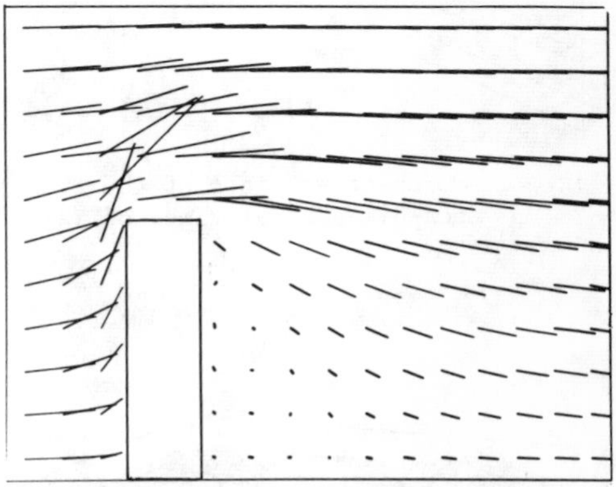

Fig. 4. Vertical plane of velocity vectors near the center of the structure in Fig. 2

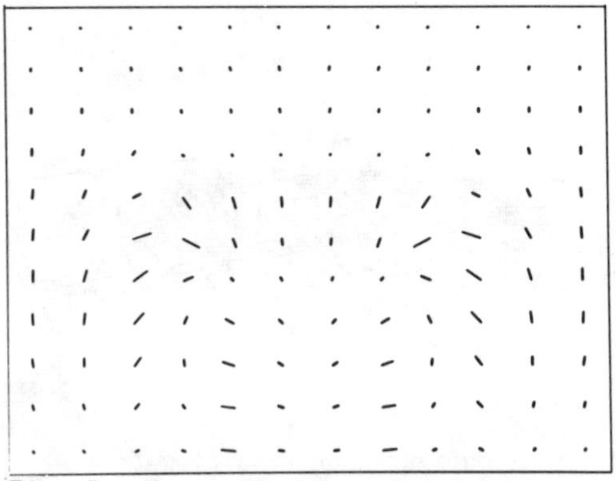

Fig. 5. Vertical plane of velocity vectors directly behind the structure in Fig. 2, as viewed from downstream

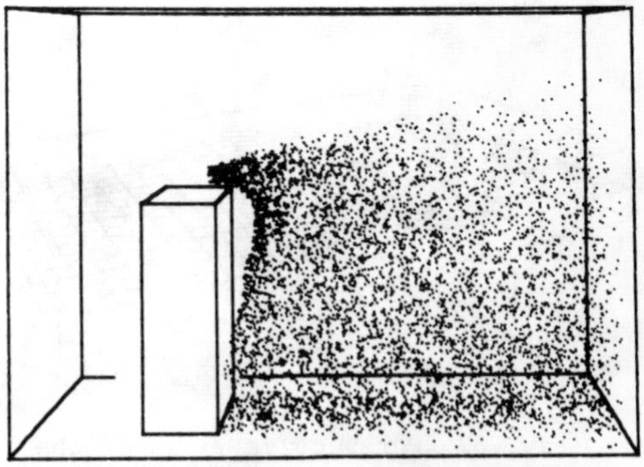

Fig. 6. Perspective view of the particle distribution in the flow field shown in Figs. 2-5

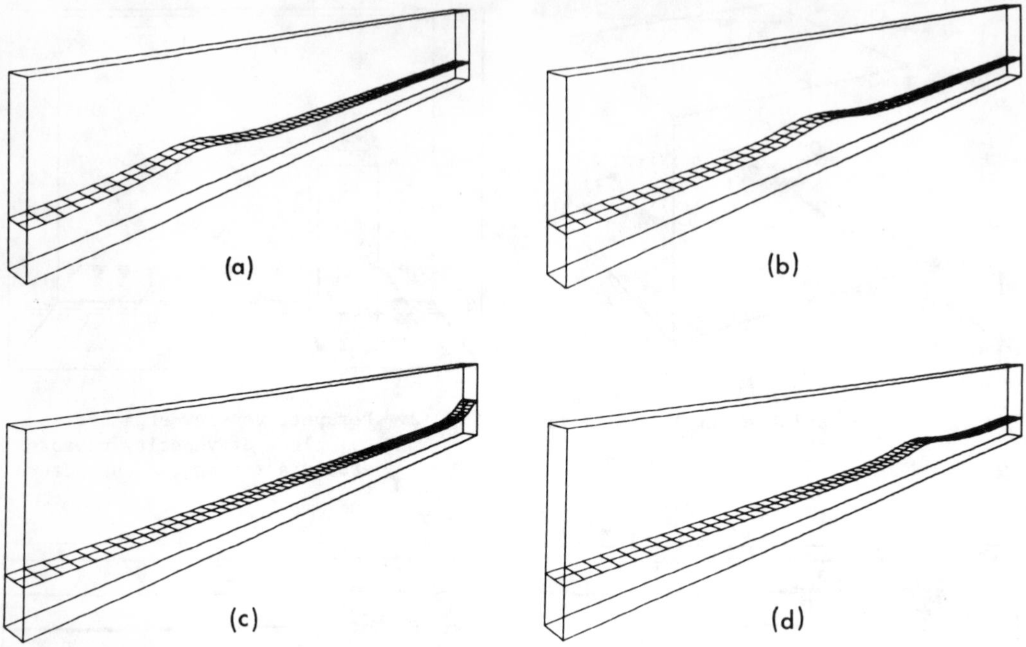

Fig. 7. Perspective views of the free surface configuration of a solitary wave at non-dimensional times of (a) 12.0, (b) 15.0, (c) 31.0, and (d) 41.0

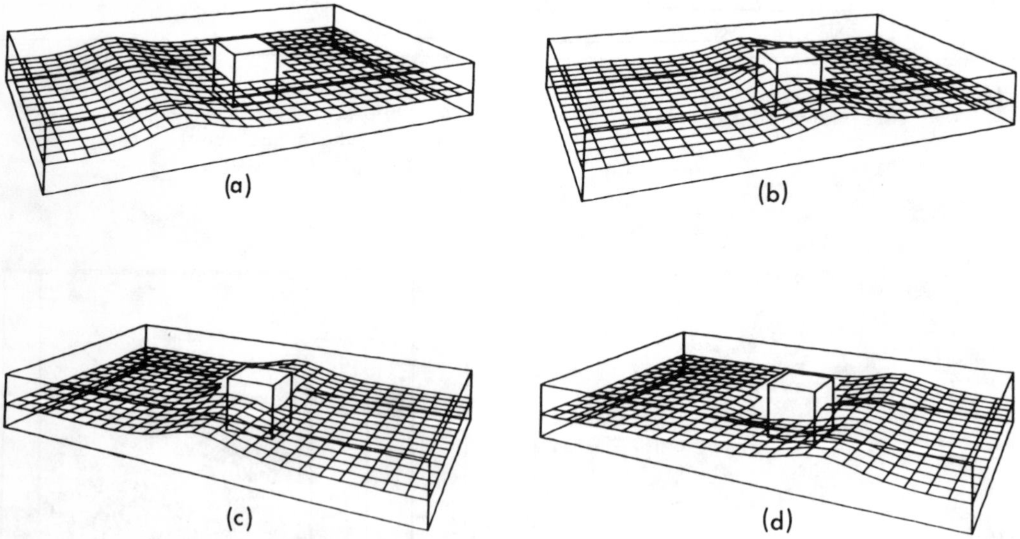

Fig. 8. Perspective views of the free surface configuration resulting from a solitary wave interacting with a rectangular structure

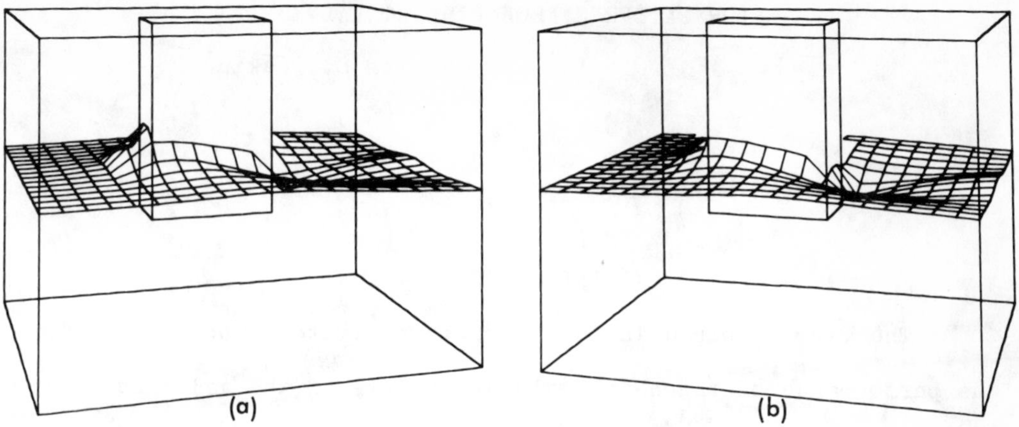

Fig. 9. Perspective views of the free surface configuration resulting from flow past a blunt body

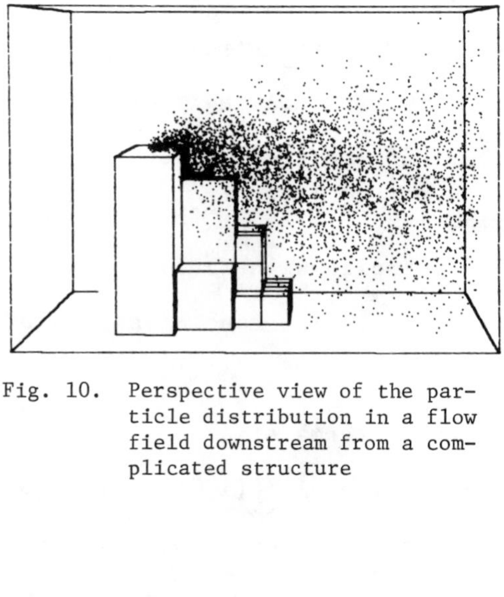

Fig. 10. Perspective view of the particle distribution in a flow field downstream from a complicated structure

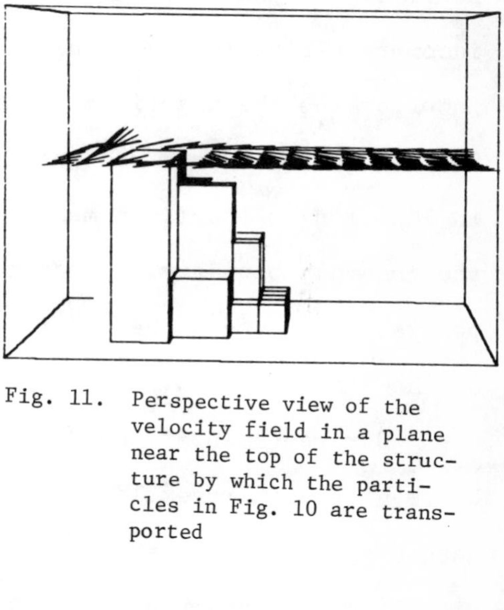

Fig. 11. Perspective view of the velocity field in a plane near the top of the structure by which the particles in Fig. 10 are transported

FLOW PATTERNS AROUND HEART VALVES

by Charles S. Peskin[1]

INTRODUCTION

The flow of blood in the heart is intimately connected with the performance of the heart valves. Points of the valve leaflet are carried along at the local fluid velocity. At the same time, these points exert forces on the fluid which significantly alter the fluid motion. The cardiac tissue, valve and heart wall, forms in essence a boundary of the fluid. We idealize this boundary as a neutrally buoyant, force-generating structure immersed in a viscous incompressible fluid. A numerical method is then introduced which incorporates the algorithm of Chorin (1969) for solving the Navier-Stokes equations on a rectangular mesh. In the present work we have, in addition to the fluid mesh, a collection of points representing the immersed boundary. These need not coincide with fluid mesh points. We connect the two representations by introducing an analog of the δ- function.

Our representation of the boundary in terms of its forces bears a certain resemblance to the work of Viecelli (1969,1971) who used a system of pressures just sufficient to prevent the fluid from violating a boundary constraint. In the present work, however, the motion of the boundary is not known in advance, and our method also includes the possibility of tangential forces.

---

[1]Albert Einstein College of Medicine, Bronx, New York, USA.
This work was performed while the author was a candidate for the degree of Doctor of Philosophy in the Sue Golding graduate division.

EQUATIONS OF MOTION

Our object here is to put the equations of motion in a form which is suitable for the construction of a numerical scheme. We remark that the term "boundary" will be used loosely here to include the possibility of a structure with finite thickness, like the heart wall. Also, all of our material boundaries will be immersed, and we regard the fluid as a whole as contained in a periodic box.

Let a dense sequence of material sample points of the immersed boundary be labeled by the index $k = 1,2, \ldots$ and let $\underline{x}_k$ be the position in space of the point k. Then $\{\underline{x}_k\}$ completely determines the configuration of the immersed boundary, the deformations of which are continuous. The state of stress can be characterized in integral form by giving the force $\underline{G}(R)$ applied by the boundary to each region R of the fluid. To define the local intensity of this force along the boundary, we introduce vectors $\underline{f}_k$ which are related to $\underline{G}(R)$ as follows:

$$\underline{G}(R) = \lim_{N\to\infty} N^{-1} \sum_{k=1}^{N} \underline{f}_k \, \delta_k(R) \tag{1}$$

where

$$\delta_k(R) = \begin{Bmatrix} 1 & \underline{x}_k \in R \\ 0 & \underline{x}_k \notin R \end{Bmatrix} = \int_R \delta(\underline{x}-\underline{x}_k)\, d\underline{x} \tag{2}$$

Note that $\mu(R)$ given by

$$\mu(R) = \lim_{N\to\infty} N^{-1} \sum_{k=1}^{N} \delta_k(R) \tag{3}$$

forms a measure on regions of the fluid. Roughly this measure is the fraction of sample points of the boundary contained in the region R. This shows that (1) is an integral and that $\underline{f}_k$ is the force per unit measure with the measure defined by (3).

Now, using (2) and interchanging the order of integration

$$\underline{G}(R) = \int_R \lim_{N\to\infty} N^{-1} \sum_{k=1}^{N} \underline{f}_k \; \delta(\underline{x}-\underline{x}_k)\, d\underline{x} \tag{4}$$

which shows that $\underline{F}(\underline{x})$ given by

$$\underline{F}(\underline{x}) = \lim_{N\to\infty} N^{-1} \sum_{k=1}^{N} \underline{f}_k \; \delta(\underline{x}-\underline{x}_k) \tag{5}$$

is the force density in the fluid at least formally.

The equation of motion of the fluid under the influence of such a force density is

$$\underline{u}_t = \mathcal{P}\,(\, - \underline{u}\cdot\nabla\, \underline{u} + \nabla^2\underline{u} + \underline{F}\,) \tag{6}$$

where $\mathcal{P}$ is the orthogonal projection onto the space of periodic, divergence-free vector fields, see for example Chorin (1969).

Under the influence of the velocity field $\underline{u}$, the equation of motion of the immersed boundary is

$$\frac{d\underline{x}_k}{dt} = \underline{u}(\underline{x}_k) = \int_{\text{fluid}} \underline{u}(\underline{x})\; \delta(\underline{x}-\underline{x}_k)\, d\underline{x} \tag{7}$$

We complete the equations of motion by specifying a stress-strain relation for the immersed boundary in the form

$$\underline{f}_k = \underline{f}_k(\, \ldots\, \underline{x}_k,\, \ldots\,) \tag{8}$$

For active boundaries like the heart wall these functions will change with time.

The equations of motion are (5) - (8). Eqs. (5) and (7) are integral transformations between boundary and fluid quantities. If p is the difference in dimensionality between fluid and boundary, then $\underline{F}(\underline{x})$ is singular like a δ-function in p dimensions. In the numerical scheme $\underline{F}$ will be of order h^{-p}, where h is the mesh width.

FINITE REPRESENTATION OF THE FLUID

Here we use the methods of Chorin (1969) and introduce a rectangular mesh of mesh width h covering our 1×1 periodic domain. Temporarily assuming that $\underline{F}_{ij}$ is known, we advance the velocity field according to

$$\begin{aligned} (I + \delta t\, Q_x(u_x^n))\ \underline{u}^* &= \underline{u}^n + \delta t\, \underline{F} \\ (I + \delta t\, Q_y(u_y^n))\ \underline{u}^{**} &= \underline{u}^* \\ \underline{u}^{n+1} &= P\, \underline{u}^{**} \end{aligned} \tag{9}$$

where Q_x and Q_y are difference operators, each in one space direction only, depending on $\underline{u}$, such that $Q_x(u_x^n) + Q_y(u_y^n)$ corresponds to $\underline{u}^n \cdot \nabla - \nabla^2$; and where P is a discrete operator corresponding to $\mathcal{P}$.

FINITE REPRESENTATION OF THE BOUNDARY: NUMERICAL STABILITY

The representation is the same as that which was described under Equations of Motion, but with N fixed such that the distance between nearest neighbors is about h/2. The finite set of functions

$$\underline{f}_k(\underline{x}_1 \ldots \underline{x}_N) \qquad k = 1,2, \ldots N \tag{10}$$

are derived from (8) for any particular N-point configuration $(\underline{x}_1 \ldots \underline{x}_N)$ first, by allowing the $\underline{x}_k$, $k \geq N$, to assume their equilibrium values; then by defining $N^{-1}\underline{f}_k$ as the force applied to the fluid by point k in the resulting configuration.

To secure numerical stability we use at each time step forces $N^{-1}\underline{f}_k^*$ which, neglecting interactions through the fluid, will be recovered at the end of the time step. These are the solution of

$$\left.\begin{aligned} \underline{x}_k^* &= \underline{x}_k^n + \delta t\, \underline{u}^n(\underline{x}_k^n) + (\delta t)^2 (9/64h^2)\, \beta\, N^{-1}\, \underline{f}_k \\ \underline{f}_k^* &= \underline{f}_k(\underline{x}_1^* \ldots \underline{x}_N^*) \end{aligned}\right\} \tag{11}$$

where β is a parameter of order 1, and where the reason for the factor $9/64h^2$ will appear in the next section.

The solution of (11) by Newton's method is greatly facilitated when the functions (10) are specialized by assuming that all of the forces are generated as tensions in straight line segments connecting specified pairs of boundary points. It is then the case that one has to compute the factors of a matrix which is symmetric, positive definite, and sparse. By appropriate numbering one can make the structure of the matrix like that shown here. One can factor such a matrix without introducing non-zero elements in the region marked 0. These regions are therefore ignored, and the computation becomes efficient.

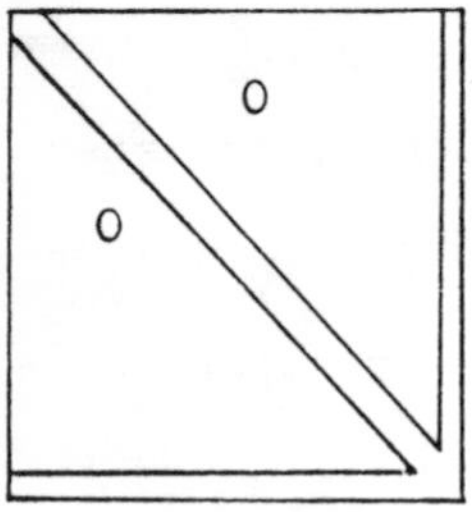

CONNECTING THE BOUNDARY AND FLUID

Points of the immersed boundary need not coincide with fluid mesh points. The required connection is made by the discrete analog of equations (5) and (7):

$$\underline{F}_{ij} = N^{-1} \sum_{k=1}^{N} D_{ij}(\underline{x}_k^n)\, \underline{f}_k^* \tag{12}$$

$$\underline{x}_k^{n+1} = \underline{x}_k^n + \delta t \sum_{ij} \underline{u}_{ij}^{n+1} D_{ij}(\underline{x}_k^n)\, h^2 \tag{13}$$

The function D_{ij} in the foregoing is constructed as follows. Let

$$\phi(r) = \begin{cases} (1 + \cos(\pi r/2))/4 & |r| \leq 2 \\ 0 & |r| > 2 \end{cases} \tag{14}$$

so that $\phi(r)$ has the form shown (next page).

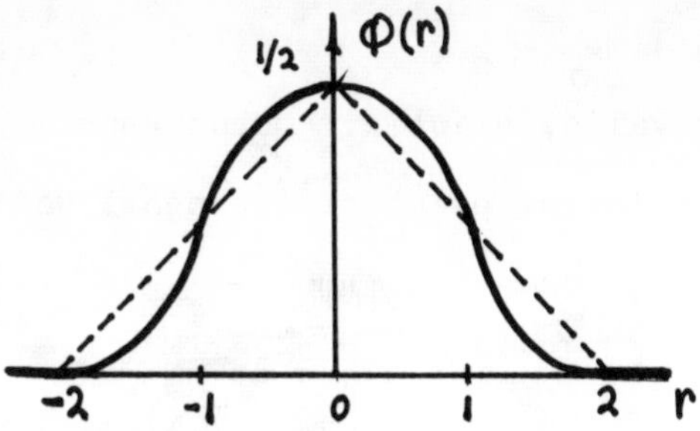

Next, with $\underline{x} = (xh, yh)$, let

$$D_{ij}(\underline{x}) = h^{-2} \, \phi(x-i) \, \phi(y-j) \qquad (15)$$

The function ϕ satisfies the following conditions:

a. $\quad \int \phi(r)\, dr = 1$

b. $\quad \phi(r) \gtrsim 0,$ and $\phi(r) = 0$ when $|r| \gtrsim 2$

c. For all r

$$\sum_{k \text{ even}} \phi(r-k) = \sum_{k \text{ odd}} \phi(r-k) = 1/2 \;\rightarrow\; \sum_{k} \phi(r-k) = 1$$

d. For all r

$$\sum_{k} \phi^2(r-k) = 3/8$$

which implies that for all r,s

$$\sum_{k} \phi^2(r-k) \geq \sum_{k} \phi(r-k)\, \phi(s-k)$$

The importance of (a) – (c) is discussed in Peskin (1972) where we used, in effect, the function ϕ given by the dotted line in the above figure. That function satisfies (a) – (c) but not (d). The advantage of (d) appears when the boundary is highly stressed; it guarantees that when force is applied by a boundary point to the fluid there will be a qualitatively reasonable relationship between the effect at that point and the effect at other points.

SUMMARY OF THE NUMERICAL SCHEME

$$(\underline{u}^n_{ij}, \underline{x}^n_k) \xrightarrow{(11)} \underline{f}^*_k \xrightarrow{(12)} \underline{F}_{ij} \xrightarrow{(9)} \underline{u}^{n+1}_{ij} \xrightarrow{(13)} \underline{x}^{n+1}_k$$

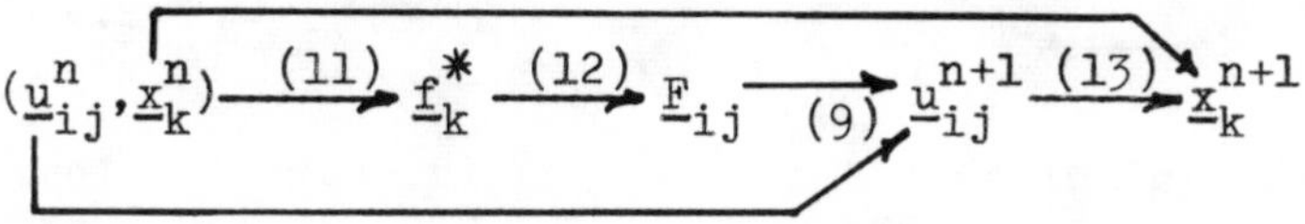

RESULTS

The chambers are muscular; the valve, elastic. Cords under low tension connect valve tips to the lowest point of the heart wall.

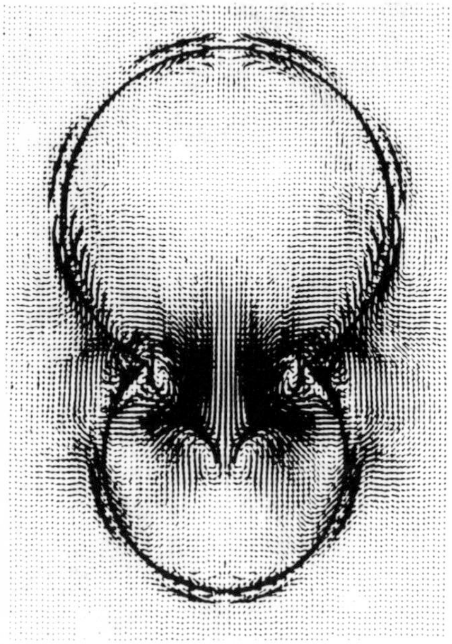

(a) The valve opening

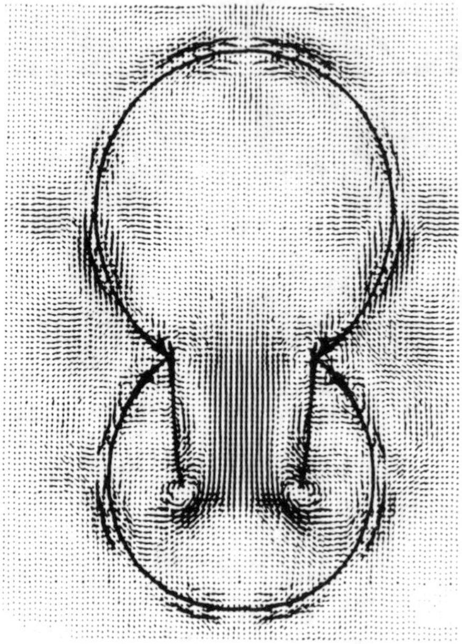

(b) Vortex formation

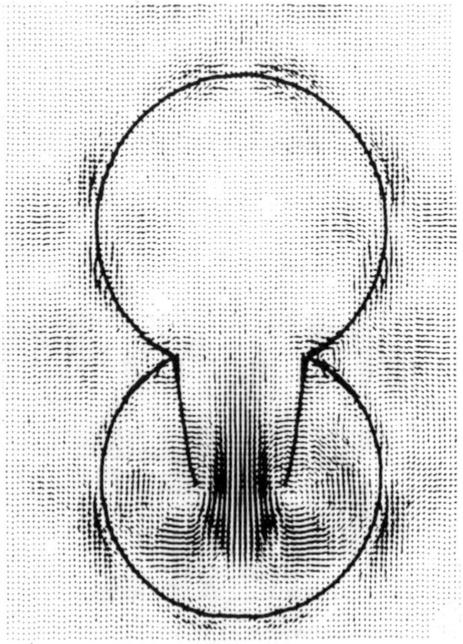

(c) Partial closure by vortex streamlines

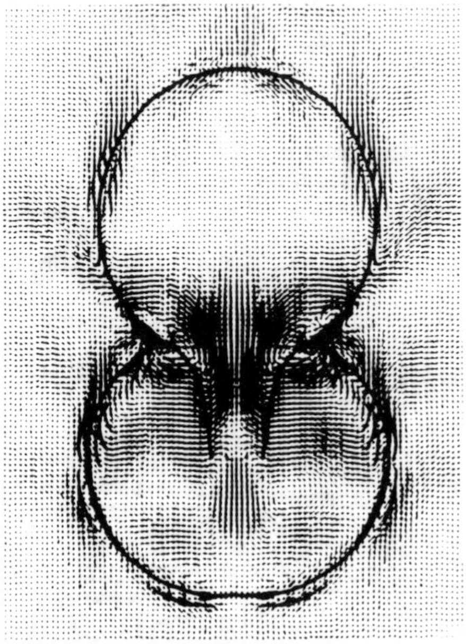

(d) Further closure by contracting lower chamber

CONCLUDING REMARKS: STRENGTHS AND LIMITATIONS OF THE METHOD

The main strength of this technique is its generality. One can specify the properties of the heart and valve essentially at will and, in effect, design a cardiac structure and observe its performance in the computer. This is a useful way to seek understanding of the natural heart and to approach the design of artificial hearts and valves.

In practice we are limited to two dimensions (in principle the technique generalizes at once to three) and also to Reynolds numbers which are low compared to those of human physiology. But the range of Reynolds number from the elephant to the dormouse is about 100:1. This encourages us to believe that the Reynolds number is not a crucial parameter in the physiology of Mammalian hearts.

ACKNOWLEDGMENTS

My advisers in this work have been A.J. Chorin and E.L. Yellin. I am very grateful to them, and also to O. Widlund, for their encouragement and for the countless discussions from which these methods have grown. This work was supported in part by the National Institutes of Health (USA): Institute of General Medical Sciences.[1] Computer time and office space were generously made available by the Courant Institute of Mathematical Sciences under contract with the United States Atomic Energy Commission.[2]

REFERENCES

Chorin, A.J. (1969) Math Comp 23: 341-353

Peskin, C.S. (1972) J Comp Phys (to appear)

Viecelli, J.A. (1969) J Comp Phys 4: 543-551

(1971) J Comp Phys 8: 119-143

1 5T5GM1674

2 AT (11-1) 3077
AT (30-1) 1480

CALCUL DE L'ECOULEMENT D'UN FLUIDE VISQUEUX COMPRESSIBLE
AUTOUR D'UN OBSTACLE DE FORME PARABOLIQUE

par Roger PEYRET
Université Paris VI (Collaborateur extérieur de l'O.N.E.R.A.)
et Henri VIVIAND
Office National d'Etudes et de Recherches Aérospatiales (ONERA) 92320 - Châtillon (France)

I - FORMULATION DU PROBLEME

De nombreuses études ont été consacrées au calcul numérique d'écoulements compressibles visqueux pour des configurations géométriques simples, telles que la plaque plane [1] à [4], un dièdre [5], une cavité rectangulaire [6], un avant-corps ou le nez d'un obstacle émoussé [7] à [10], l'écoulement de culot [11] à [15], sans mentionner les écoulements monodimensionnels. En ce qui concerne le calcul de l'écoulement complet autour d'un obstacle fini, seul le cas du cylindre en écoulement transitoire semble avoir été traité [16].

Le travail présenté ici constitue la deuxième phase d'une étude consacrée au calcul numérique de l'écoulement stationnaire d'un fluide visqueux compressible autour d'obstacles paraboliques. Dans la première phase [17], nous nous sommes limités au cas d'un obstacle parabolique infini ; nous considérons maintenant le cas d'un obstacle fini formé de deux arcs de paraboles cofocales (Fig. 1).

Cet obstacle est placé dans un écoulement supersonique uniforme d'un gaz parfait à chaleurs spécifiques constantes. On admet que les coefficients de viscosité et de conductibilité thermique sont constants, et on se place dans le cas où la paroi est à température fixée.

La résolution numérique des équations de Navier-Stokes est effectuée dans un système de coordonnées paraboliques (ξ, η) reliées aux coordonnées cartésiennes (x, y) par :

$$(1) \qquad x = \frac{1}{2}(\xi^2 - \eta^2) \quad , \quad y = \xi\eta$$

La paroi latérale de l'obstacle a pour équation $\eta = \eta_0$, avec $0 \leq \xi \leq \xi_0$, et le culot est défini par $\xi = \xi_0$, $0 \leq \eta \leq \eta_0$ (Fig. 1).

Les équations de Navier-Stokes stationnaires peuvent s'écrire sous la forme condensée :

$$(2) \qquad \mathbb{F}_\xi + \mathbb{G}_\eta = \frac{1}{Re}\mathbb{W} + \mathbb{S}$$

où $\mathbb{F}, \mathbb{G}, \mathbb{S}$ sont des vecteurs dont les 4 composantes s'expriment en fonction de la masse volumique ρ, de la température T, de la pression $p = \rho T$, et des quantités U, V reliées aux composantes u, v de la vitesse dans le repère (ξ, η) par :

$$U = hu = \xi u + \eta v$$
$$V = hv = -\eta u + \xi v$$

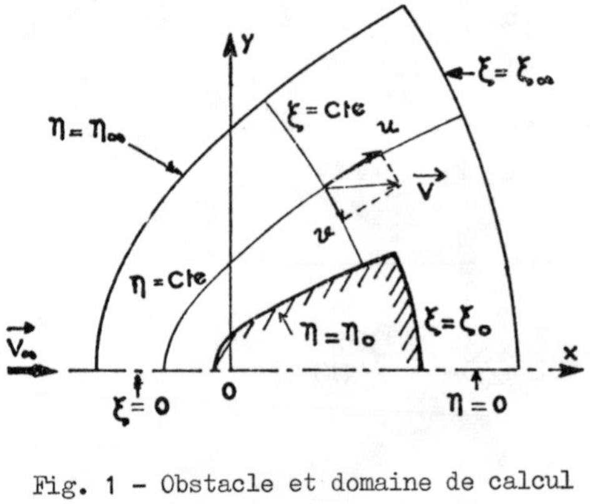

Fig. 1 - Obstacle et domaine de calcul

où $r = \sqrt{\xi^2 + \eta^2}$, et où u , v sont les composantes cartésiennes de la vitesse. Le terme dissipatif W contient outre les inconnues U , V, T, leurs dérivées premières et secondes. Les expressions complètes des vecteurs F, G , S et W sont données dans [17].

Les variables thermodynamiques p, ρ , T et la vitesse ont été rendues adimensionnelles à l'aide des valeurs correspondantes à l'infini. Les distances sont rapportées à une longueur de référence L pour l'instant arbitraire, qui sert à définir le nombre de Reynolds Re , basé en outre sur la vitesse à l'infini.

Le domaine de calcul est limité par différentes paraboles du réseau qui sont les parois ($\xi = \xi_0$; $\eta = \eta_0$), l'axe de symétrie ($\xi = 0$; $\eta = 0$), et deux frontières artificielles ($\eta = \eta_\infty$; $\xi = \xi_\infty$). Les conditions aux limites sont les suivantes :

a) sur la parabole $\eta = \eta_\infty$ qui représente l'infini amont, on impose des conditions d'écoulement uniforme, soit :

(3) $$U = \xi \quad , \quad V = -\eta_\infty \quad , \quad \rho = T = 1$$

b) sur la parabole $\xi = \xi_\infty$ qui limite le domaine de calcul à l'aval, on écrit des conditions en partie arbitraires. On a constaté que la condition utilisée en [17], à savoir l'annulation de dérivées par rapport à ξ , entraînait l'apparition d'oscillations dans le sillage remontant progressivement jusqu'au culot. On a alors utilisé une simple extrapolation linéaire de U, V, ρ et T , qui s'est avérée satisfaisante.

c) à la paroi, la vitesse est nulle et la température est donnée :

(4) $$U = V = 0 \quad , \quad T = T_p$$

d) les conditions de symétrie sur l'axe Ox deviennent, en amont de l'obstacle ($\xi = 0$, $\eta_0 \leqslant \eta \leqslant \eta_\infty$) :

(5) $$U = 0 \quad , \quad V_\xi = \rho_\xi = T_\xi = 0$$

et dans le sillage ($\eta = 0$, $\xi_0 \leqslant \xi \leqslant \xi_\infty$) :

(6) $$V = 0 \quad , \quad U_\eta = \rho_\eta = T_\eta = 0$$

II - RESOLUTION NUMERIQUE

Schéma aux différences - La solution des équations stationnaires (2) est obtenue comme limite, lorsque le temps t croît indéfiniment, de la solution d'équations instationnaires associées que nous notons :

(7) $$f_t + F_\xi + G_\eta = \frac{1}{Re} W + S$$

Dans [17], nous avions pris pour f le terme exact f_e (éq. 8) correspondant aux équations de Navier-Stokes instationnaires. Les schémas considérés nécessitent dans ce cas un pas de temps Δt proportionnel à ρ ; or, la masse volumique devenant très petite au voisinage du culot, cette formulation n'est plus adaptée au cas présent. Nous avons alors choisi un terme f artificiel (éq. 9) de façon à avoir un Δt indépendant de ρ

et optimum pour tous les points du domaine de calcul :

(8) $$f_e = \begin{pmatrix} h^2 \rho \\ h^2 \rho U \\ h^2 \rho V \\ \frac{h^2 p}{\gamma M_\infty^2} + \frac{\gamma-1}{2} \rho (U^2+V^2) \end{pmatrix} , \quad (9) \quad f = \begin{pmatrix} \rho \\ U \\ V \\ T/(\gamma M_\infty^2) \end{pmatrix}$$

Quoiqu'il en soit, ici comme en [17], la discrétisation n'est pas consistante dans l'état transitoire avec (7). Cette discrétisation est choisie de façon à améliorer la stabilité et la convergence vers l'état stationnaire ; plus précisément, pour une équation modèle avec une seule variable d'espace :

(10) $$\varphi_t = \mathcal{L} \varphi \quad , \quad \mathcal{L} \varphi \equiv \frac{1}{Re} \varphi_{\xi\xi} - A \varphi_\xi \qquad (A = Cte)$$

on écrit :

(11) $$\frac{1}{\Delta t}\left(\varphi_i^{n+1} - \varphi_i^n \right) = \frac{1}{Re\, \Delta\xi^2} \left(\varphi_{i+1}^n - 2\varphi_i^n + \varphi_{i-1}^{n+1} \right) - \frac{A}{2\Delta\xi} \left(\varphi_{i+1}^n - \varphi_{i-1}^{n+1} \right)$$

où $\varphi_i^n = \varphi(\xi_i, t_n)$, $\xi_i = (i-1)\Delta\xi$, $t_n = n\,\Delta t$. Ce schéma est consistant, à l'état stationnaire, avec l'équation $\mathcal{L}\varphi = 0$ et l'erreur est $O(\Delta\xi^2)$. Ce procédé, déjà utilisé en [18] est équivalent à la méthode de Gauss-Seidel avec relaxation pour la résolution de $\mathcal{L}\varphi = 0$.

La condition $\Delta t < Re\, \Delta\xi^2 (1 + |A| Re\, \Delta\xi/2)^{-1}$ entraîne la stabilité du schéma (11) en même temps qu'elle assure, si $A > 0$, que l'état stationnaire est obtenu plus rapidement avec (11) qu'avec le schéma explicite classique ; si $A < 0$, cette dernière propriété nécessite de plus $|A|\, \Delta\xi < 2\, Re^{-1}$. Dans ce dernier cas, on aurait intérêt à inverser le rôle des points i-1 et i+1 de façon à éviter la contrainte $|A|\, \Delta\xi < 2\, Re^{-1}$.

Les équations (7) sont discrétisées par un schéma du type (11). L'étude de la stabilité a été faite en [17] en négligeant l'effet des dérivées premières devant celui des dérivées secondes ; le résultat, valable seulement pour Re assez petit, est le suivant :

(12) $$\Delta t < \min. \left\{ \frac{3}{4} Re\, \Delta^2 , \frac{Re\, Pr}{\gamma} \Delta^2 \right\} , \quad \Delta^2 = \frac{\Delta\xi^2\, \Delta\eta^2}{\Delta\xi^2 + \Delta\eta^2}$$

Ce critère est deux fois moins restrictif que celui qui correspond au schéma explicite classique.

Traitement du voisinage de la paroi - On utilise un maillage rectangulaire uniforme de pas $\Delta\xi, \Delta\eta$ tel que $\xi_i = (i-1)\Delta\xi$, $i = 1$ à I, $\eta_j = (j-1)\Delta\eta$, $j = 1$ à J. La paroi latérale $\eta = \eta_0$ et le culot $\xi = \xi_0$ appartiennent aux lignes du maillage et correspondent respectivement à $j = j_P$ et $i = i_P$.

Les conditions aux limites énoncées au § I sont discrétisées de façon classique, les dérivées étant approchées par des différences décentrées à trois points. Une difficulté se présente lorsqu'on écrit l'équation de quantité de mouvement sur la ligne voisine de la paroi (soit $j = j_P + 1$ ou $i = i_P + 1$) car la discrétisation du gradient de pression transversal avec le schéma centré utilisé au point courant fait intervenir la valeur de la pression et donc de la masse volumique sur cette paroi. Or, la détermination de la masse volumique à la paroi est un problème délicat ; en particulier l'utilisation de l'équation de continuité instationnaire avec différences décentrées conduit à une évolution monotone non convergente de la masse volumique en fonction du temps, croissante au point d'arrêt, décroissante sur le culot. On a donc recherché des méthodes ne faisant pas intervenir la masse volumique à la paroi.

La solution proposée dans [17] consiste à ne pas utiliser l'équation de quantité de mouvement transversale sur la ligne voisine de la paroi, mais à calculer la composante transversale de la vitesse à l'aide de l'équation de continuité stationnaire écrite à la paroi, soit $(V_\eta)_{i,j_P} = 0$ et $(U_\xi)_{i_P,j} = 0$. Ces dérivées, discrétisées par un schéma décentré du second ordre, donnent :

$$V_{i,j_P+1} = \frac{1}{4} V_{i,j_P+2} \quad , \quad U_{i_P+1,j} = \frac{1}{4} U_{i_P+2,j} \tag{13}$$

Cette méthode, qui s'est révélée satisfaisante dans le cas de la parabole infinie [17], ne peut être appliquée aux points $(i_P, j_P+1), (i_P+1, j_P)$ voisins de l'arête du culot car les dérivées de la vitesse ne sont pas définies à l'arête. Par suite, les vitesses U_{i_P+1,j_P} et V_{i_P,j_P+1} ont été calculées par une opération de moyenne à partir de 4 points voisins.

D'autre part, du fait des forts gradients existant au voisinage de l'arête, la précision d'une différence décentrée à trois points, bien que théoriquement du second ordre, devient illusoire dans ce voisinage. On a constaté que les résultats pouvaient être améliorés par l'emploi d'une autre méthode, précise au premier ordre seulement, et qui peut être appliquée à tous les points voisins des parois.

Dans cette deuxième méthode, on utilise, comme au point courant, l'équation de quantité de mouvement transversale pour obtenir la vitesse sur la ligne voisine de la paroi, mais on discrétise le gradient de pression par une différence avancée à deux points ne faisant pas intervenir la pression à la paroi.

III - RESULTATS

Des applications numériques ont été faites pour un obstacle défini par $\xi_0 = 1$ et $\eta_0 = 0{,}5$. La longueur l de cet obstacle, rapportée à L, est donc $\frac{1}{2}(\xi_0^2 + \eta_0^2) = \frac{5}{8}$, et le nombre de Reynolds basé sur l est $Re_l = \frac{5}{8} Re$. On a pris $\Delta\xi = \Delta\eta = 0{,}05$ $\xi_\infty = \eta_\infty = 2$ (I = J = 41), $i_P = 21$, $j_P = 11$. Les paramètres de l'écoulement sont : $M_\infty = 2$, $\gamma = 1{,}4$, $Pr = 0{,}72$, $T_P = 1{,}2$.

Un premier calcul a été effectué à $Re = 40$ ($Re_l = 25$) en partant d'un état initial arbitraire. Ce calcul a été arrêté lorsque

$$\bar{R} = \underset{k}{\mathrm{Max}} \left\{ \underset{i,j}{\text{valeur moyenne}} \left(\left| f_{i,j}^{n+1} - f_{i,j}^{n} \right| \right)_k \right\} \leq 1{,}5 . 10^{-5}$$

où k refère à la $k^{ième}$ composante de f (éq. 9). Notons que ce maximum est obtenu pour $k = 1$, et que $\Delta t = 0{,}51.10^{-2}$. Le résidu local est partout voisin de cette valeur moyenne sauf dans les zones à forts gradients où il atteint sa valeur maximale $R = \underset{k}{\mathrm{Max}} \left\{ \underset{i,j}{\mathrm{Max}} \left(\left| f_{i,j}^{n+1} - f_{i,j}^{n} \right| \right)_k \right\}$ qui vaut alors $2{,}5.10^{-4}$. Précisons que le calcul a d'abord été effectué en utilisant la première méthode de traitement du voisinage de la paroi, et que les résultats obtenus ont été améliorés, dans une phase finale, par application de la seconde méthode.

Un deuxième calcul a été effectué à $Re = 80$ ($Re_l = 50$) en partant des résultats précédents, et en utilisant alors tout au long la seconde méthode de traitement de la paroi. Le pas de temps est le même que pour $Re = 40$. Ce calcul a été arrêté lorsque $\bar{R} \leq 5.10^{-6}$; on a alors $R = 6{,}5.10^{-5}$.

Pour $Re = 40$, compte tenu des modifications effectuées au cours du calcul (2800 cycles de temps) sans revenir à l'état initial, on ne peut évaluer le nombre minimal d'itérations effectivement nécessaire pour atteindre l'état de convergence cité plus haut. Par contre, pour $\mathcal{R}e = 80$, 1400 cycles de temps ont été nécessaires pour obtenir $\overline{\mathcal{R}} = 5.10^{-6}$. Il faut cependant noter que, au bout de 1000 cycles, on avait $\overline{\mathcal{R}} = 2,4.10^{-5}$ et $\mathcal{R} = 3.10^{-4}$, et que les résultats n'ont pas varié de façon significative compte tenu de la précision du calcul. Le temps de calcul sur machine IBM 360-50 est de 21 minutes pour 100 cycles de temps.

L'obstacle et le maillage sont représentés sur la figure 2 qui montre aussi les positions de la couche choc et de la ligne sonique. Les figures 3 à 7 montrent des profils, à ξ fixé, du nombre de Mach M, de la composante u de la vitesse, de la température T et de la masse volumique ρ . On peut distinguer sur ces profils les régions caractéristiques de l'écoulement : couche de choc, couche visqueuse à la paroi ou sillage, séparées par une zone à gradient plus faible.

On a placé sur les profils à $\xi = 0$ (axe de symétrie amont) les valeurs théoriques données par les relations de Rankine-Hugoniot ; on remarquera la bonne concordance avec les résultats du calcul.

Les oscillations qui apparaissent sur la face aval du choc ou encore au voisinage de l'arête du culot (pour les profils de ρ seulement) pourraient être fortement diminuées, comme en [17], par l'utilisation d'un maillage plus fin.

On a représenté sur la figure 8 le champ des directions du vecteur vitesse dans la région du culot pour $\mathcal{R}e_\ell = 50$. Cette figure met en évidence une zone de recirculation qui n'existait pas pour $\mathcal{R}e_\ell = 25$.

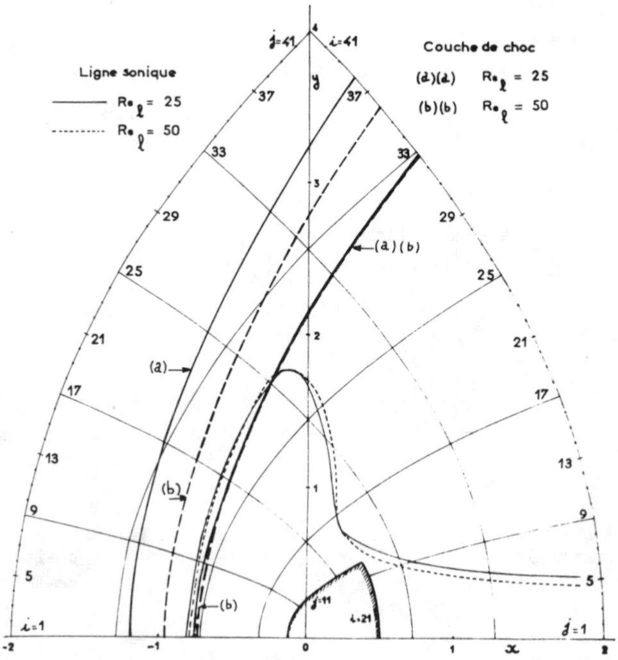

Fig. 2 - Couche de choc et ligne sonique

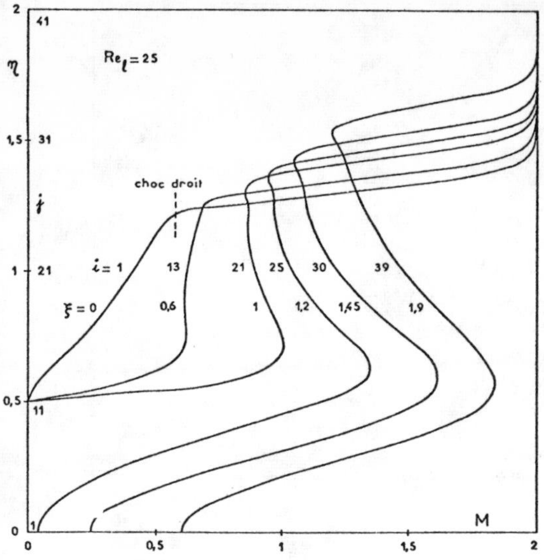

Fig. 3 - Profils de nombre de Mach à $Re_l = 25$

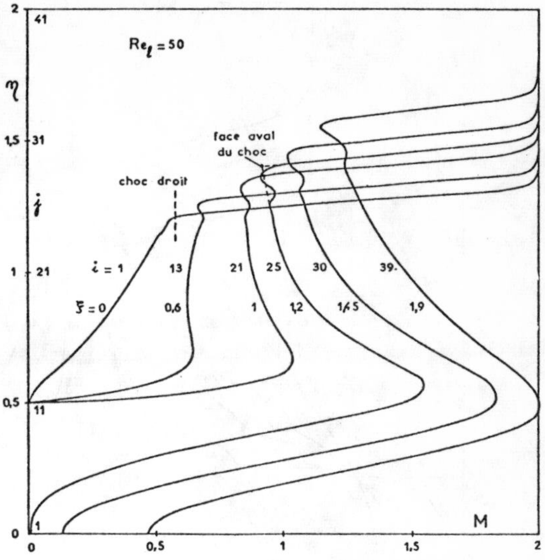

Fig. 4 - Profils de nombre de Mach à $Re_l = 50$

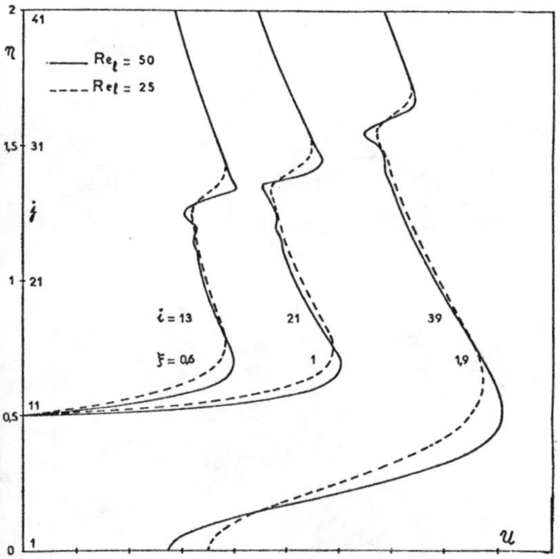

Fig. 5 - Profils de la composante u de la vitesse

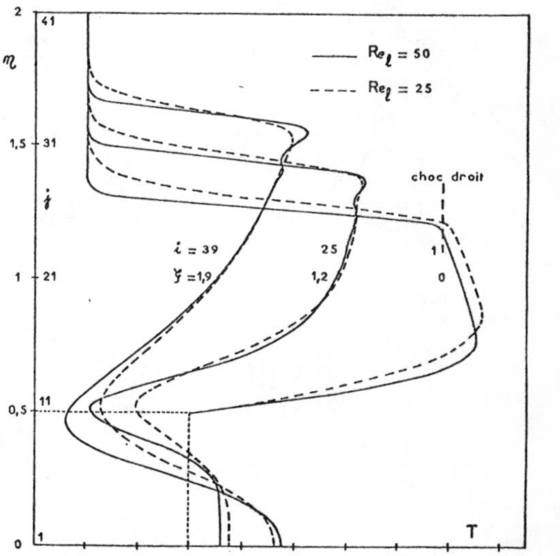

Fig. 6 - Profils de température

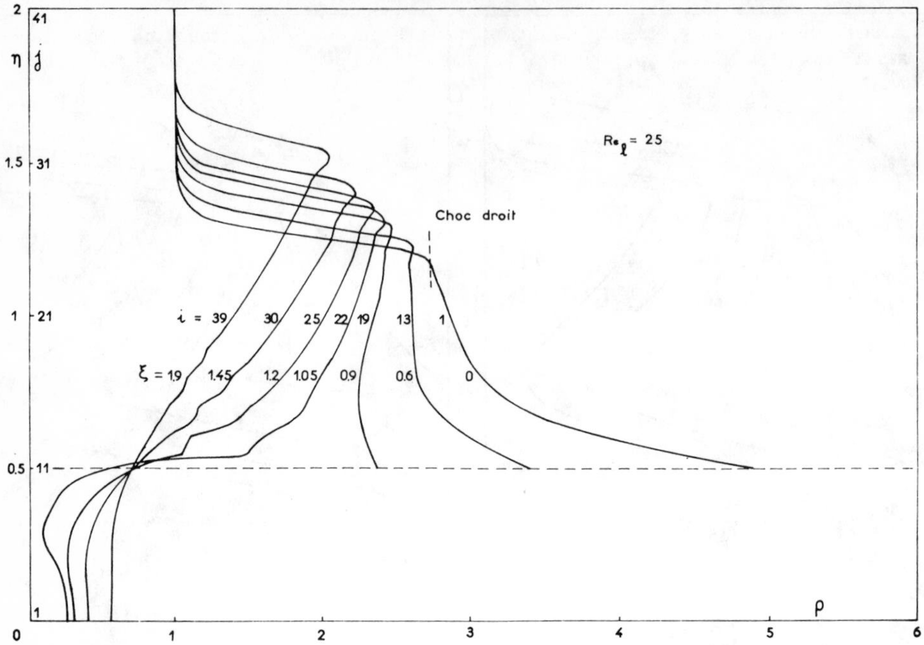

Fig. 7 - Profils de masse volumique à $Re_l = 25$

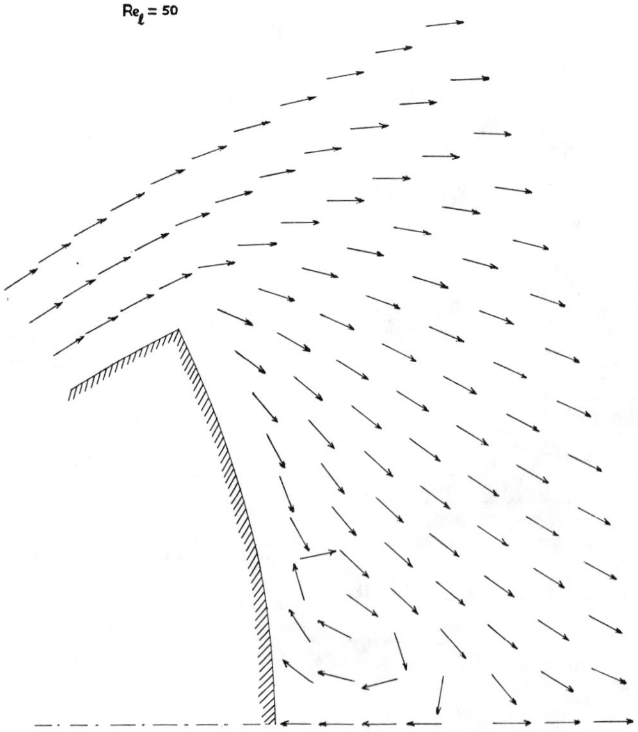

Fig. 8 - Champ des directions des vitesses à $Re_l = 50$ au voisinage du culot

IV - CONCLUSIONS

Nous avons présenté les premiers résultats obtenus au cours de calculs de l'écoulement complet d'un fluide visqueux compressible autour d'un obstacle présentant un culot. Ces résultats sont relatifs à des nombres de Reynolds encore petits et des calculs sont en cours pour atteindre de plus grandes valeurs du nombre de Reynolds.

Par ailleurs, différents perfectionnements sont actuellement étudiés : maillage resserré dans les zones de forts gradients, dilatation de coordonnée dans la direction aval, choix optimal du terme β en particulier par la considération d'échelles de temps différentes selon les composantes (cf. [18]). Signalons aussi que parallèlement au schéma étudié ici, des résultats préliminaires ont été obtenus au moyen du schéma proposé en [19] dans le cas de l'obstacle parabolique infini et qu'on a constaté un très bon accord avec les résultats présentés en [17].

REFERENCES

[1] Thommen, H.U., ZAMP 17, 369 - 384 (1966)

[2] Kurzrock, J.W., and Mates, R.E., AIAA paper 66-30 (1966)

[3] Butler, T.D., Physics of Fluids 10, 6, 1205-1215 (1967)

[4] Mac Cormack, R.W., Proc. 2nd Int. Conf. on Numerical Methods in Fluid Dynamics, 151-163, Univ. Calif. Berkeley (Sept. 1970)

[5] Brailovskaya, I.Yu., Fluid Dynamics 2, 3, 49-55 (1967)

[6] Polezhaev, V.I., Fluid Dynamics 2, 2, 21-27 (1967)

[7] Moretti, G., and Salas, M.D., AIAA paper 69-139 (1969)

[8] Molodtsov, V.K., U.S.S.R. Comput. Math. and Math. Physics 9, 5, 320-329 (1969)

[9] Molodtsov. V.K., and Tolstych, A.J., Proc. Session on Numerical Methods in Gas-dynamics, 2nd Int. Colloq. on Gasdynamics of Explosions and Reacting Systems, vol. I, 37-54, Novossibirsk (Août 1969)

[10] Pavlov, B.M., ibid., 55-66

[11] Myshenkov, V.I., ibid, 67-82

[12] Allen, J.S., and Cheng, S.I., Physics of Fluids 13, 1, 37-52 (1970)

[13] Roache, P.J., and Mueller, T.J., AIAA J. 8, 3, 530-538 (1970)

[14] Brailovskaya, I. Yu., Dokl. Akad. Nauk, SSSR, 197, 3, 542-544 (1971)

[15] Ross, B.B., and Cheng, S.I., Proc. 2nd Int. Conf. on Numerical Methods in Fluid Dynamics, 164-169, Univ. of Calif. Berkeley (Sept. 1970)
AIAA paper 72-115 (1972)

[16] Scala, S.M., and Gordon, P., AIAA J. 6, 5, 815-822 (1968)

[17] Peyret, R., et Viviand, H., La Rech. Aérosp. 1972-3, 123-131 (Mai-Juin 1972)

[18] Fortin, M., Peyret, R., et Temam, R., Journ. Mécanique, 10, 3, 357-390 (1971)

[19] Mac Cormack, R.W., AIAA paper 69-354 (1969)

SOLUTIONS NUMERIQUES DES EQUATIONS DE NAVIER-STOKES POUR LES ECOULEMENTS EN COUCHES VISQUEUSES

K. ROESNER+ et Şt.N. SAVULESCU++

RESUME

On construit des solutions numériques des équations de Navier-Stokes en partant d'une formulation intégro-différentielle de celles-ci, adaptée pour les écoulements en couches visqueuses incompressibles. Par un procédé itératif on obtient des profils de vitesse et de pression pour l'écoulement visqueux à la parois d'une plaque plane, dont la première approximation est la couche limite de Blasius.

INTRODUCTION

Le but de ce travail est d'obtenir des informations sur les modifications apportées par les termes des équations complètes de Navier-Stokes sur les solutions de première approximation données par la théorie de la couche limite. Afin d'évaluer numériquement l'influence de ces termes, on les présente dans une nouvelle forme mathématique par les opérations suivantes:

- la transformation des coordonnées du type von-Mises normalisée [1]
- l'intégration le long de la normale à la parois, l'introduction des conditions aux limites et l'obtention d'une formulation intégro-différentielle [1] , adaptée pour les écoulements près des parois.

A l'aide de cette formulation, la vitesse parallèle à la parois est exprimée sous la forme d'une série finie, dont le premier terme représente l'approximation couche limite. Les autres termes, provenant de la contribution des équations complètes de Navier-Stokes, sont exprimés par des différences entre des quantités rendues typiques à l'aide d'une double intégrale normalisée. Le processus itératif part de l'approximation couche limite et se déroule ultérieurement en tenant compte de tous les termes supplémentairs, de certaines conditions imposées aux profils de vitesse, ainsi que de la variation

+ Inst. für Angew. Math. der Albert-Ludwigs-Universität und Inst. für Angew. Math. und Mech. der DFVLR, Freiburg i. Br.

++ Inst. de Mécanique des Fluides, Bucarest, et Inst. für Angew. Math. der Albert-Ludwigs-Universität, Freiburg

temporelle qui en résulte moyennant une durée de référence T.
Pour l'écoulement de Blasius on a calculé, en premier approche, la modification des profils de vitesse, due seulement au mécanisme perturbateur des équations complètes de Navier-Stokes.

PHYSIQUE DES COUCHES VISQUEUSES

En généralisant la notion de couche limite, nous définissons les couches visqueuses par les propiétés suivantes:

- Le domaine de l'écoulement visqueux est compris entre une surface donnée (y=0) - dite PAROIS - et une surface arbitraire $\delta(x,z,t)$ - dite EPAISSEUR - assurant la continuité de la vitesse et de la pression,
- Le champ des vitesses $\vec{V}(x,y,z,t) = \vec{V}_P + \vec{j}v$ permet toujours de distinguer un écoulement principal $\vec{V}_P := \vec{i}u + \vec{k}w$, parallèle à la parois(coordonnées x,z), avec $|\vec{V}_P| \gg v$ et dont les gradients $\vec{V}_{P/y}$ sont aussi très importants,
- Les distributions locales suivant y (x_1, z_1, t_1 étant fixes)-dites PROFILS- de la vitesse $\vec{V}$ et de la pression p sont des fonctions continues, bornées, uniquement déterminées par rapport à y,
- Le mouvement du fluide incompressible (ϱ = const.), de viscosité ν = const. est régi par les équations de Navier-Stokes, écrites sous la suivante forme non-dimensionelle:

$$(\vec{V}_P)_{yy} = Re_T\,(\vec{V}_P)_t + Re\left[(\vec{V}\nabla)\vec{V}_P + \nabla p - \vec{F}\right] - (\vec{V}_P)_{xx} - (\vec{V}_P)_{zz}\,,$$

$$(1)\qquad p_y = -\frac{Re_T}{Re}v_t + (\vec{V}\nabla)v + \frac{1}{Re}\Delta v + F^{(y)}\,,$$

$$v_y = -u_x - w_z\,,$$

où $Re := \dfrac{LV_r}{\nu}$, $Re_T := \dfrac{L^2}{\nu T}$, L, V_r, T étant les grandeurs de référence.

Les solutions de ces équations, en valeurs instantanées, doivent satisfaire au conditions aux limites:

A LA PAROIS (y=0) : $\vec{V} = \vec{V}_o(x,z,t)$, $p = p_o(x,z,t)$,

A L'EPAISSEUR ($y = \delta(x,z,t)$) :

$$\vec{V}_P = \vec{V}_{P_e}(x,z,t)\,,\ \vec{\lambda}\cdot\vec{V}_{P/y} = q_{v_e}(x,z,t)\,,$$

$$(p + \frac{\varrho}{2}v^2)_{/y} = q_{p_e}(x,z,t)$$

aussi qu'aux conditions initiales specifiées pour chaque problème.
La <u>fig. 1.</u> montre le schéma physique d'une couche visqueuse et le système des coordonnées Oxyz, choisi de telle manière que $u_e = w_e = U_\infty \frac{\sqrt{2}}{2}$.

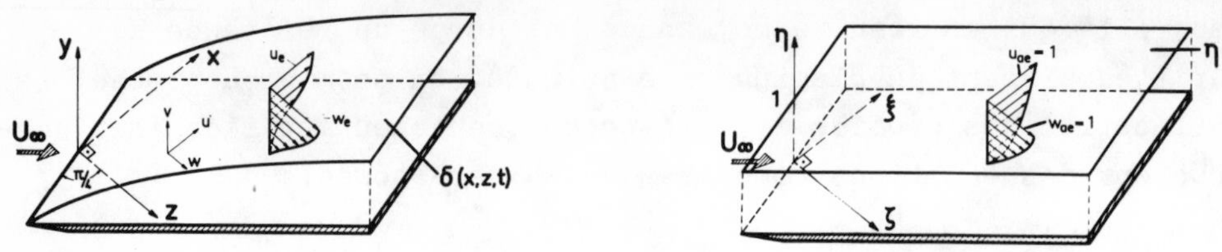

Fig. 1. Le schéma physique et les systèmes des coordonnées

LES EQUATIONS TRANSFORMEES ET LA FORMULATION INTEGRO-DIFFERENTIELLE

On a montré [1] qu'une transformation de y dans η est une généralisation de la transformation von-Mises. On introduit le système des coordonnées

$$(2) \qquad \begin{pmatrix} x \\ y \\ z \\ t \end{pmatrix} \longrightarrow \begin{pmatrix} \xi = x \\ \eta = \dfrac{\int_0^y u\,dy'}{\int_0^\delta u\,dy'} = \dfrac{\psi}{\psi_\delta} \\ \zeta = z \\ \tau = t \end{pmatrix}, y = \psi_\delta \int_0^\eta \frac{d\eta'}{u}$$

qui, tout en assurant les conditions de transformation ponctuelle, permet la transformation inverse.

L'opérateur intégral non-linéaire [1] :

$$(3) \qquad (f)^+ = \frac{\int_0^\eta \left[\int_1^{\eta'} f\, d\eta'' \right] d\eta'}{\int_0^1 \left[\int_1^{\eta'} f\, d\eta'' \right] d\eta'} \equiv \frac{F^+}{F_1^+} \quad ; \quad 0 \le \eta \le 1 \, ,$$

- exprimé par le symbole $(\ )^+$ - possède la propriété de contraction pour une classe assez large des fonctions f(u,v,w). L'étude numérique de cet opérateur montre que pour beaucoup de fonctions $(f)^+ \sim \eta(2-\eta)$. Cette observation a été utilisée dans le cas du système (1).

Ensuite on écrit le système (1) sous la forme suivante:

$$\begin{aligned}(Du^2)_{\eta\eta} = & - Re(\psi_\delta^2)_\xi \eta u_\eta + 2 Re\, \psi_\delta y_\eta \left(p + \frac{u^2}{2}\right)_\xi + 2 Re_T \cdot \psi_\delta (y_\eta u_\tau - y_\tau u_\eta) + \\ & + 2 Re \cdot \psi_\delta \left(w y_\eta u_\zeta - u_\eta \frac{\partial}{\partial \zeta} \int_0^\eta w y_\eta \, d\eta' \right) - 2 Re \cdot \psi_\delta y_\xi p_\eta + \\ & + 2\psi_\delta \left(2 y_\xi u_{\xi\eta} + y_{\xi\xi} u_\eta - y_\eta u_{\xi\xi} + 2 y_\zeta u_{\zeta\eta} + y_{\zeta\zeta} u_\eta - y_\eta u_{\zeta\zeta} \right) + \\ & + (u^2 D_\eta)_\eta + 2 Re \cdot \psi_\delta \left[(v - u y_\xi - w y_\zeta)_{\eta=0} u_\eta + y_\eta F^{(x)} \right] \\ \equiv & \; f_{u_B} + \sum_{1}^{7} {}^{i} f_{u_i} \; ,\end{aligned}$$

$$(Duw)_{\eta\eta} = -Re(\psi_\delta^2)_\xi \eta w_\eta + 2Re\cdot\psi_\delta y_\eta \left(p+\frac{w^2}{2}\right)_\zeta + 2\,Re_T\cdot\psi_\delta(y_\eta w_\tau - y_\tau w_\eta) +$$
$$+ 2Re\cdot\psi_\delta\left(\psi_\delta w_\xi - w_\eta \frac{\partial}{\partial\xi}\int_0^\eta w y_\eta\, d\eta'\right) - 2\,Re\cdot\psi_\delta y_\zeta p_\eta +$$
$$+ 2\psi_\delta\left(2y_\xi w_{\xi\eta} + y_{\xi\xi} w_\eta - y_\eta w_{\xi\xi} + 2y_\zeta w_{\zeta\eta} + y_{\zeta\zeta} w_\eta - y_\eta w_{\zeta\zeta}\right) +$$
$$+ (uwD_\eta + Dwu_\eta - Duw_\eta)_\eta + 2Re\cdot\psi_\delta\left[(v-uy_\xi-wy_\zeta)_{\eta=0} w_\eta + y_\eta F^{(z)}\right]$$
$$\equiv f_{w_B} + \sum_{1}^{7} {}_i f_{w_i} ,$$

$$\left(p+\frac{v^2}{2}\right)_\eta = \left[(uy_\xi + wy_\zeta)v_\eta - y_\eta(uv_\xi + wv_\zeta)\right] - \frac{Re_T}{Re}(y_\eta v_\tau - y_\tau v_\eta) - y_\eta F^{(y)}$$
$$+ \frac{1}{Re}\left[(\eta_y D v_\eta)_\eta - 2y_\xi v_{\xi\eta} - y_{\xi\xi} v_\eta + y_\eta v_{\xi\xi} - 2y_\zeta v_{\zeta\eta} - y_{\zeta\zeta} v_\eta + y_\eta v_{\zeta\zeta}\right]$$
$$\equiv \sum_{1}^{7} {}_i f_{p_i} ,$$

$$v_\eta = (uy_\xi + wy_\zeta)_\eta - (\psi_\delta)_\xi - (y_\eta w)_\zeta \equiv \sum_{1}^{3} {}_k g_k ,$$

$$D := 1 + y_\xi^2 + y_\zeta^2 .$$

Les conditions aux limites, pour le problème de la plaque plane, s'écrivent (fig. 1.) :

$$\eta = 0 : \quad u = v = w = 0 \quad ; \quad p = p_0(\xi,\zeta,\tau) \; ;$$
(4) $$\eta = 1 : \quad u_e = w_e = \frac{\sqrt{2}}{2} U_\infty \quad ; \quad u_\eta = w_\eta = 0 \; ; \quad \left(p + \frac{v^2}{2}\right)_\eta = 0 \; ;$$
$$\xi + \zeta = 0 : \quad \psi_\delta = 0 \; ; \quad (\ldots)_{\xi,\zeta,\tau} = 0$$

En intégrant deux fois les deux premières équations par rapport à η, en y introduisant les conditions aux limites (4) ainsi que les notations de (3) on obtient finalement le système:

$$u^2 = \frac{1}{D}\left[(Du^2)_{\eta=1}(f_{u_B})^+ + \sum_{1}^{7} {}_i F_{u_{i_1}}^+ \left\{(f_{u_i})^+ - (f_{u_B})^+\right\}\right]$$
$$=: \Phi_1(u, v, w, p, \psi_\delta; Re, Re_T) ,$$

$$\psi_\delta^2 = -\frac{1}{Re}\int_{-\zeta}^{\xi} \frac{(Du^2)_{\eta=1} - \sum_{1}^{7} {}^{i} F_{u_{i1}}^{+}}{(\eta u_\eta)_{\eta=1}^{+}} d\xi'$$

$$=: \phi_2 \, (u, v, w, p, \psi_\delta \, ; \, Re, Re_T) \quad ,$$

(5)

$$v = u y_\xi + w y_\zeta - \eta (\psi_\delta)_\xi - \frac{\partial}{\partial\zeta}\int_0^\eta w y_\eta \, d\eta'$$

$$=: \phi_3 (u, v, w, p, \psi_\delta) \, ,$$

$$p = -\frac{v^2}{2} + \sum_{1}^{7} {}^{i} \int_0^\eta f_{p_i} \, d\eta' =: \phi_4 \, (u, v, w, p, \psi_\delta ; Re, Re_T),$$

$$w = \frac{1}{Du}\left[(Duw)_{\eta=1} (f_{w_B})^+ + \sum_{1}^{7} {}^{i} F_{w_{i1}}^{+} \left\{ (f_{w_i})^+ - (f_{w_B})^+ \right\} \right] =: \phi_5 \, .$$

On considère ces relations comme des identités qu'on doit vérifier par des solutions obtenues après un processus itératif, en commencant par des solutions de première approximation moyennant le Re_T qui satisfasse aux conditions de compatibilité:

$$u^2 \geq 0 \quad ; \quad \psi_\delta^2 \geq 0 \, .$$

TRAITEMENT NUMERIQUE DU SYSTEME (5)

La méthode numérique utilisée est représentée schématiquement dans le diagramme du flux de la fig. 2. Le programme, écrit en FORTRAN V, détermine d'abord Re, Re_T qui assurent la condition de compatibilité $u^2 \geq 0$; $\psi_\delta^2 \geq 0$. Sitôt q'une grandeur a été déterminée, les autres qui en dépendent, sont à nouveau calculées. On a imposé une limite des erreurs moyennant une intégrale suivant η, prise sur les valeurs absolues des itérations successives, entre $\eta = 0$ et $\eta = 1$.

Le programme présenté ici est écrit pour un réseau de 6 points (voir le fig. 3.),en tenant compte de la capacité de l'UNIVAC 1106 de l'Université de Freiburg i. Br.
La machine fournit les résultats numériques concernant les distributions de la vitesse u(η), v(η), w(η) , de la pression p(η), ainsi que y(η) sous la forme des diagrammes.

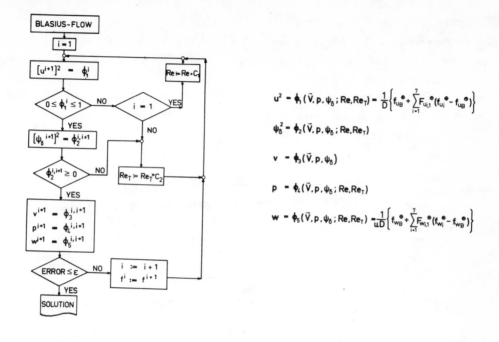

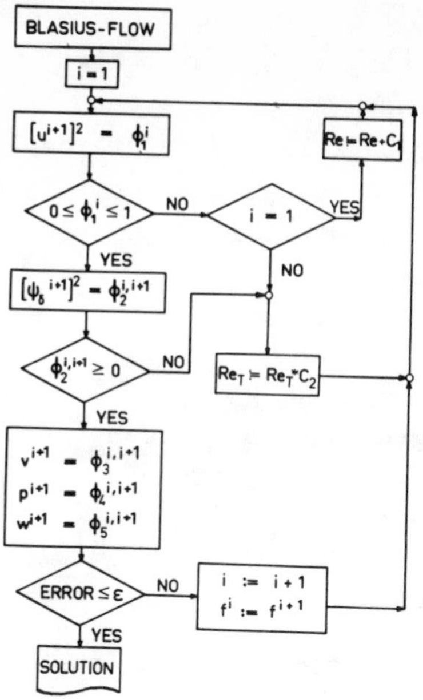

$$u^2 = \phi_1(\vec{V}, p, \psi_\delta ; Re, Re_T) = \frac{1}{D}\left\{f_{u_B}^{\oplus} + \sum_{i=1}^{7} F_{u_{i,1}}^{\oplus}(f_{u_i}^{\oplus} - f_{u_B}^{\oplus})\right\}$$

$$\psi_\delta^2 = \phi_2(\vec{V}, p, \psi_\delta ; Re, Re_T)$$

$$v = \phi_3(\vec{V}, p, \psi_\delta)$$

$$p = \phi_4(\vec{V}, p, \psi_\delta ; Re, Re_T)$$

$$w = \phi_5(\vec{V}, p, \psi_\delta ; Re, Re_T) = \frac{1}{uD}\left\{f_{w_B}^{\oplus} + \sum_{i=1}^{7} F_{w_{i,1}}^{\oplus}(f_{w_i}^{\oplus} - f_{w_B}^{\oplus})\right\}$$

<u>Fig. 2.</u> Diagramme du flux numérique et le système des équations

LES RESULTATS ET LEUR INTERPRETATION

Du grand nombre des résultats numériques obtenus nous présentons ici seulement quelques uns, comme étant les plus significatifs pour la méthode d'intégration numérique proposée.

Il faut tout d'abord souligner le charactère de premier approche en ce qu'on a exploré un nombre réduit des points et que l'erreur d'itération a eu un taux élevé. Mais en dépit de cela, le calcul de tous les termes des équations complètes Navier-Stokes pour les couches visqueuses a révelé des aspects nouveaux par rapport à l'approximation couche limite. En premier lieu, les solutions n'existent, suivant notre méthode, que pour certaines valeurs couplées de Re, Re_T. En maintenant fixe Re on a cherché la compatibilité de la solution en variant Re_T. De cette manière on présente maintenant les distributions de $U(y), v(y), W(y), p(y)$ pour le couple des paramètres $Re=10^3$, $Re_T=340$. Ce couple Re, Re_T correspond approximativement au stage de la transition dite de la stabilité linéaire tridimensionnelle, en ce que le profil Blasius $U(y)$ n'est pas profondément modifié, une variation temporelle d'une fréquence $f = (Re_T.V_r)/(Re.L)$ de l'ordre de grandeur des oscillations Tollmien-Schlichting y apparaît, W(y) indique la formation des tourbillons longitudinaux et p(y) lui aussi manifeste un changement de signe mentionné dans la litérature de la transition [1] . Pour trouver d'autre couples Re, Re_T , espéciallement dans le domainedes grandes valeurs de Re ($\approx 10^5$) il faut améliorer certains détails de la stratégie numérique.

REMERCIMENTS

Les auteurs tiennent à exprimer leur gratitude à Deutsche Forschungsgemeinschaft, qui leur a donné la possibilité d'effectuer ce travail.

REFERENCES

[1] Săvulescu, Şt. N. Tranziţia de la scurgerea laminară la cea turbulentă (Transition from laminar to turbulent flow) Editura Academiei Republicii Socialiste România, Bucureşti - 1968

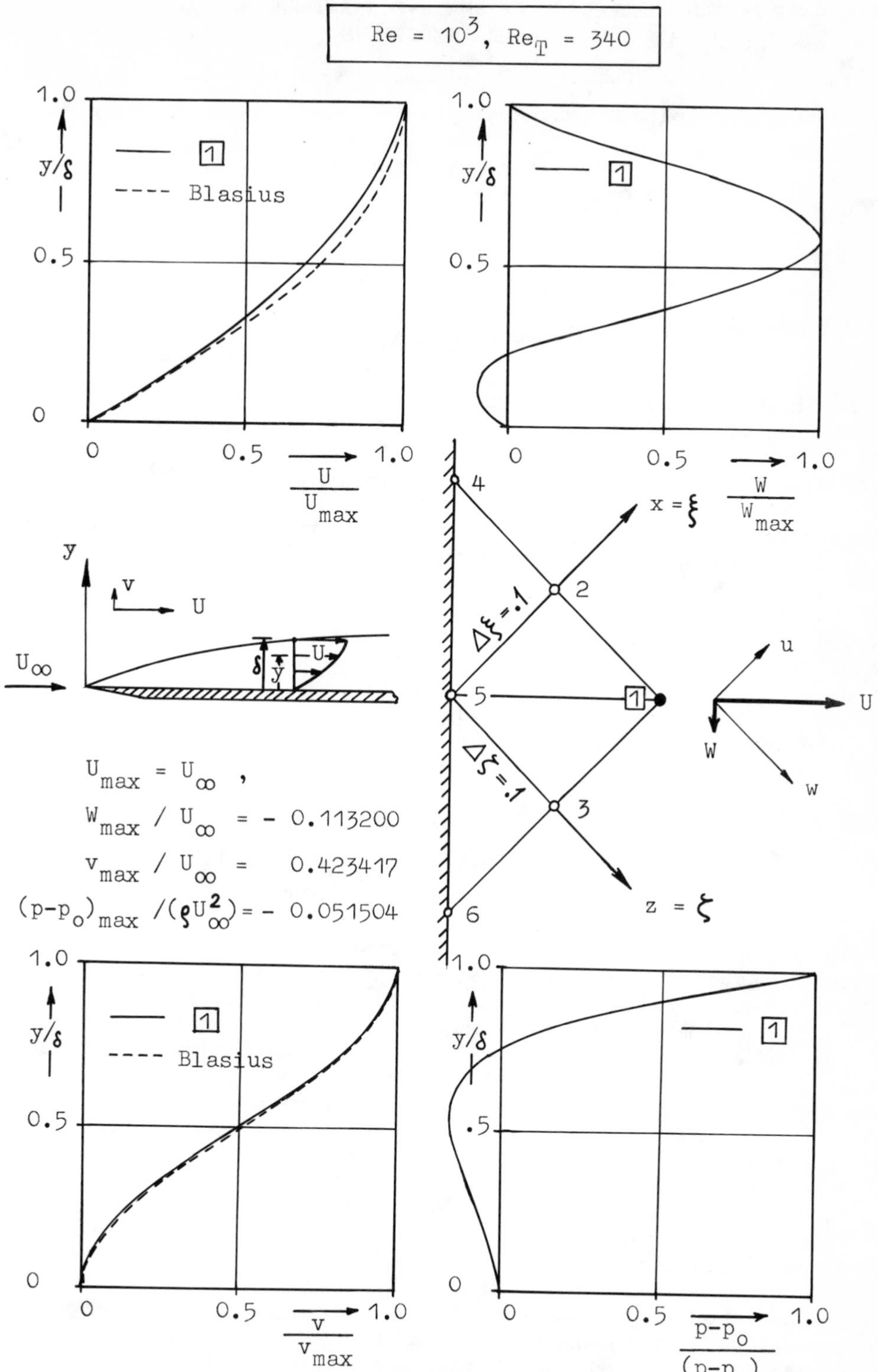

Fig. 3. Les résultats numériques

NUMERICAL SIMULATION OF SMALL-SCALE THERMAL CONVECTION IN THE ATMOSPHERE

Richard C. J. Somerville
Courant Institute of Mathematical Sciences, New York University
and
Goddard Institute for Space Studies-NASA
2880 Broadway, New York, N.Y. 10025, U.S.A.

ABSTRACT

Non-hydrostatic convection is studied by numerical integration of a Boussinesq system in three space dimensions and time. To simulate cloud convection, parameterized effects of latent heat and small-scale turbulence have been included. In this paper, the results are compared, in simplified cases without the parameterizations, with cell structure observed in Rayleigh-Bénard laboratory convection experiments in air. At a Rayleigh number of 4000, the numerical model successfully simulates the experimentally observed evolution, including some prominent transient features, of the flow from a randomly perturbed conductive initial state to a final state of large-amplitude, steady, two-dimensional rolls. At a Rayleigh number of 9000, the model reproduces the experimentally observed unsteady equilibrium of vertically coherent oscillatory waves superimposed on rolls. In both cases, good quantitative agreement with laboratory data is obtained.

INTRODUCTION

Recently it has become clear that several important aspects of even the simplest examples of thermal convection are intrinsically time-dependent and three-dimensional in space. For example, in Rayleigh-Bénard convection between horizontal surfaces maintained at different constant temperatures, several flow regimes are distinguishable experimentally (Krishnamurti, 1970). The simplest of these is the steady two-dimensional roll, which occurs at moderately supercritical values of the Rayleigh number, R, as theoretically predicted (Schlüter, Lortz, and Busse, 1965; Busse, 1967). At larger values of R, however, unsteady and three-dimensional flows occur, and, at very large R, turbulence is found. The availability of good laboratory data, the simple specifications of the problem, and the transition to turbulence through a sequence of successively more complicated flow regimes, make Rayleigh-Bénard convection an excellent test problem for numerical methods. In the present study, a numerical model being developed to simulate small-scale moist convection in the atmosphere is applied to the simpler problem of non-linear Rayleigh-Bénard convection in the laboratory. The numerical results are then compared with experimental data, with the dual aim of studying Rayleigh-Bénard convection and of testing the model more thoroughly than will be possible when it is applied to the much less well observed phenomena of cloud-scale convection in the atmosphere.

EQUATIONS AND BOUNDARY CONDITIONS

The foundation of the present model of small-scale atmospheric convection is the Boussinesq approximation to the Navier-Stokes equation. To this have been added parametric representations of the dynamic and thermodynamic effects of phase changes of water, conservation equations and boundary conditions for water substance, a non-linear eddy viscosity, and an extension of the basic Boussinesq system to the more general anelastic system. When these modifications are removed, the classical Boussinesq system for dry shallow convection, as traditionally used in studies of Rayleigh-Bénard convection (Chandrasekhar, 1961), is recovered.

In this paper, we present tests of the model performed by making these simplifications and then comparing the numerical results with laboratory observations (Willis and Deardorff, 1970) of Rayleigh-Bénard convection in a thin layer of air. Results and details of the more complete model will be given in a later paper.

The equations and notation are standard and have been used in many previous studies of Rayleigh-Bénard convection (e.g., Lipps and Somerville, 1971).

The fluid is characterized by a mean density ρ_o, coefficient of thermal expansion α, coefficient of kinematic viscosity ν, and coefficient of thermometric conductivity κ, all assumed constant. The fluid is contained between horizontal conducting surfaces separated by a distance d and maintained at different constant temperatures, the lower surface being the warmer. The temperature difference is denoted by $\Delta T > 0$. We take Cartesian coordinates with x and y in the horizontal, z positive upwards, and an origin on the lower surface. A gravity vector of magnitude g, directed downward, represents the only body force.

In the Boussinesq approximation, the density ρ is constant except in the buoyancy term, where

$$\rho = \rho_o(1 - \alpha T'), \tag{1}$$

where T' is a departure from the mean temperature. We employ dimensionless variables, scaling all lengths by d, time by $d^2\kappa^{-1}$, temperature by ΔT, and pressure by $\rho_o\kappa^2 d^{-2}$. Then denoting the unit vertical vector by $\underline{k}$, pressure by p, temperature by T, and the velocity vector by $\underline{v}$, the dimensionless equations for conservation of mass, momentum, and thermodynamic energy are

$$\nabla \cdot \underline{v} = 0 \tag{2}$$

$$\left(\frac{\partial}{\partial t} + \underline{v} \cdot \nabla\right) \underline{v} = -\nabla p + \sigma R T \underline{k} + \sigma \nabla^2 \underline{v}, \tag{3}$$

$$\left(\frac{\partial}{\partial t} + \underline{v} \cdot \nabla\right) T = \nabla^2 T . \tag{4}$$

The parameters are the Rayleigh number

$$R = g\alpha\Delta T d^3 \kappa^{-1} \nu^{-1}$$

and the Prandtl number $\sigma = \nu\kappa^{-1}$.

So that a quantitative comparison will be possible between the numerical results and experiments performed in a container which is bounded at top and bottom by rigid surfaces, we choose boundary conditions as follows:

$$\underline{v} = 0 \quad \text{and} \quad T = 1 \quad \text{on } z = 0 \tag{5}$$

$$\underline{v} = 0 \quad \text{and} \quad T = 0 \quad \text{on } z = 1 \tag{6}$$

Because the experimental tank has a horizontal extent much larger than its depth, we employ a computational domain with this property and use cyclic (periodic) boundary conditions at the side walls. In the calculations to be reported in this paper, the horizontal extent in the x direction is 6 times the depth, and in the y direction it is 4.9 times the depth. This domain, which was also used by Lipps and Somerville (1971), is about the largest consistent with adequate resolution and computer capacity.

The initial state consists of the conductive solution to (2) - (6),

$$\underline{v} = 0, \quad T = 1 - z \tag{7}$$

plus a random perturbation of small amplitude added to the temperature field.

NUMERICAL METHOD

An adaptation of Chorin's (1968) method has been used to solve the above system. This method solves the equations in their fully time-dependent form and is not to be confused with Chorin's (1967) "artificial compressibility" method for steady flows, which the present author has used to treat convection in rotating coordinates (Somerville, 1971) as well as the conventional two-dimensional, steady Rayleigh-Bénard problem (Willis, Deardorff, and Somerville, 1972).

A very worthwhile modification due to Thirlby (1970) has been incorporated. This consists in the replacement of the Samarskii (1963) alternating direction implicit scheme, used by Chorin (1968), by the less economical but more accurate three-dimensional Douglas-Peaceman-Rachford scheme (Douglas, 1962).

For the calculations reported below, the number of grid points in the x, y, and z directions were 44, 36, and 17, respectively, at R = 4000 and 48, 40, and 21 at R = 9000. The dimensionless time step was 0.00273 at R = 4000 and 0.00175 at R = 9000. The integration of a case in three space dimensions, from the initial state given above to a final steady or equilibrium unsteady state, typically requires about 1000 time steps, which, with a Fortran code, consumes about ten hours of IBM 360/95 computer time.

RESULTS

We now present the results of two numerical integrations of the system described above, both of which are for air (Prandtl number = 0.7). The specifications of the two cases differ only in the value of the Rayleigh number, R. In the first case, R = 4000, and in the second, R = 9000. Both cases are illustrated by contour plots of the vertical velocity field in the horizontal plane at the mid-level of the fluid ($z = \frac{1}{2}$). The abscissa in each plot is x, and the ordinate is y. The heavy lines are the zero isotachs which separate ascending and descending regions. The contour interval, in units of the dimensionless vertical velocity, is 5 in Figs. 1 - 5, and 8 in Fig. 6.

Figs. 1 - 5 are a time sequence showing the evolution of the flow at R = 4000. This case, except for a different random initial temperature perturbation, is identical in specification to a case integrated with an entirely different numerical method by Lipps and Somerville (1971). It is thus interesting (and gratifying) to note that the final state (Fig. 5) is one of steady rolls with wavelength 3 times the fluid depth, which is the result achieved in the earlier integration and is also consistent with experiment (Willis, Deardorff, and Somerville, 1972). The two integrations differed in the details of the transient evolution to the steady state, however, which may partially be due to the different random initial temperature perturbation. In particular, the present integration converged to the steady state about three times faster than did the previous integration. Because the initial state is not realistic, it is not possible to compare the evolution time to experiment.

The present case is particularly interesting in that the details of the transient evolution correspond quite closely with a description of an experimental phenomenon observed by Willis and Deardorff (1970, p.670):

> "In regions where the rolls are strongly curved (which occur even within a rectangular convection chamber), it seems that the roll or cell diameter often becomes too large for equilibrium to be maintained. Then a new cell appears as an expanding blob at the region of maximum curvature of a roll edge or in the centre of the larger cell. Invariably the new cell fails to remain symmetrically within the adjacent rolls or larger cell, and migrates to one side where it disturbs neighboring rolls until a new quasi-steady pattern is obtained."

Figs. 1 - 4 apparently reproduce this complicated process quite faithfully.* In Fig. 1, the right-hand side of the flow is tending to

*At the conference, computer-generated motion pictures of the evolution of this case and the following one were shown.

rolls, but on the left a large cell of descending air has become "too large for equilibrium to be maintained." In Fig. 2, 150 time steps later, "a new cell appears as an expanding blob...in the center of the larger cell." This rising plume then "migrates to one side where it disturbs neighboring rolls." Because of the periodic side boundary conditions, the plume may be seen breaking out of the descending cell at the left side of Fig. 3 and affecting the neighboring rolls on the right side of the figure. The plume then also breaks through the other side of the cell, and the resulting "new quasi-steady pattern" appears in Fig. 4. The line of cells on the left side then coalesces to create the final steady state of rolls shown in Fig. 5. Thus the numerical simulation seems to have captured in considerable detail the transient three-dimensional evolution by which the flow "solves" the preferred mode question en route to a two-dimensional steady state.

In Fig. 6, the vertical velocity pattern in an unsteady quasi-equilibrium state near the end of the integration at R = 9000 is shown. Diagonally oriented rolls (which are of course permitted by the periodic side boundary conditions) predominate, and non-stationary waves are superimposed upon them. Only the pattern at the mid-level ($z = \frac{1}{2}$) is shown. The wave form, however, is essentially independent of depth. This oscillatory modulation of the rolls is the phenomenon observed in experiments on convection in air at R = 9000 by Willis and Deardorff (1970), which has been treated theoretically by Busse (1972) and independently simulated numerically by F. Lipps (unpublished). An intercomparison of the experimental, theoretical, and numerical findings is in progress, and preliminary results are encouraging. The dimensionless period of the waves in this case is 0.18. Their dimensionless wavelength is 2.6 (the experimental value is 2.5), and the wavelength of the rolls themselves is 3.6 (the experimental value is 3.5). The vertical coherency is also observed experimentally.

CONCLUSION

The numerical model, applied to Rayleigh-Benard convection in air, has simulated time-dependent three-dimensional flows with a considerable degree of realism. Thus, when the model is extended to moist atmospheric convection, some confidence in its dynamical foundation is justified.

ACKNOWLEDGMENTS

All of the computer programming and much of the analysis has been done by Maja Broman, without whose help this work could not have been completed.

Support has been provided at the Goddard Institute for Space Studies by a National Research Council - National Aeronautics and Space Administration Senior Postdoctoral Research Associateship.

REFERENCES

Busse, F.H., J. Math. Phys., 46, 140-150 (1967).

Busse, F.H., J. Fluid Mech., 52, 97-112 (1972).

Chandrasekhar, S., Hydrodynamic and Hydromagnetic Stability, Clarendon, Oxford (1961).

Chorin, A.J., J. Comp. Phys., 2, 12-26 (1967).

Chorin, A.J., Math. Comp., 22, 745-762 (1968).

Douglas, J., Numerische Math., 4, 41-63 (1962).

Krishnamurti, R., J. Fluid Mech., 42, 295-307 (1970).

Lipps, F.B., and Somerville, R.C.J., Phys. Fluids, 14, 759-765 (1971).

Samarskii, A.A., U.S.S.R. Comp. Math. and Math. Phys., 5, 894-926 (1963).

Schlüter, A., Lortz, D. and Busse, F.H., J. Fluid Mech., 23, 129-144 (1965).

Somerville, R.C.J., Geophys. Fluid Dyn., 2, 247-262 (1971).

Thirlby, R., J. Fluid Mech., 44, 673-693 (1970).

Willis, G.E. and Deardorff, J.W., J. Fluid Mech., 44, 661-672 (1970).

Willis, G.E., Deardorff, J.W. and Somerville, R.C.J., J. Fluid Mech., 54, 351-367 (1972).

Fig. 1. Vertical velocity at R = 4000 after 350 time steps

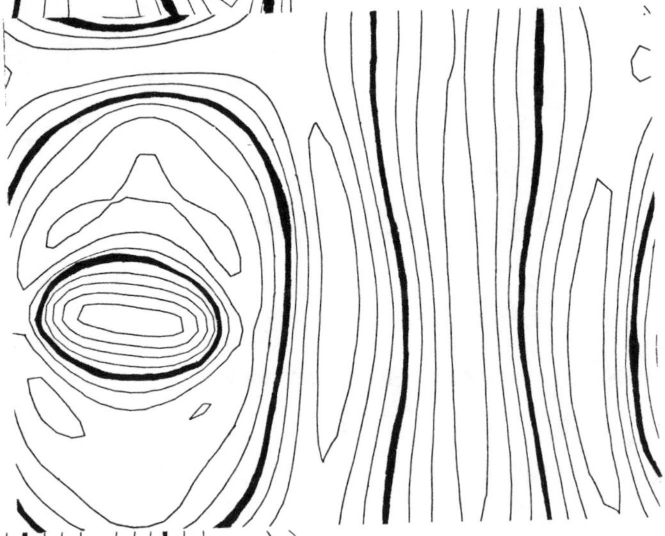

Fig. 2. Vertical velocity at R = 4000 after 500 time steps.

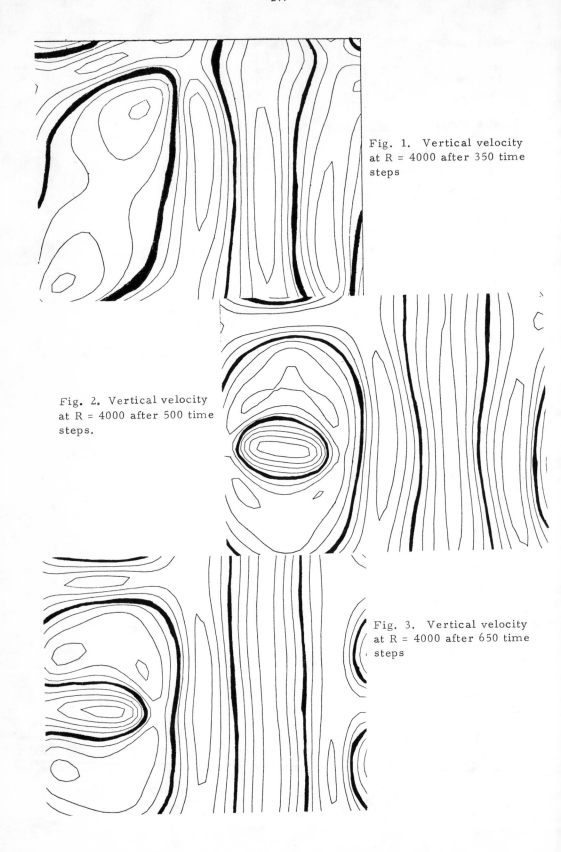

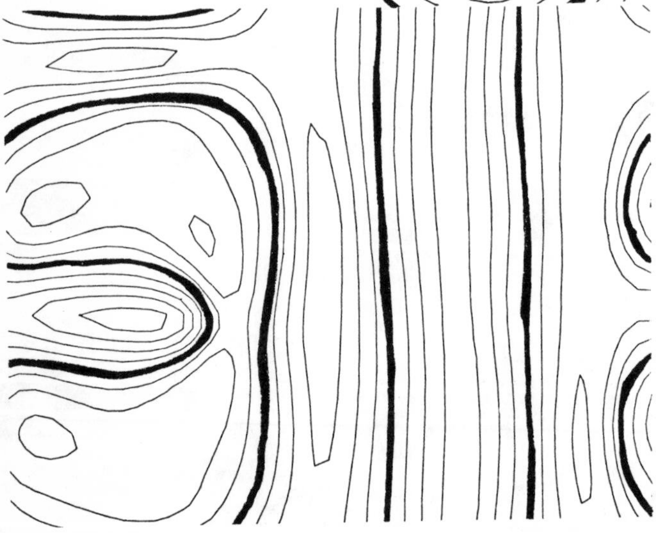

Fig. 3. Vertical velocity at R = 4000 after 650 time steps

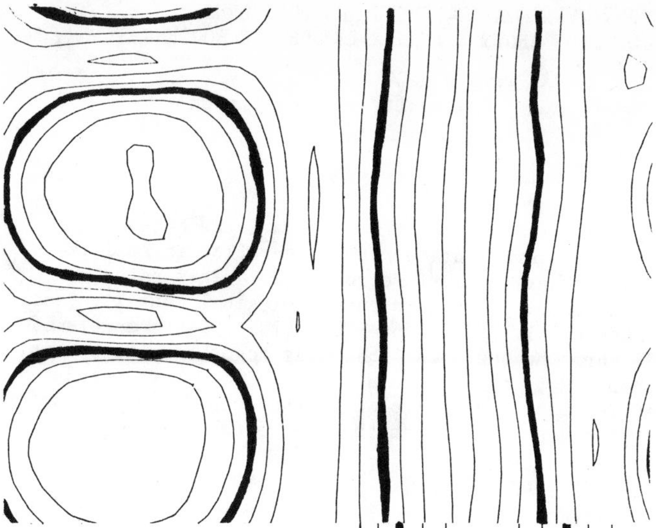

Fig. 4. Vertical velocity at R = 4000 after 800 time steps

Fig. 5. Vertical velocity at R = 4000 after 1100 time steps. The solution is steady.

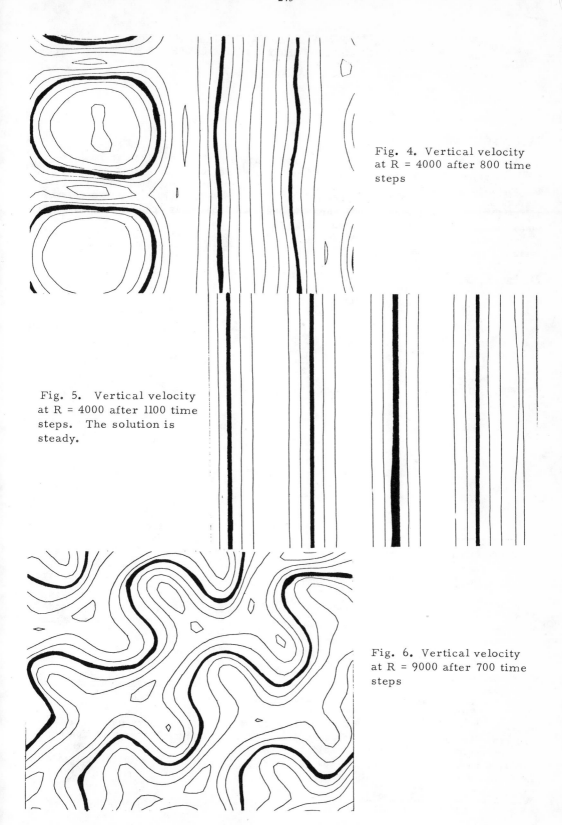

Fig. 6. Vertical velocity at R = 9000 after 700 time steps

SOME COMPARISONS BETWEEN MIXING-LENGTH AND TURBULENT ENERGY EQUATION MODELS OF FLOW ABOVE A CHANGE IN SURFACE ROUGHNESS

P. A. Taylor
Department of Oceanography, University of Southampton

1. Introduction

The aim of this study is simply to make some direct comparisons between several different theoretical models of turbulent boundary-layer flow in applications to steady flow above an abrupt change in surface roughness. This situation has received considerable attention recently in a micrometeorological context, (see reviews by Laikhtman (1970), Panchev et al. (1971) and Plate (1971)). The models of Taylor (1969) and Peterson (1969) form the basis for the study together with a model based on hypotheses used by Glushko (1965) and Novikova (1969). They have been programmed using, as far as possible, the same finite difference schemes and computations have been made for two typical cases of roughness-change flow.

2. Basic Equations and Boundary Conditions

With the usual turbulent boundary-layer approximations (see Reynolds (1968)), no pressure gradient and neglecting viscous shear stresses the two basic equations to be incorporated in all of the models are the horizontal momentum equation,

$$U\frac{\partial U}{\partial x} + W\frac{\partial U}{\partial z} = \frac{\partial \tau}{\partial z} \tag{1}$$

and the continuity equation,

$$\frac{\partial U}{\partial x} + \frac{\partial W}{\partial z} = 0 \tag{2}$$

Here U and W are the mean velocity components as in Fig. 1 while $\tau = -\overline{u'w'}$ is the kinematic Reynolds shear stress.

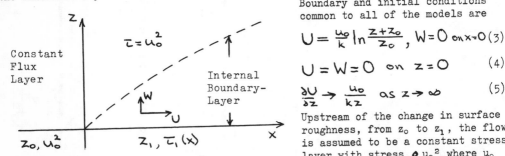

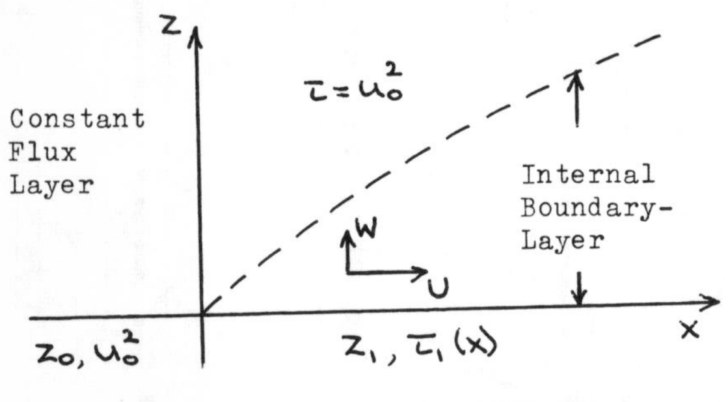

Fig. 1. The Internal Boundary-Layer

Boundary and initial conditions common to all of the models are

$$U = \frac{u_0}{k}\ln\frac{z+z_0}{z_0}, \quad W = 0 \text{ on } x = 0 \tag{3}$$

$$U = W = 0 \quad \text{on } z = 0 \tag{4}$$

$$\frac{\partial U}{\partial z} \rightarrow \frac{u_0}{kz} \quad \text{as } z \rightarrow \infty \tag{5}$$

Upstream of the change in surface roughness, from z_0 to z_1, the flow is assumed to be a constant stress layer with stress ρu_0^2 where u_0 is the surface friction velocity.

To close the system of equations (1) and (2) we need additional information relating τ and the velocity field. One simple way is to use a mixing-length model. Other methods involve the turbulent energy equation which, with appropriate boundary-layer approximations we write in the form

$$U\frac{\partial E}{\partial x} + W\frac{\partial E}{\partial z} = \tau\frac{\partial U}{\partial z} + \frac{\partial}{\partial z}\left(K_E\frac{\partial E}{\partial z}\right) - \varepsilon \tag{6}$$

Here $E = \frac{1}{2}\overline{q^2} = \frac{1}{2}\overline{u'_i u'_i}$ is the turbulent kinetic energy divided by density and ε is the rate of dissipation of turbulent energy. The diffusion of turbulent energy $\left(-\frac{\partial}{\partial z}(\overline{q^2 w'}) - \frac{1}{\rho}\frac{\partial}{\partial z}(\overline{w'p'})\right)$ is assumed here to be a gradient diffusion process with eddy diffusion coefficient K_E. The hypothesis for energy diffusion used by Bradshaw et al. (1967) does not appear to be readily applicable to the problems we are considering. We now describe the features of the individual models used in the intercomparisons.

a) ML - a mixing-length model

This model is essentially the same as that described in Taylor (1969). The finite difference scheme has been changed to use implicit representation of the diffusion terms; the results are virtually identical. The system of equations is closed by postulating the relationship

$$\tau^{1/2} = l_M \frac{\partial U}{\partial z} \tag{7}$$

where l_M, the mixing-length for momentum, is assumed to be of the form

$$l_M = k(z + z_i) \tag{8}$$

Here k is von Karman's constant (taken as 0.4) and $z_i(x)$ is a local value of the roughness length (z_o for $x < o$ and z_1 for $x > o$).

b) GL - a Glushko-type model

In this model we adapt the hypotheses used by Glushko (1965) (see also Beckwith and Bushnell (1968)).to the internal boundary-layer problem. Similar work is described by Novikova (1969) with a different treatment of the mixing-lengths and incorporating thermal effects.

Mean and turbulent quantities are related by assuming

$$\tau = K_M \frac{\partial U}{\partial z} \tag{9}$$

where

$$K_M = E^{1/2} l_G \tag{10}$$

and

$$l_G = k^*(z + z_i) \tag{11}$$

We take the constant $k^* = k/\sqrt{\lambda}$ where λ is the equilibrium constant stress layer value of the ratio τ/E. In the turbulent energy equation we assume $K_E = K_M$ and represent the dissipation by

$$\varepsilon = E^{3/2} / l_{DG} \tag{12}$$

where the "dissipation length", l_{DG} is given by

$$l_{DG} = k_D^*(z + z_i) \tag{13}$$

with $k^*_D = k/\lambda^{3/2}$

The values of the constants k^*, k^*_D are chosen to be consistent with a local balance of production and dissipation of turbulent energy in an equilibrium constant stress layer with the usual logarithmic velocity profile in which $\tau = u_o^2 = \lambda E$.

The numerical value used initially for λ is 0.16 as suggested by Peterson (1969). We would thus have, for an equilibrium constant flux layer, $\tau = 0.16E = 0.08\overline{q}^2$ which, while it appears to be a reasonable value based on micrometeorological evidence, is certainly much lower than that found in "wind-tunnel" boundary layers. Bradshaw et al. (1968) assume $\tau = 0.15\overline{q}^2$ in their model of "flat-plate" boundary-layers; this is in good agreement with values given by Hinze (1959). The results of tests using different values of λ will be given later.

The initial and boundary conditions on E are $E = u_o^2/\lambda$ on $x = 0$; $E \rightarrow u_o^2/\lambda$ as $z \rightarrow \infty$ and $\frac{\partial E}{\partial z} = 0$ on $z = 0$ (i.e. no flux of turbulent energy through the ground).

c) EP - Peterson's (1969) model

The fundamental difference between this and the previous models is that the mean flow and turbulent stress are no longer directly related by the use of an eddy viscosity. In place of this the assumption is made that

$$\tau = \lambda E \tag{14}$$

throughout the entire flow region.

The turbulent energy equation then plays the central role in this model. The assumptions made within it are that

$$K_E = K_M = \tau / \frac{\partial U}{\partial z} \tag{15}$$

and

$$\varepsilon = \tau^{3/2} / l_D \quad \text{with} \quad l_D = k(z + z_i) \tag{16}$$

Equation (16) is equivalent to Peterson's hypothesis with $l_D = kz$ since here the lower boundary condition is to be applied on $z = 0$. The value of λ is again initially taken as 0.16.

The boundary conditions proposed by Peterson for the turbulent energy equation which he writes in the form

$$U\frac{\partial \tau}{\partial x} + W\frac{\partial \tau}{\partial z} = \lambda\tau\frac{\partial U}{\partial z} + \frac{\partial}{\partial z}\left(\frac{\tau}{\partial U/\partial z}\frac{\partial \tau}{\partial z}\right) - \lambda\tau^{3/2}/k(z+z_i) \qquad (17)$$

after substituting for E are;

$$\tau = u_0^2 \text{ on } x = 0 \quad ; \quad \tau \to u_0^2 \text{ as } z \to \infty$$

$$\text{and, on } z = 0 \text{ for } x > 0, \qquad \tau = \left[kz_1 (\partial U/\partial z)_0\right]^2 \qquad (18)$$

This latter condition is essentially a requirement that the velocity profile near the ground is of the form

$$U = \frac{\tau_1^{1/2}}{k}\ln\left(\frac{z+z_1}{z_1}\right) \qquad (19)$$

and in practice is used in this way. $\tau_1(x)$ is the surface kinematic shear stress for $x>0$.

d) PM - the modified Peterson model

The only internal modification made here to the EP model is in the expression for the coefficient of eddy diffusivity for turbulent energy. In place of (15) we use

$$K_E = k(z+z_i)\tau^{1/2} \qquad (20)$$

This assumption has also been made by Shir (1972) in a recent analysis which includes dynamic pressures effects and uses a modified form for the dissipation and mixing lengths.

This was found to be necessary in order to use an alternative surface boundary condition,

$$\frac{\partial \tau}{\partial z} = 0 \quad \text{on } z = 0 \qquad (21)$$

The EP model was found to be inherently unstable when this condition was imposed.

Further details of the models and of the finite difference schemes are given in an appendix (unpublished) to this paper which the author will be pleased to supply on request.

3. Qualitative Features of the Models

The models considered are all of a relatively simple boundary-layer type. Only a single "turbulence" equation has been used and the length scales involved have all been specified a priori. This is in contrast to developments of the von Karman approach with $l \propto \psi/\partial\psi/\partial z$ where ψ is some quantity associated with the velocity field.

The main difference between the models considered is that ML and GL use an eddy viscosity concept in the momentum equation while EP and PM assume instead that $\tau \propto E$. This leads to corresponding mathematical differences in the nature of the governing equations. The ML model gives rise to a single parabolic second order partial differential equation (PDE), essentially in U, to be solved simultaneously with the continuity equation which plays a secondary role in this context. The GL model gives two simultaneous parabolic PDEs together with continuity while the EP and PM models involve a single parabolic PDE (the energy equation) to be solved simultaneously with two first order equations, i.e. momentum and continuity. In the case of the EP and PM models the momentum equation is playing a somewhat secondary role. A gradient diffusion process is assumed for turbulent energy and so we do not have systems of a hyperbolic nature as in the work of Bradshaw et al. (1968).

4. Some results

The two cases considered are a) $M = -5$ and b) $M = +3$ where $M = \ln\frac{z_0}{z_1}$. Case a) corresponds to flow from a relatively smooth to a rough surface while b) is the rough to smooth case. In these initial tests we take $\lambda = 0.16$ in the EP, PM and GL models. Figures 2 and 5 show the downstream variation in (surface shear stress)$^{\frac{1}{2}}$ for cases a) and b) respectively. The mixing length model predicts the most extreme stress values in both cases. For M = +3 we were unable to obtain any results using the EP model. The system of equations appeared to be inherently unstable whatever grid spacings were used both with the basic finite difference scheme and a similar explicit scheme. The instability occurred near the surface very close to the roughness change, a region where any model of the type discussed here will be unrealistic. Peterson (1969) presents results for rough to smooth cases but does not include the region close to the change.

In case a) the choice of surface boundary condition affects the Peterson-type models (EP and PM) considerably near the change in roughness but well downstream the results from EP, PM and (not shown) a PM model with boundary condition (18), are very close. Results from the GL model are generally closer to the ML model than to the Peterson type models.

Velocity profiles at particular downstream locations are compared in Figs. 3 and 6 and shear stresses in Figs. 4 and 7. A major difference between the models' predictions is the presence of pronounced inflection in the EP and PM velocity profiles which is absent in the ML model and present to a much lesser extent in the GL results. While there is considerable evidence of inflection points in observed profiles (see Peterson (1969)) similar effects can be caused by thermal or orographic effects. Another noticeable difference between the models is that the depth of the modified or internal boundary layer as indicated by the shear stress and velocity profiles is greatest in the ML model and least in the Peterson model. Some recent comparisons with data from the Risö tower in Denmark (see Petersen and Taylor (1972)) indicate that internal boundary layer depths predicted by the Peterson model are too low. The shape of the Risö profiles appear to be reasonably close to that predicted by the GL model although direct comparisons have not yet been made.

Figures 2 to 7 represent the basic results of the intercomparisons which are intended mainly to illustrate the differences and similarities between the models. In view of the discrepancy in the values of λ chosen by Peterson (0.16) and Shir (0.22) for the atmosphere and Bradshaw et al. for the "wind tunnel" (0.30) we also ran some numerical experiments with different values of λ using the EP and GL models. Some results with M = -5 are shown in Figs. 8, 9 and 10. As velocities are scaled w.r.t. u_o a higher value of λ corresponds to lower turbulent intensities in the upstream flow and vice versa. We observe that higher values of λ correspond to slightly more pronounced modifications to the flow in the internal boundary layer; i.e. higher stresses and deeper internal boundary layers. This is because the "extra" turbulent energy generated by the roughness change (for $M<0$) represents a greater proportion of the energy when the upstream, or "background" level is lower. The effects of changing λ are more pronounced in the EP model than in the GL case where the velocity profiles are almost indistinguishable by $\frac{x}{z_1} = 10^3$.

5. Conclusions

The basic conclusion that can be drawn is that there are quite considerable differences between the predictions of the ML and GL models on one hand and the EP and PM on the other. Taking a higher value for λ in the EP model reduces these differences a little. Observations indicate that we should expect slight inflection in the velocity profiles downwind of a change in surface roughness. This is absent from the ML results and weak in the GL results. On the other hand the Risö observations suggest that the EP and PM models predictions of the depth of the modified layer is too low. This could be increased by increasing the value of λ but some justification for this would be required.

In carrying out the computations described here we found that the methods used were apparently unconditionally stable for the ML model, reasonably stable for the GL model, less so for the PM and much less so for the EP models. Indeed for M = +3 our EP model appears to be inherently unstable. Times required for computations with the EP model tended to be about 10 to 20 times those required for ML as a result of the restrictions that had to be placed (by trial and error) on the step sizes in the x-direction. So far we have been unsuccessful in determining the exact nature and cause of the instabilities that developed with too large a step size.

Given the task of choosing one of these four models for application to this and similar problems we would recommend the GL model, in part simply because it gives results in about the middle of the range. Another reason is that it is more readily generalised to problems of the boundary-layer type where zeros of shear stress and velocity shear could occur but where we would not expect zeros of turbulent energy or diffusion coefficients. The obvious examples are in studies of pipe or channel flow where a generalisation of mixing-length or Peterson type models would involve infinite values for mixing-length or zero values for λ at the centre-line. A more balanced and informed view must await detailed comparisons with experimental results.

References

BECKWITH I.E. and BUSHNELL D.M. 1968. 'Calculation of mean and fluctuating properties of the incompressible turbulent boundary layer', in Proceedings of the 1968 AFOSR-IFP-STANFORD conference on Computation of Turbulent Boundary Layers.

BRADSHAW P., FERRISS D.H. and ATWELL N.P. 1967. 'Calculation of boundary-layer development using the turbulent energy equation', J. Fluid Mech. 28, pp. 593-616.

GLUSHKO G.S. 1965. 'Turbulent boundary layer on a flat plate in an incompressible fluid', Izv. Akad. Naut. SSSR, Ser. Mek. 4, pp. 13-23; Trans. MinofAv TIL/T 5664.

HINZE J.O. 1959. 'Turbulence', McGraw-Hill, New York.

LAIKHTMAN D.L. 1970. Physics of the Atmospheric Boundary Layer, 2nd edition. GIMIZ, Leningrad (in Russian).

NOVIKOVA S.P. 1969. 'On the problem of transformation of the meteorological elements', Meteorologia i Gidrologia 12, pp. 89-93.

PANCHEV S., DONEV E. and GODEV N. 1971. 'Wind profile and vertical motions above an abrupt change in surface roughness and temperature', Boundary-Layer Meteorol. 2, pp. 52-63.

PETERSEN E.L. and TAYLOR P.A., 1972. 'Some comparisons between observed wind profiles at Risö and theoretical predictions for flow over inhomogeneous terrain'. To appear.

PETERSON E.W. 1969. 'Modifications of mean flow and turbulent energy by a change in surface roughness under conditions of neutral thermal stability', Quart J. Roy. Meteorol. Soc. 95, pp. 561-575.

PLATE E.J. 1971. Aerodynamic Characteristics of Atmospheric Boundary-Layers. AEC Critical Review Series, US Atomic Energy Commission.

REYNOLDS W.C. 1968. 'A morphology of the prediction methods' in Proceedings of the 1968 AFOSR-IFP-STANFORD on Computation of Turbulent Boundary-Layers.

SHIR C.C. 1972. 'A numerical computation of air flow over a sudden change in surface roughness', J. Atmos. Sci. 29, pp. 304-310.

TAYLOR P.A. 1969. 'On wind and shear stress profiles above a change in surface roughness', Quart. J. Roy. Meteorol. Soc. 95, pp. 77-91.

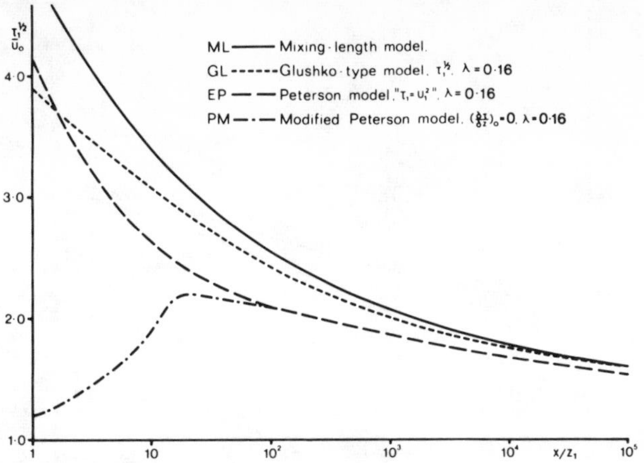

Fig. 2. Downstream variation in (surface shear stress)$^{1/2}$; M = -5

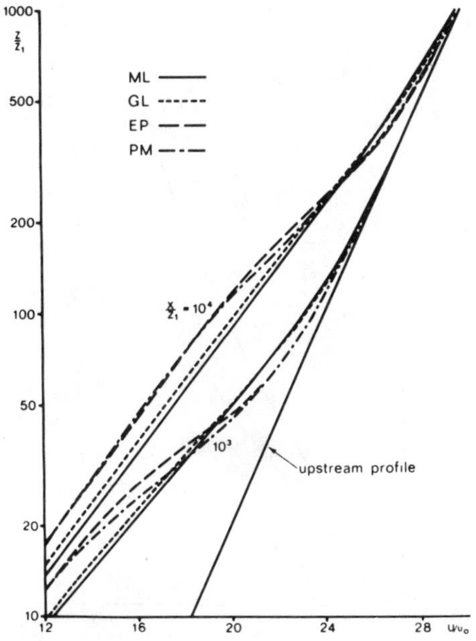

Fig. 3. Velocity profiles at $\frac{x}{z_1} = 10^3, 10^4$; M = -5

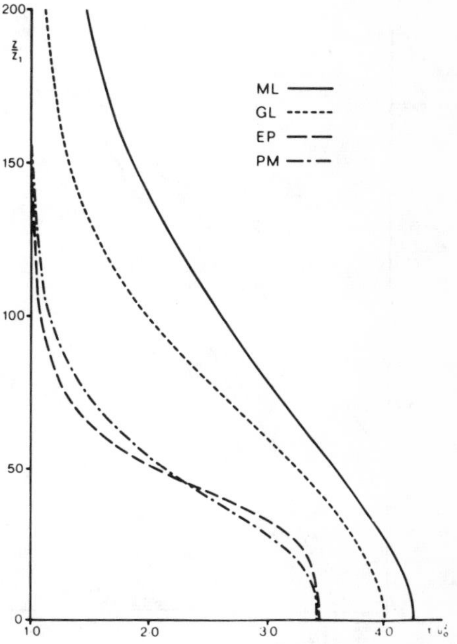

Fig. 4. Shear stress profiles at $\frac{x}{z_1} = 10^3$; M = -5

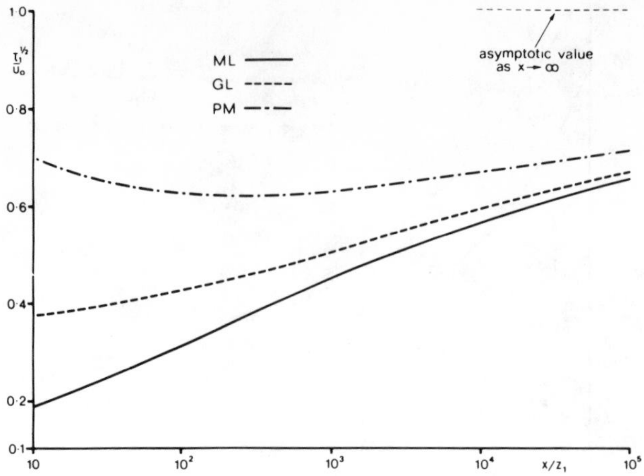

Fig. 5. Downstream variation in (surface shear stress)$^{1/2}$; M = +3

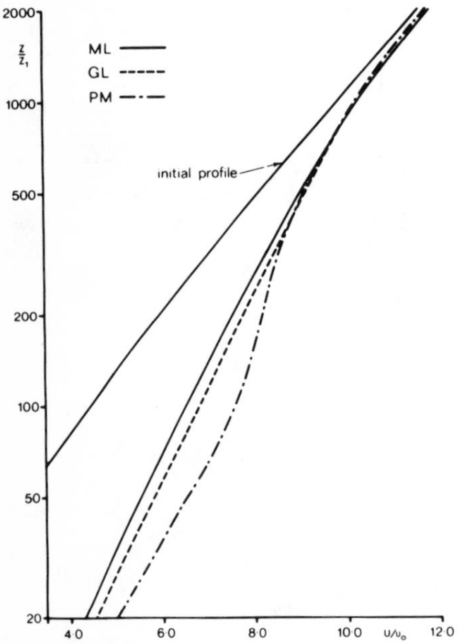

Fig. 6. Velocity profiles at $\frac{x}{z_1}=10^4$; M = +3

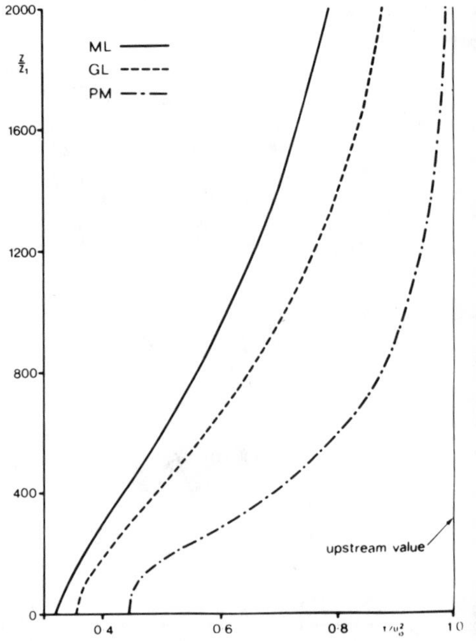

Fig. 7. Shear stress profiles at $\frac{x}{z_1}=10^4$; M = +3

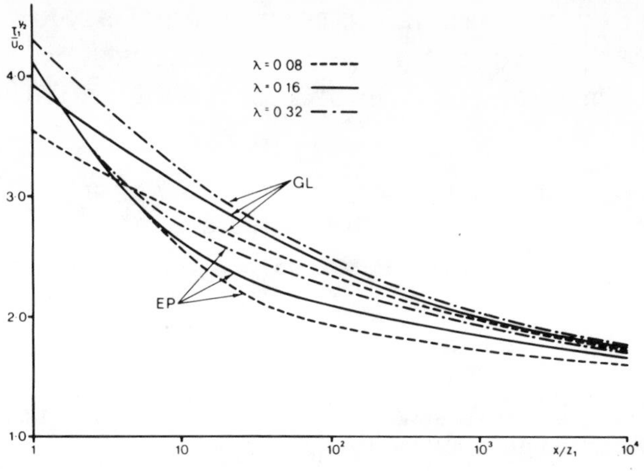

Fig. 8. Downstream variations in (surface shear stress)$^{1/2}$ for different values of λ ; M = -5

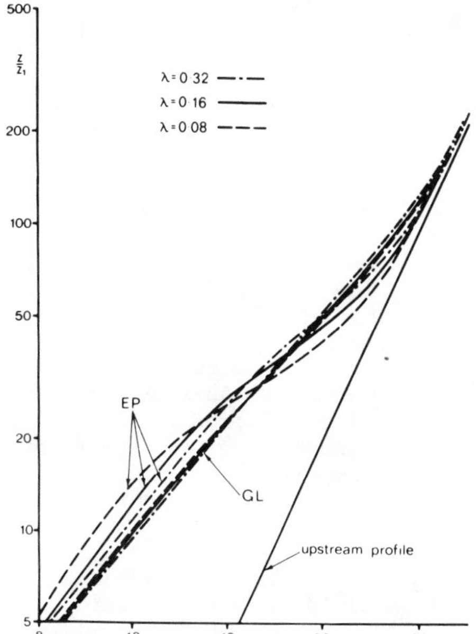

Fig. 9. Velocity profiles at $\frac{x}{z_1}$=10^3 using EP and GL models with different values of λ ; M = -5

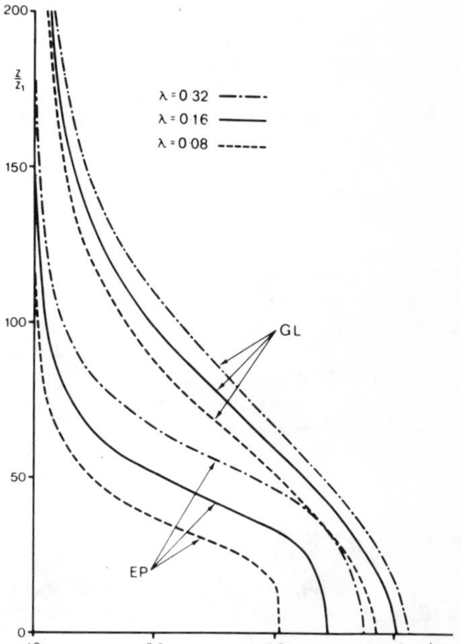

Fig. 10. Shear stress profiles at $\frac{x}{z_1}$=10^3 with different values of λ; M = -5

NUMERICAL SOLUTION OF THE UNSTEADY NAVIER-STOKES EQUATIONS IN CURVILINEAR COORDINATES: THE HYPERSONIC BLUNT BODY MERGED LAYER PROBLEM*

by

Keith J. Victoria and George F. Widhopf

The Aerospace Corporation

Los Angeles, California 90009

INTRODUCTION

The use of the Navier-Stokes (N-S) equations to solve hypersonic low Reynolds/high Knudsen number flow problems has been conjectured and, indeed, such calculations have been carried out by many investigators in the last decade.** In most cases, order of magnitude physical arguments have been utilized to argue the validity of a continuum description of the flow in the layer near a body surface. However, the specific question of whether a solution of the N-S equations provides an accurate description of the flow structure near a body is still unresolved. The resolution of this question is of interest since a continuum approach, at present, is more adaptable to solving complicated fluid-physics problems than the non-continuum method. As a consequence, knowledge of the limits of the N-S equations in describing this type of flow phenomena is of general interest.

The question of the limits of applicability of the N-S equations to transitional flows was directly addressed by Vogenitz and Takata who obtained solutions for the shock layer flow surrounding a blunt body using the Monte Carlo (M. C.) molecular simulation technique originated by Bird. Comparisons were included in that paper between the molecular simulation results and corresponding thin layer continuum solutions obtained by Levinsky and Yoshihara and H. K. Cheng. These comparisons showed substantial differences between the flow field variables in the shock layer region of the flow field as well as in the diffuse shock structure itself. The inability of the N-S equations to properly describe the shock structure was not surprising since many investigators (e.g., Liepmann, et al. 1962, 1964) have shown that the N-S equations do not adequately describe the structure of strong planar shock waves ($M \gtrsim 2$). However, it was not clear whether the disagreement in the shock layer was due to a breakdown in the applicability of the N-S stress-strain model and the Fourier heat conduction law at these conditions, or to the inaccuracy of the thin shock layer equations used to generate the continuum results. Hence, this study was undertaken to provide an accurate continuum solution in order to clarify the issue.

The initial problem chosen was the primary one discussed by Vogenitz and Takata, that is, the flow about a sphere in a Mach 10 flow at a freestream Reynolds number based on nose radius of 152. The corresponding Knudsen number is 0.10 and for purposes of discussion this is termed to be a transitional flow regime. A steady numerical solution of the N-S equations was obtained for these flow conditions by using the leap-frog/Dufort-Frankel difference approximation and relaxing the solution in time. A description of the numerical technique as well as the results obtained in this study, together with comparisons with corresponding M. C. and thin layer solutions, are included in the following sections.

---

* This work was supported by the Advanced Research Projects Agency under SAMSO Contract F04701-71-C-0172.

** A listing of some of these references is included in Dellinger's paper.

MATHEMATICAL FORMULATION

PARTIAL DIFFERENTIAL EQUATIONS

In this study, the compressible N-S equations were employed to model the flow. The generalized coordinate system used in this study is depicted in Figure 1. The nondimensional axisymmetric N-S equations written in this curvilinear coordinate system are

Continuity:

$$\frac{\partial \rho}{\partial t} + \frac{1}{h}\left\{\frac{\partial \rho u}{\partial x} + \kappa \rho v\right\} + \frac{\partial \rho v}{\partial y} + \frac{1}{r}\left\{\rho u \sin\beta + \rho v \cos\beta\right\} = 0 \qquad (1)$$

x-Momentum:

$$\rho\frac{\partial u}{\partial t} + \frac{1}{h}\left\{\rho u\frac{\partial u}{\partial x} + \kappa\rho u v + \frac{\partial p}{\partial x}\right\} + \rho v\frac{\partial u}{\partial y} = \frac{1}{h}\left\{\frac{\partial}{\partial x}\tau_{xx} + 2\kappa\tau_{xy}\right\}$$

$$+ \frac{\partial}{\partial y}\tau_{xy} + \frac{1}{r}\left\{(\tau_{xx} - \tau_{\phi\phi})\sin\beta + \tau_{xy}\cos\beta\right\} \qquad (2)$$

y-Momentum:

$$\rho\frac{\partial v}{\partial t} + \frac{1}{h}\left\{\rho u\frac{\partial v}{\partial x} - \kappa\rho u^2\right\} + \rho v\frac{\partial v}{\partial y} + \frac{\partial p}{\partial y} = \frac{1}{h}\left\{\frac{\partial}{\partial y}\tau_{xy} + \kappa(\tau_{yy} - \tau_{xx})\right\}$$

$$+ \frac{\partial}{\partial y}\tau_{yy} + \frac{1}{r}\left\{\tau_{xy}\sin\beta + (\tau_{yy} - \tau_{\phi\phi})\cos\beta\right\} \qquad (3)$$

Energy:

$$\rho\frac{\partial e}{\partial t} + \frac{1}{h}\rho u\frac{\partial e}{\partial x} + \rho v\frac{\partial e}{\partial y} + \gamma(\gamma-1)M_\infty^2\, p\left\{E_{xx} + E_{yy} + E_{\phi\phi}\right\}$$

$$= -\frac{\gamma}{Pr}\left[\frac{1}{h}\left\{\frac{\partial}{\partial x}q_x + \kappa q_y\right\} + \frac{\partial}{\partial y}q_y + \frac{1}{r}\left\{q_x\sin\beta + q_y\cos\beta\right\}\right] \qquad (4)$$

$$+ \gamma(\gamma-1)M_\infty^2\left\{\tau_{xx}E_{xx} + \tau_{yy}E_{yy} + \tau_{xy}E_{xy} + \tau_{\phi\phi}E_{\phi\phi}\right\}$$

Constituitive:

Stress-Strain Relationship:

$$\tau_{xx} = \mu(2 + s)E_{xx} + s(E_{yy} + E_{\phi\phi}) \quad ; \quad s = -2/3$$

$$\tau_{yy} = \mu(2 + s)E_{yy} + s(E_{xx} + E_{\phi\phi}) \qquad (5)$$

$$\tau_{xy} = \mu E_{xy}$$

Fourier Heat Conduction Law:

$$q_x = -\frac{k}{h}\frac{\partial T}{\partial x} \quad ; \quad q_y = -k\frac{\partial T}{\partial y} \tag{6}$$

Viscosity Law:

$$\mu = \mu(T) = T^{\omega} \quad ; \quad \omega = 1/2 \tag{7}$$

State:

$$p = (\rho\ T)/\left(\gamma M_{\infty}^2\right) \quad ; \quad e = T \quad ; \quad k = \mu \tag{8}$$

Definitions:

Strains:

$$E_{xx} = \frac{1}{h}\left\{\frac{\partial u}{\partial x} + v\kappa\right\} \quad ; \quad E_{yy} = \frac{\partial v}{\partial y}$$

$$E_{xy} = \frac{\partial u}{\partial y} + \frac{1}{h}\left\{\frac{\partial v}{\partial x} - u\kappa\right\} \quad ; \quad E_{\phi\phi} = \frac{1}{r}\left\{u \sin\beta + v\cos\beta\right\} \tag{9}$$

Coordinate Metrics:

$$h(x, y) = 1 + \kappa(x)\ y \quad ; \quad r(x, y) = z(x) + y \cos\beta(x) \quad ;$$

$$\kappa(x) = -d\beta/dx \tag{10}$$

Here ρ, u, v, e, T, and p are density, streamwise component of velocity, normal component of velocity, internal energy, temperature and pressure, respectively. The remaining quantities μ, k, γ, R, M and Pr are viscosity, thermal conductivity, ratio of specific heats, gas constant, Mach number and Prandtl number, respectively. The spatial variables, x and y, are defined in Figure 1 and t denotes the temporal variable. Dimensional variables are denoted by a star superscript and reference quantities are taken to be freestream density, ρ_{∞}^*, velocity, U_{∞}^*, temperature, T_{∞}^*, and nose radius, $R_n^* = 1/\kappa^*(0)$. The variables appearing in the differential equations have been nondimensionalized with respect to their freestream values, where the reference pressure, length and time are defined as: $p_{\infty}^* = \rho_{\infty}^* U_{\infty}^{*2}$, $L_{\infty}^* = \mu_{\infty}^*/(\rho_{\infty}^* U_{\infty}^*)$ and $t_{\infty}^* = \mu_{\infty}^*/(\rho_{\infty}^* U_{\infty}^{*2})$.

This set of partial differential equations subject to the boundary conditions described in the following section were solved numerically using the following finite difference approximations.

FINITE DIFFERENCE EQUATIONS

The difference form of the partial differential equations is obtained by employing the leap-frog/Dufort-Frankel approximation. In this approximation the time and the streamwise spatial derivatives take the form

$$f(x,y,t) = f(j\Delta x, k\Delta y, n\Delta t) = f^n_{jk}$$

$$\bar{f}^n_{jk} = (f^{n+1}_{jk} + f^{n-1}_{jk})/2$$

$$\left(\frac{\partial f}{\partial t}\right)^n_{jk} = \frac{\Delta_n f^n_{jk}}{\Delta t} = \frac{1}{2\Delta t}\left\{f^{n+1}_{jk} - f^{n-1}_{jk}\right\}$$

$$\left(\frac{\partial f}{\partial x}\right)^n_{jk} = \frac{\Delta_j f^n_{jk}}{\Delta x_j} = \frac{1}{(1+R_j)R_j\Delta x_j}\left\{R_j^{\,2}\left(f^n_{j+1k} - f^n_{jk}\right) - \left(f^n_{j-1k} - f^n_{jk}\right)\right\} \tag{11}$$

$$\left(\frac{\partial f}{\partial x}\right)^n_{j+1k} = \frac{\overset{*}{\Delta}_j f^n_{j+1k}}{\Delta x_j} = \frac{1}{(1+R_j)R_j\Delta x_j}\left\{(1+R_j)^2\left(f^n_{j+1k} - \bar{f}^n_{jk}\right) - \left(f^n_{j+1k} - f^n_{j-1k}\right)\right\}$$

$$\left(\frac{\partial f}{\partial x}\right)^n_{j-1k} = \frac{\overset{*}{\Delta}_j f^n_{j-1k}}{\Delta x_j} = \frac{1}{(1+R_j)R_j\Delta x_j}\left\{(1+R_j)^2\left(\bar{f}^n_{jk} - f^n_{j-1k}\right) - R_j^{\,2}\left(f^n_{j+1k} - f^n_{j-1k}\right)\right\}$$

where $R_j = \Delta x_{j-1}/\Delta x_j$. Thus, using these difference approximations, the dependent variables $(\rho, u, v, T)^{n+1}_{jk}$ can be calculated explicitly in terms of the flow quantities at the previous two time levels.

The consistency and accuracy of the finite difference approximation can be established by application to the model equations

$$\frac{\partial u}{\partial t} + \frac{\partial f}{\partial x} = \frac{\partial \sigma}{\partial x} \quad ; \quad \sigma = \epsilon(x)\frac{\partial u}{\partial x} \quad ; \quad f = u^2/2 \tag{12}$$

The corresponding finite difference equations are

$$\left(u^{n+1}_j - u^n_j\right)\left(1+\Lambda_j\right) - \left(u^{n-1}_j - u^n_j\right)\left(1-\Lambda_j\right) + 2\tau_j\Delta_j f^n_j = 2\tau_j\Delta_j\sigma^n_j$$

$$\begin{aligned} \sigma^n_{j+1} &= \epsilon_{j+1}\,\overset{*}{\delta}_j u^n_{j+1}/\Delta x_j \\ \sigma^n_{j-1} &= \epsilon_{j-1}\,\overset{*}{\delta}_j u^n_{j-1}/\Delta x_j \\ \sigma^n_j &= \epsilon_j\,\Delta_j\,u^n_j/\Delta x_j \end{aligned} \tag{13}$$

where

$$\tau_j = \Delta t/\Delta x_j \quad \text{and}$$

$$\begin{aligned} \overset{*}{\delta}_j u^n_{j+1} &= \left\{(1+R_j)^2\left(u^n_{j+1} - u^n_j\right) - \left(u^n_{j+1} - u^n_{j-1}\right)\right\}/\left\{(1+R_j)R_j\right\} \\ \overset{*}{\delta}_j u^n_{j-1} &= \left\{(1+R_j)^2\left(u^n_j - u^n_{j-1}\right) - R_j^{\,2}\left(u^n_{j+1} - u^n_{j-1}\right)\right\}/\left\{(1+R_j)R_j\right\} \end{aligned} \tag{14}$$

$$\Lambda_j = \theta_j\left(R_j^2\,\epsilon_{j+1} + \epsilon_{j-1}\right)/R_j^2 \quad ; \quad \theta_j = \tau_j/\Delta x_j$$

The differential equation actually solved can be simply reconstructed by use of Taylor's series expansion of the finite difference equations. The result is

$$\left(\frac{\partial u}{\partial t}\right)_j^n + \Lambda_j \left\{ \left(\frac{\partial^2 u}{\partial t^2}\right)_j^n \frac{\Delta t}{2!} + \left(\frac{\partial^4 u}{\partial t^4}\right)_j^n \frac{(\Delta t)^3}{4!} + \ldots \right\} + \left(\frac{\partial^3 u}{\partial t^3}\right)_j^n \frac{(\Delta t)^2}{3!} + \ldots$$

$$+ \left(\frac{\partial f}{\partial x}\right)_j^n + R_j \left(\frac{\partial^3 f}{\partial x^3}\right)_j^n \frac{(\Delta x_j)^2}{3!} + \ldots$$

$$= \left(\frac{\partial \sigma}{\partial x}\right)_j^n + 2\epsilon_j \left(\frac{\partial^3 u}{\partial x^3}\right)_j^n \frac{(1-R_j)\Delta x_j}{3!} \tag{15}$$

$$+ \left\{ R_j \left(\frac{\partial^3 \sigma}{\partial x^3}\right)_j^n - 2R_j \left[\frac{\partial}{\partial x}\left(\epsilon \frac{\partial^3 u}{\partial x^3}\right)\right]_j^n \right.$$

$$\left. + \frac{\left(R_j^2 + R_j + 1\right)}{2} \epsilon_j \left(\frac{\partial^4 u}{\partial x^4}\right)_j^n \right\} \frac{(\Delta x_j)^2}{3!} + \ldots$$

Hence, if the second order temporal derivative is small compared with the first order temporal derivative for $\Delta t \Lambda_j/2 = 0(1)$, or $\Delta t \Lambda_j/2 \to 0$ in the limit of vanishing spatial stepsize, the differential equation solved is formally second order accurate, temporally. Stability of the calculation of the (linearized about an order 1 mean flow) unsteady difference equation demands that $\Delta t/\Delta x \leq 1$ ($\Delta t^*/\Delta x^* \leq 1/U^*$) thus, for calculation near the maximum allowable timestep, $\Delta t \Lambda_j/2 = 0(\epsilon_j)$. Numerical experiments performed on unsteady diffusion problems have indicated that temporal consistency is relatively easy to obtain. Formal second order spatial accuracy is maintained for variable mesh spacing, $R_j \neq 1$, if $(1 - R_j) = 0(\Delta x_j)$. For equal mesh spacing, $R_j = 1$, second order accuracy is assured.

The governing equations (Eqs. 1 - 10) were written in the expanded form shown as a result of the following consideration. When the equations were written in conservative form, it was impossible to calculate an accurate u velocity profile near the axis of symmetry, $x = 0$. The troublesome term was discovered to be the derivative $\partial(r\rho u^2)/\partial x$ in the x-momentum equation. As $x \to 0$, $r(x, y) \to 0$ as well as $u \sim x$, hence, this derivative has the form $\partial(xu^2)/\partial x$. The finite difference approximation to a derivative of this type is

$$\frac{\Delta_j (xu^2)_j^n}{\Delta x} = \frac{\partial(xu^2)}{\partial x} + \frac{\partial^3 (xu^2)}{\partial x^3} \frac{(\Delta x)^2}{3!} + \ldots = 3x^2 + (\Delta x)^2 \tag{16}$$

for $u \sim x$. Hence, when $x \sim \Delta x$, the error generated is 1/3 the derivative being approximated. To eliminate this error, the derivative was expanded with use of the continuity equation to the form $r\rho u(\partial u/\partial x)$. The finite difference approximation of this type of derivative, modeled as before is

$$x_j u_j^n \frac{\Delta_j u_j}{\Delta x} = xu \left\{ \frac{\partial u}{\partial x} + \frac{\partial^3 u}{\partial x^3} \frac{(\Delta x)^2}{3!} + \ldots \right\} = x^2 \tag{17}$$

for $u \sim x$, i.e., the truncation error is reduced to zero and the correct representation is recovered. This modification significantly improved the accuracy of the u velocity calculation near the symmetry line. The entire equation set was then expanded to the form shown.

Due to an instability in the calculation of density, which developed in the freestream portion of the flow field after long calculation times, the normal derivative in the continuity equation was recombined to conservative form. It is believed that this (very slow) instability is associated with the non-linearity of the equation set and the neutrally stable difference scheme used and was triggered by very small numerical error in the freestream. A mechanism for partially overcoming this instability is believed to be associated with the added (second order) term in the finite difference representation of the recombined derivative, i.e.,

$$\frac{1}{(hr)_{jk}} \frac{\Delta_k (hr\rho v)^n_{jk}}{\Delta y} = \left[1 + \frac{\kappa \cos\beta}{hr} (\Delta y)^2\right] \frac{\partial \rho v}{\partial y} + \frac{1}{hr}(\kappa r + h\cos\beta)\,\rho v$$

$$+ \frac{1}{hr}(\kappa r + h\cos\beta) \frac{\partial^2 \rho v}{\partial_y^2} \frac{(\Delta y)^2}{2} + S$$

$$\frac{1}{(hr)_{jk}} \left[(hr)_{jk} \frac{\Delta_k (\rho v)^n_{jk}}{\Delta y} + (\kappa r + h\cos\beta)_{jk} (\rho v)^n_{jk}\right] = \frac{\partial \rho v}{\partial y} \tag{18}$$

$$+ \frac{1}{hr}(\kappa r + h\cos\beta)\,\rho v + S$$

where S is additional second order error. The first change (coefficient of the derivative) does not alter the structure of the continuity equation, but the second change is a (second order) diffusion term in y on the right-hand side of the continuity equation (as $v \rightarrow -\sin\beta$). This (albiet) small change stabilized the previously unstable calculation for the density for 3800 cycles (≈ 45 characteristic times). However, even though the shock layer properties converged, the freestream calculation again became unstable. This was overcome by averaging the density in the inhomogeneous term in the continuity equation between the n+1 and n-1 time levels. This change renders the continuity equation unconditionally stable while introducing a second order time inconsistency. The steady state solution was unaffected except for the important fact that the solution was now stable in the freestream as well. This was checked by running the solution to 8000 cycles by which time the entire flow field was converged to six significant figures.

STABILITY OF THE NUMERICAL EQUATIONS

The linear stability characteristics of the numerical scheme used are well known for calculation of one, two and three dimensional flows with respect to coordinate systems which have zero curvature. For axisymmetric flows with curvature, this is not so, since determination of even linear stability characteristics is a formidable task. Two criteria were found for the set of linearized equations in the limit of zero curvature and vanishing mean flow in the x-direction and either $r(x, y) = r(x) \rightarrow 0$ or $r(x) \rightarrow \infty$. These are

$$\Delta t \le \frac{M_\infty \Delta y_{min}}{M_\infty v + \sqrt{T}\left[1 + 1.44(\Delta y_{min}/\Delta x_{min})^2\right]^{1/2}} \qquad r \rightarrow 0 \tag{19}$$

$$\Delta t \le \frac{M_\infty \Delta y_{min}}{M_\infty v + \sqrt{T}\left[1 + (\Delta y_{min}/\Delta x_{min})^2\right]^{1/2}} \qquad r \rightarrow \infty \tag{20}$$

The first of these is the most restrictive and the actual time steps used in the present computations were determined by letting $v \rightarrow 1$ and evaluating the temperature at the stagnation condition. Although approximate, this expression was adequate to determine a stable time step for the numerical calculations performed. The equality sign was used for the numerical calculations.

BOUNDARY CONDITIONS

The boundary conditions for the numerical problem were expressed in the following forms:

Solid Boundary: $(y = 0)$:

No slip conditions were applied at the wall since at cold wall conditions it has been shown (Hayes and Probstein) that slip effects are negligible with respect to their influence on the flow field away from the wall. The wall temperature was taken to be the same as the freestream and the wall density was computed from the unsteady continuity equation applied on the wall. The finite difference equations are of the form

$$(\rho v)_{j0}^{n+1} = - (\rho v)_{j2}^{n-1}$$

$$\Delta_n \rho_{j1}^{n} = - \frac{\Delta t}{2\Delta y_1} \frac{1}{z(x_j)} \left\{ (hr\rho v)_{j2}^{n} - (hr\rho v)_{j0}^{n} \right\}$$

where $h_{j0} = 1 - \kappa \Delta y_1$ and $r_{j0} = z(x_j) - \Delta y_1 \cos \beta_j$.

The solution for ρv was analytically continued outside the physical domain (one row of mesh points into the wall) by a second order extrapolation in space and time of the form

$$f_{j0}^{n+1} = 2 f_{j1}^{n} - f_{j2}^{n-1} + 0 \left[(\Delta y)^2, \Delta y \Delta t, (\Delta t)^2 \right]$$

where f represents a general function. This method allowed the use of centered spatial differencing on the wall, was stable and produced the (required) zero spatial gradient at the wall in the steady limit.

Upstream Boundary: $(y = \Delta = \text{constant})$:

Freestream properties were imposed, i.e., $\rho = T = 1$, $u = \cos \beta$ and $v = - \sin \beta$, where the magnitude of Δ was chosen large enough to completely enclose the diffuse shock wave.

Outflow Plane: $(x = x_{max})$:

All variables were computed by integrating the field equations. Centered spatial differencing was used with a closure scheme similar to that used for computing wall density (using the x coordinate and time).

Stagnation Line: $(x = 0)$:

On the stagnation line, the flow variables and their derivatives have the following properties which can be derived from series expansions of the variables about $x = 0$:

$$F(0, y, \phi) = F(0, y, \phi+\pi)$$

$$\frac{\partial F}{\partial x}(0, y, \phi) = -\frac{\partial F}{\partial x}(0, y, \phi+\pi)$$

where $F = \rho$, $u/\cos\beta$, $v/\sin\beta$, and T. These properties were used to develop the following extrapolation equation to determine the flow variables on the stagnation line

$$F^n_{1k} = \left\{(1+R_2)^2 F^n_{2k} - (R_2)^2 F^n_{3k}\right\}/(1+2R_2)$$

where $R_2 = \Delta x_1/\Delta x_2$. This procedure proved to be both stable and accurate.

INITIAL CONDITIONS

The initial conditions for all x stations were chosen to be the thin layer stagnation line solution profiles of Levinsky and Yoshihara. This procedure is not essential since the flow can be started from uniform conditions with the wall conditions serving as the driving functions.

NUMERICAL RESULTS

The flow about a spherical nosetip at hypersonic flow conditions of $M_\infty = 10$, $Re_\infty = R^*_n/L^*_\infty = 152$ $(K_n = 0.1)$ was the initial problem solved using the numerical technique described. The outflow plane for this calculation was located 73.8^o from the stagnation line. The solution was obtained for a monatomic gas with the properties $\gamma = 5/3$, $\mu = T^{1/2}$, $Pr = 2/3$ and zero bulk viscosity.

The steady state flow field was essentially converged after the flow was relaxed for a characteristic time, $U^*_\infty t^*/R^*_n = 10$ (i.e., ten times the time required for a particle traveling at the freestream velocity to traverse a distance of one sphere radius). The converged solution was checked for global conservation of mass, momentum and total energy using an integration region bounded by a line of constant y in the freestream, the stagnation line, the body and four different downstream positions (lines of constant x). The error in the balances for all conservative quantities, quoted in terms of percentage of freestream inflow, was less than one percent at the x station nearest the stagnation line and monatonically decreased to less than one-half of one percent at the x station nearest the outflow boundary.

Shown in Figures 2 - 5 are the nondimensional distributions of density, the two velocity components and the temperature on the stagnation line of the sphere. Here, each variable has been nondimensionalized with respect to its freestream counterpart. In each of these figures, the results of both the thin layer and the Monte Carlo solutions for the corresponding flow case are also shown. The N-S solutions of Levinsky and Yoshihara and H. K. Cheng are plotted as one solution (designated THIN LAYER) due to their close agreement. The agreement between the present results and the Monte Carlo solution is very good in the region where the Navier-Stokes equations are expected to be valid. This region can be defined using the results obtained by Liepmann, et al. (1962, 1964), where the ability of the N-S equations to describe the structure of planar shock waves was studied. There it was determined that for reasonable convergence of the Chapman-Enskog series, values of the expansion parameter τ/p (ratio of shear stress to pressure) are limited by approximately 0.2. Thus, this parameter was calculated along each body normal to determine the approximate range of validity of the N-S equations in this particular case. Here, the stress, τ, was taken to be the stress in the normal direction, τ_{yy}. A plot of variation of τ/p at the two stations is included in the insert in Figure 2. In view of the results of Liepmann, et al., a value of $|\tau/p| = 0.2$ was selected and the corresponding location has been indicated on each figure. As can be seen, the N-S and M. C. solution are in very good agreement within this region. In fact, the region of agreement is more extensive than that determined using $|\tau/p| = 0.2$ as a criteria. The comparison with the thin layer results is not as good, with substantial disagreement in the region of the layer downstream of the location of the point where $|\tau/p| = 0.2$ as well as upstream of this point. This is best exemplified by the differences in the density and the streamwise velocity distribution as shown in Figures 2 and 4.

Shown in Figures 6 - 9 are the distributions of flow field variables at a station farther downstream on the hemisphere ($\pi/2 - \beta = 73.8^{o}$). At this station only Monte Carlo distributions were available for comparison purposes. Good agreement is noted for all the flow variables in the shock layer region. Again, breakdown occurs upstream of the point where $|\tau/p| = 0.2$.

CONCLUSIONS

The Navier-Stokes solution of this study agrees with the Monte Carlo result in the shock layer region within the statistical scatter of the Monte Carlo calculation. Thus, it appears that the origin of the disagreement between the thin layer Navier-Stokes solution and the molecular simulation solution in the shock layer region is in the thin layer approximations and not in the Navier-Stokes stress-strain model and Fourier heat conduction law. It should be pointed out, however, that at this transitional flow condition ($K_n = 0.10$) the Monte Carlo solution shows that the temperature is not in equilibrium. Thus, for this particular flow condition, the molecular simulation technique is more appropriate.

The study also revealed interesting results regarding the numerical computation procedure in addition to the physical results. In particular, it was found that application of the leap-frog/Dufort-Frankel finite difference approximation to the continuity equation in curvilinear coordinates can lead to numerical instabilities. A simple time-averaging of the density in the inhomogeneous term in the continuity equation eliminated these instabilities. With this modification, it was possible to converge the numerical solution (cycle to cycle) for all flow variables to six significant figures.

Global conservation of mass, momentum and total energy can be accurately maintained with use of the non-conservative form of the Navier-Stokes equations. These quantities were found to be conserved to within one percent of their respective freestream inflow values.

REFERENCES

Bird, G. A., "Shock Wave Structure in a Rigid Sphere Gas," Rarefied Gas Dynamics, edited by J. H. deLeeuw, Supplement 3, Vol. I, 1965, p. 216.

Cheng, H. K., "The Blunt-Body Problem in Hypersonic Flow at Low Reynolds Number," Cornell Aero Lab Report No. AF-1285-A-10, June 1963.

Dellinger, T. C., "Computation of Nonequilibrium Merged Stagnation Shock Layers by Successive Accelerated Replacement," AIAA Preprint No. 69-655, June 16 - 18, 1969.

Hayes, W. D. and Probstein, R. F., Hypersonic Flow Theory, Academic Press, New York, 1959, p.p. 375 - 395.

Levinsky, E. S. and Yoshihara, H., Hypersonic Flow Research edited by F. R. Riddell, Academic Press, New York, 1962, p. 81.

Liepmann, H. W., Narasimha, R. and Chahine, M. T., "Structure of a Plane Shock Layer," Physics of Fluids, Vol. 5, No. 11, November 1962.

Liepmann, H. W., Narasimha, R. and Chahine, M. T., "Theoretical and Experimental Aspects of the Shock Structure Problem," Proc. of the 11th International Congress of Applied Mechanics, Munich, Germany, 1964, Springer-Verlag, edited by Henry Görtler.

Vogenitz, F. W. and Takata, G. Y., "Monte Carlo Study of Blunt Body Hypersonic Viscous Shock Layers," TRW Systems Group Report No. 06488-6470-RO-00, September 1970.

ACKNOWLEDGMENTS

The authors would like to acknowledge the constructive criticism and suggestions given by Prof. Toshi Kubota of the California Institute of Technology and Drs. Frank L. Fernandez and Thomas D. Taylor of The Aerospace Corporation throughout the course of this research. Special mention is also due to Mrs. Leila Jennings of The Aerospace Corporation who constructed a very efficient computer code to carry out the numerical calculations and to Dr. William S. Helliwell of The Aerospace Corporation for suggestions pertaining to the numerical technique.

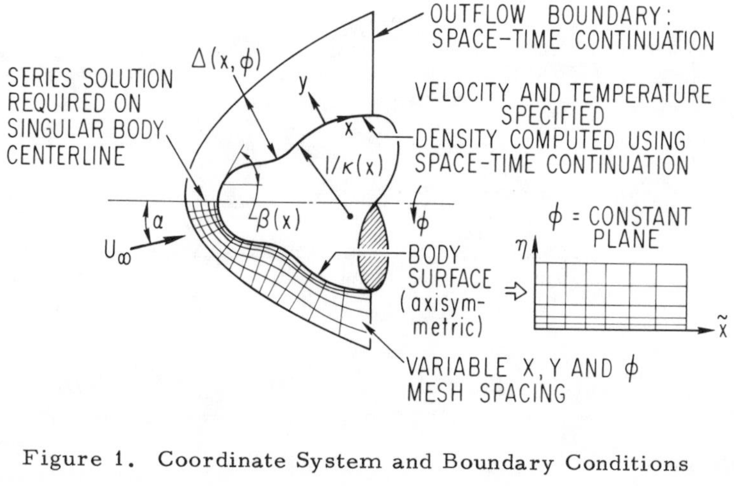

Figure 1. Coordinate System and Boundary Conditions

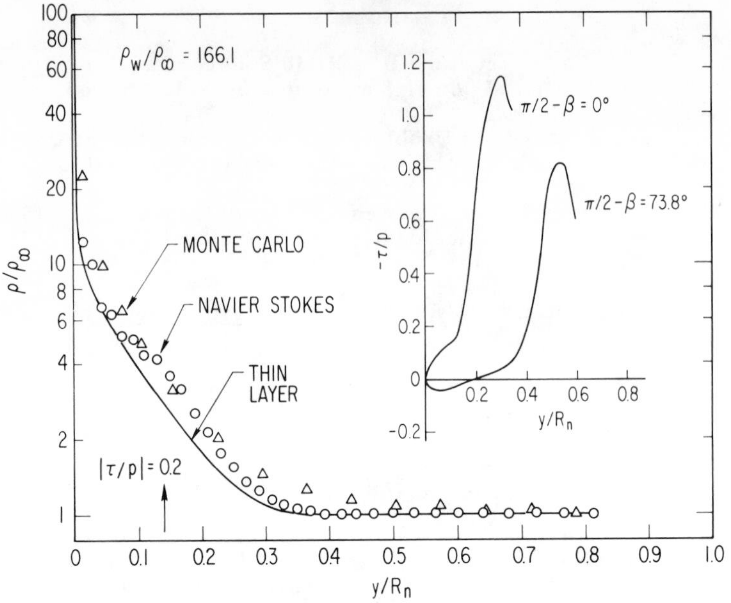

Figure 2. Stagnation Line Density

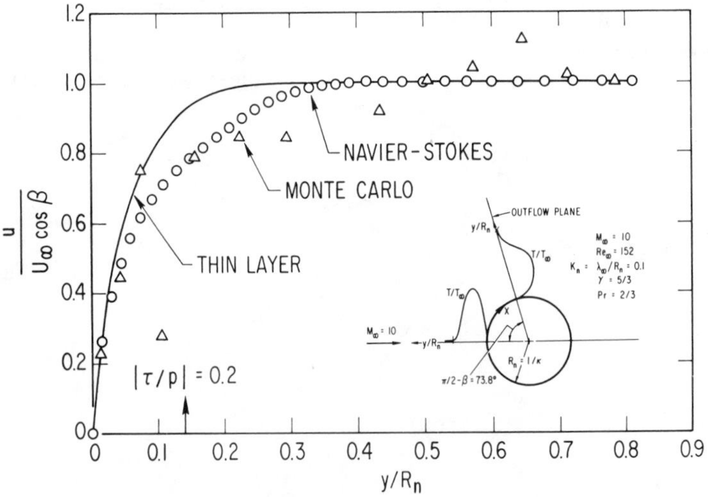

Figure 3. Stagnation Line Tangential Velocity

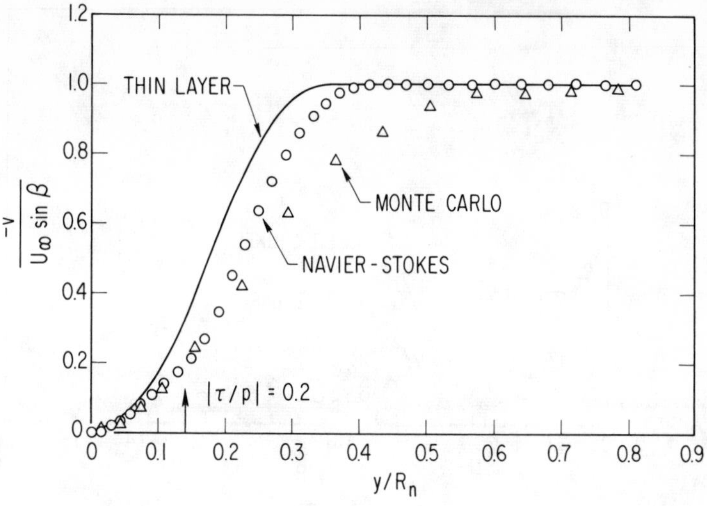

Figure 4. Stagnation Line Normal Velocity

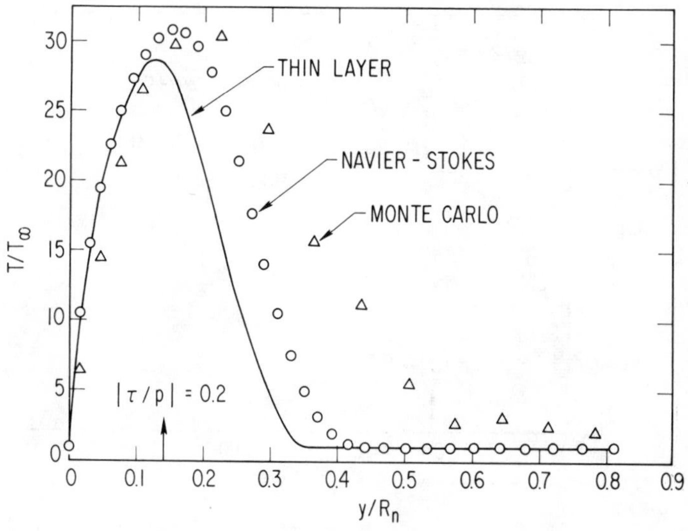

Figure 5. Stagnation Line Temperature

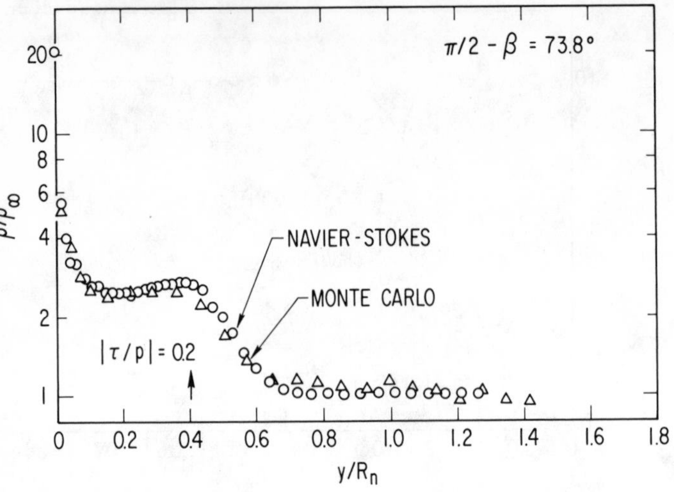

Figure 6. Outflow Plane Density

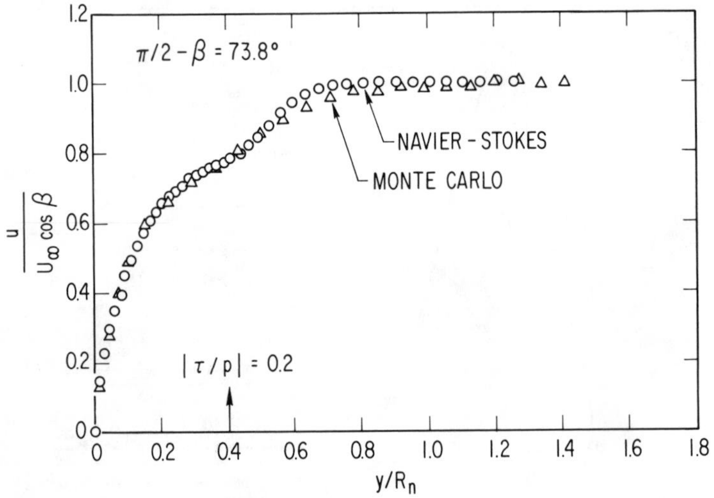

Figure 7. Outflow Plane Tangential Velocity

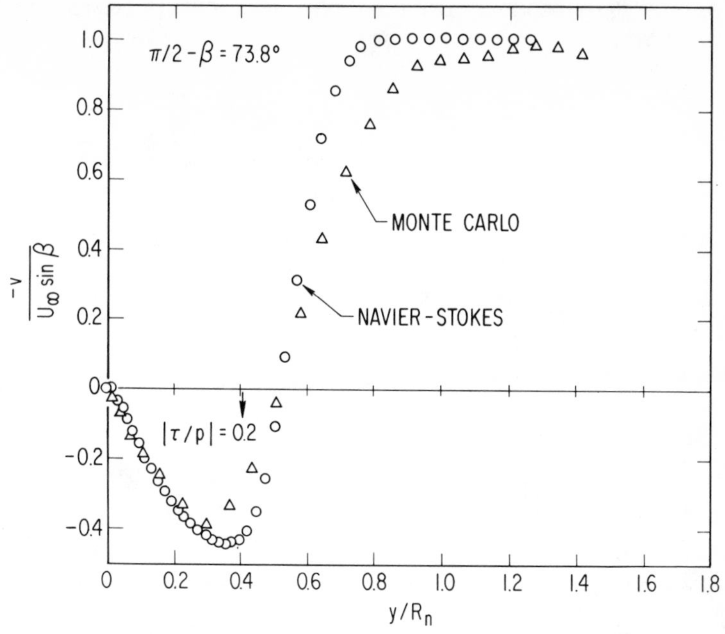

Figure 8. Outflow Plane Normal Velocity

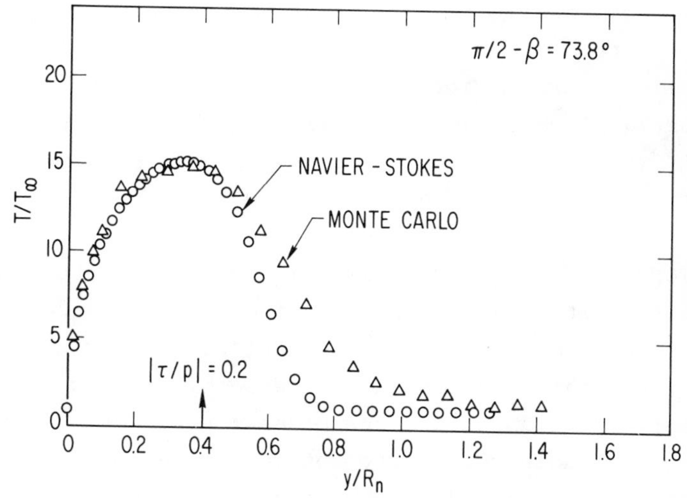

Figure 9. Outflow Plane Temperature

PROBLEME DE MECANIQUE DES FLUIDES, NON LINEAIRES, STATIONNAIRES EN METEOROLOGIE

Par WEILL ALAIN
Laboratoire de météorologie dynamique.
FRANCE

Résumé : On compare dans l'étude de deux problèmes fondamentaux de perturbations d'obstacle dans des fluides stratifiés, la méthode de discrétisation explicite et les solutions qualitatives. On montre la difficulté de l'analyse numérique en présence de discontinuités et la nécessité d'optimiser les solutions analytiques.

INTRODUCTION

Il est d'un très grand intérêt météorologique de connaître la structure des ondes d'obstacle, sans doute responsables de la génération d'une partie de la turbulence en ciel clair. Nous sommes intéressés par les solutions stationnaires lorsqu'elles existent, cette hypothèse étant justifiée par la plupart des observations (GERBIER N.-BERANGER M. 1961), (KUETTNER J.P. 1958), (LILLY D.K. 1971). Nous étudions des mouvements d'échelle horizontale d'une centaine de kilomètres au maximum (mesoscale). Les études faites sur ce sujet (QUENEY P. 1947), (LONG R.R.1953), (BRETHERTON F.P. 1966), (VERGEINER I. 1971), ont été faites avec l'hypothèse des petits mouvements, ce qui conduisait à des solutions linéarisées, mais l'amplitude de ces solutions augmentant avec l'altitude, il a paru nécessaire de reprendre l'ensemble de ces problèmes, en tenant compte de l'amplitude des ondes : problèmes généralement non linéaires. On étudie successivement les perturbations adiabatiques, stationnaires, d'amplitude finie, dans l'écoulement d'une couche plane et verticale d'un fluide stratifié, incompressible, d'étendue horizontale infinie, abordant un obstacle d'amplitude finie. On suppose le fluide non visqueux, et on néglige la force de Coriolis. ζ étant la perturbation lagrangienne, sans dimension d'une ligne de courant par rapport à sa position d'équilibre, nous sommes amenés à envisager les deux modèles A et B.

MODELE A : (Vent horizontal constant, masse volumique donnée par la relation $\rho = \rho_0 \exp - sz$, où ρ_0 est une constante, Z l'altitude, s la stabilité statique du fluide, constante positive) : à l'infini amont de l'obstacle au temps t = 0, écoulement sur un obstacle en forme de courbe de Lorentz.

MODELE B : Courant de Couette à l'infini amont de l'obstacle au temps t = 0 : le vent horizontal $\bar{u} = \pm|\gamma z|$ pour $X = \pm\infty$, le signe + correspond à $-H \leq Z \leq 0$, le signe - pour $0 \leq z \leq H$ où $\pm$H est la hauteur du domaine choisi en amont, de part et d'autre de Z = 0 ; nous avons même répartition de la masse volumique que pour A, Z est l'altitude, X l'abscisse, γ est une constante positive.

Dans les deux cas, nous avons choisi un système de coordonnées cartésiennes orthonormées, bidimensionel.

1. ETUDE du MODELE A

Considérons l'équation aux dérivées partielles, sans dimension, solution :

$$\frac{\partial^2 \zeta}{\partial z^2} + \frac{\partial^2 \zeta}{\partial X^2} + S\left(\left(\frac{\partial \zeta}{\partial Z}\right)^2 + \left(\frac{\partial \zeta}{\partial X}\right)^2 - 2\frac{\partial \zeta}{\partial Z}\right) + \zeta = 0 \qquad 1.1$$

équation dite de LONG
où S est la stabilité statique du fluide (ici sans dimension) : constante et positive, avec les conditions aux limites suivantes :
$\zeta = \zeta_0$, $\frac{\partial \zeta}{\partial z} = \frac{\partial \zeta_0}{\partial Z}$ connues au sol, solutions de l'équation linéarisée : $\zeta = \frac{\alpha a^2}{a^2 + X^2}(\cos Z - X \sin Z)$, ($\alpha$ et a paramètres positifs) ; ce qui élimine la condition à la limite supérieure.

1.1. METHODE de RESOLUTION NUMERIQUE

Nous transformons l'équation aux différences finies pour des mailles rectangulaires de tailles inégales horizontalement (PANOV J. 1951), (VERONIS G. 1971).

$$1.2 \quad \frac{\partial \zeta_{I,J}}{\partial X} \simeq \frac{\zeta_{I+1,J} - \left(\frac{h_I}{h_{I-1}}\right)^2 \zeta_{I-1,J} - \left(1 - \left(\frac{h_I}{h_{I-1}}\right)^2\right) \zeta_{I,J}}{h_I \left(1 + \frac{h_I}{h_{I-1}}\right)} - \frac{h_I\, h_{I-1}}{3!} \frac{\partial^4 \zeta_{I,J}}{\partial X^4}$$

$$1.3 \quad \frac{\partial^2 \zeta_{I,J}}{\partial X^2} \simeq \frac{2\left(\zeta_{I+1,J} + \left(\frac{h_I}{h_{I-1}}\right)\zeta_{I-1,J} - \left(1 + \left(\frac{h_I}{h_{I-1}}\right)\right)\zeta_{I,J}\right)}{h_I\, h_{I-1} \left(1 + \frac{h_I}{h_{I-1}}\right)} - \frac{\left(h_I - h_{I-1}\right)^3}{3} \frac{\partial^3 \zeta_{I,J}}{\partial X^3}$$

pour les différentielles verticales, on prend les différences finies classiques centrées

$$1.4 \quad \frac{\partial \zeta_{I,J}}{\partial Z} \simeq \frac{\zeta_{I,J+1} - \zeta_{I,J-1}}{2K} \; , \quad \frac{\partial^2 \zeta_{I,J}}{\partial Z^2} \simeq \frac{\zeta_{I,J+1} - 2\zeta_{I,J} + \zeta_{I,J-1}}{K^2}$$

où I et J sont les coordonnées d'un point du réseau choisi, h_I est le pas horizontal et k le pas vertical.

On construit un programme de résolution explicite, qui nous permet de progresser de point en point, tout en respectant les conditions aux limites : $\zeta_{0\,I,J}$ et $\zeta_{0\,I,J+1}$ connues au voisinage immédiat du sol.

La méthode utilisée permet de progresser dans le calcul des valeurs de ζ sur ordinateur, avec peu de points en mémoire, mais avec un grand degré de résolution.

On calcule ζ sur l'ensemble de points de grille, de façon à ce que les écarts entre les ζ calculés par les deux méthodes soient des infiniment petits : le nombre d'itérations pour avoir ζ spatialement stable en un point, peut varier de 1 à 1000; le temps de calcul total pour 200 X 20 points ne dépasse pas 10mn sur calculateur C.D.C. 6400 L'infini amont a été remplacé par une troncature du domaine, loin de l'obstacle, en choisissant des valeurs numériques de ζ petites, conformes aux solutions linéarisées.

On peut voir les solutions pour un obstacle large sur la figure 1.1

1.2. SOLUTION ANALYTIQUE

Dans le cas d'un obstacle très large, on peut faire l'approximation $\frac{\partial}{\partial X} \ll \frac{\partial}{\partial z}$, ce qui donne à résoudre l'équation aux dérivées partielles :

$$1.5 \qquad \frac{\partial^2 \zeta}{\partial z^2} + S\left(\left(\frac{\partial \zeta}{\partial z}\right)^2 - 2\,\frac{\partial \zeta}{\partial z}\right) + \zeta = 0$$

Dans ce cas, en utilisant la méthode de variations des constantes puisque S est petit ($S = \frac{s\,\bar{u}}{\sqrt{g s}}$ où $\bar{u}$ est la vitesse de l'écoulement, g la gravité, s la stabilité statique dimensionnée en $[L^{-1}]$), donc en utilisant la méthode de (DAVIS H.T. 1963), (CUNNINGHAM W.J. 1963), (MINORSKY N. 1947), (MILNE W.E. 1953), on obtient une solution qualitative de la forme.

$$1.6 \qquad \zeta \simeq (\exp Sz/D) \left\{ \frac{\cos\left(z + \mathrm{Log}\left(\frac{3/5\,Sz + C}{D}\right)^{\frac{1}{3}}\right)}{\left(\frac{3/5\,Sz + C}{D}\right)^{2/3}} \right\}$$

D et C sont des fonctions de X.

Cette solution bien que qualitative, permet de tracer les surfaces équiphases, calculer les longueurs d'ondes verticales, et montre que les termes quadratiques ne permettent pas de surmonter dans l'atmosphère, le terme d'amplification dû à la décroissance de la masse volumique avec l'altitude : on doit ajouter un terme dissipatif lorsque les gradients de vitesse deviennent importants.

On peut comparer sur la figure (1.2.) l'ensemble de la solution analytique et des travaux expérimentaux de D.K. LILLY à Boulder, qui montre une grande ressemblance entre les résultats théoriques et expérimentaux : tout au moins dans les traits principaux.

a) augmentation de l'amplitude avec Z.

b) même configuration des zones turbulentes.

c) même structure ondulatoire.

$$\begin{pmatrix} a = 1 \\ \alpha = 0,2 \end{pmatrix} \quad 3 = \frac{\alpha a^2}{a^2 + x^2} \quad \frac{\partial 3}{\partial z} = \frac{-\alpha a^2 x}{a^2 + x^2} \quad \bar{u} = 10 \text{ m s}^{-1} \quad s = 1,4 \ 10^{-3}$$

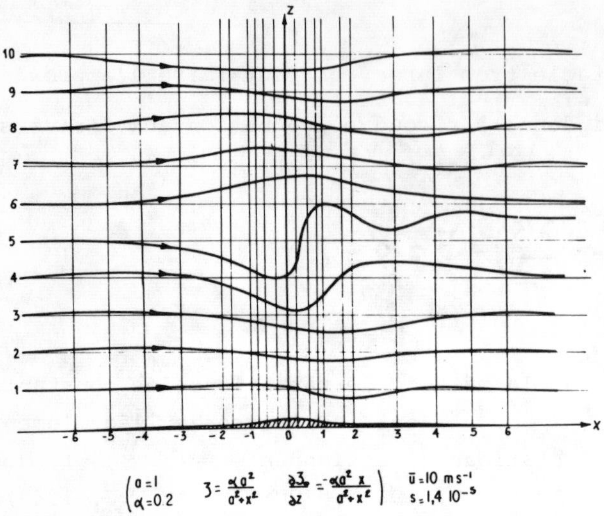

Fig. 1.1. Etude numerique. Ondes d'obstacle pour un obstacle large

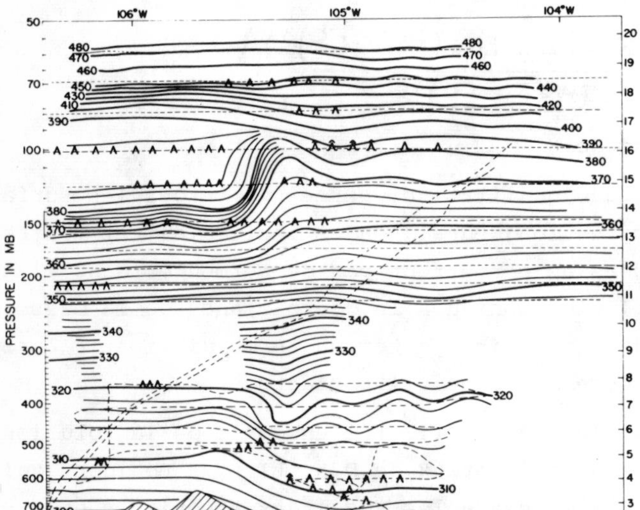

Fig. 1.2. Ondes d'obstacle étude expérimentale par D.K. LILLY à Boulder (Colorado

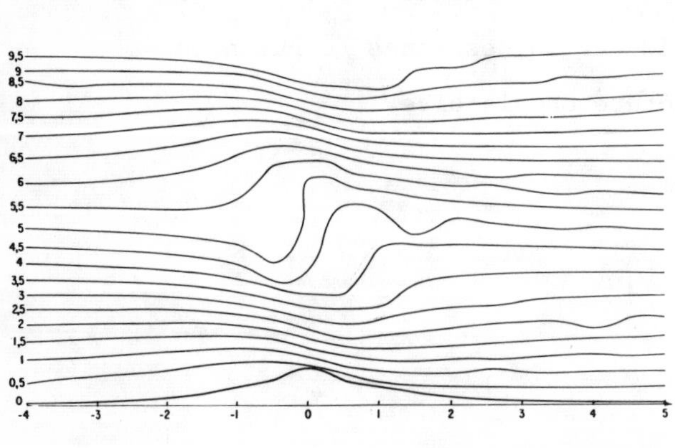

Fig. 1.2. Ondes d'obstacle: étude analytique en amplitude finie (lignes de courant)

2. ETUDE du MODELE B

Considérons l'équation aux dérivées partielles sans dimension, solution

2.1. $$\frac{\partial^2 \zeta}{\partial x^2} + \frac{\partial^2 \zeta}{\partial z^2} + \left(1 - \frac{1}{z-\zeta}\right)\left(\left(\frac{\partial \zeta}{\partial x}\right)^2 + \left(\frac{\partial \zeta}{\partial z}\right)^2 - 2\frac{\partial \zeta}{\partial z}\right) + \frac{R\,\zeta}{(z-\zeta)^2} = 0$$

où $R = \frac{g s}{\gamma^2}$ est le nombre de Richardson, supposé $> \frac{1}{4}$ conformément à la stabilité linéaire (CHANDRASEKHAR S. 1961).

2.1. RESOLUTION NUMERIQUE

Utilisons la même méthode que pour le modèle A, mais comme malheureusement nous ne pouvons pas faire l'hypothèse de non perturbation à l'infini amont, nous ne pouvons progresser que par un seul chemin, si bien que dans le domaine où le vent change de sens, apparaissent des zones tourbillonnaires uniquement dues à l'impossibilité de nous approcher de la zone où le vent s'annule, et seule semble-t-il une méthode, à partir des conditions initiales, nous permettra de résoudre numériquement ce problème.

2.2. SOLUTION ANALYTIQUE du PROBLEME

Si nous supposons l'obstacle beaucoup plus large que haut, $\frac{\partial}{\partial x} \ll \frac{\partial}{\partial z}$, comme en plus on étudie le problème pour $\bar{Z} = Z - \zeta$ petit : $1 \ll \frac{1}{\bar{Z}}$ et nous sommes donc amenés à résoudre l'équation

2.2 $$\bar{Z}^2 \frac{\partial^2 \bar{Z}}{\partial z^2} + \bar{Z}\left(\frac{\partial \bar{Z}}{\partial Z}\right)^2 + \bar{Z}(R-1) - RZ = 0$$

où $\bar{Z}$ est l'altitude de la position d'équilibre du système au temps $t = 0$
En posant $\bar{Z} = \frac{\partial z}{\partial \lambda}$ et en dérivant par rapport à λ
on obtient l'équation paramétrique linéaire

2.3 $$\frac{d^3 \bar{Z}}{d \lambda^3} + (R-1)\frac{d \bar{Z}}{d \lambda} - R\bar{Z} = 0$$

nous avons donc des solutions paramétriques du problème $Z(\lambda)$, $\bar{Z}(\lambda)$, $\zeta(\lambda)$. Nous avons pu déduire un grand nombre de propriétés des solutions, et avons même généralisé le problème à deux variables. (WEILL A. 1971). On met en évidence un système de tourbillons dont le nombre et la forme varient en fonction du nombre de Richardson Fig. (2.1, 2.2). Le problème ultérieur sera d'expliquer les problèmes énergétiques dans la zone de transition. Cette étude est une approche déterministe aux problèmes de la turbulence.

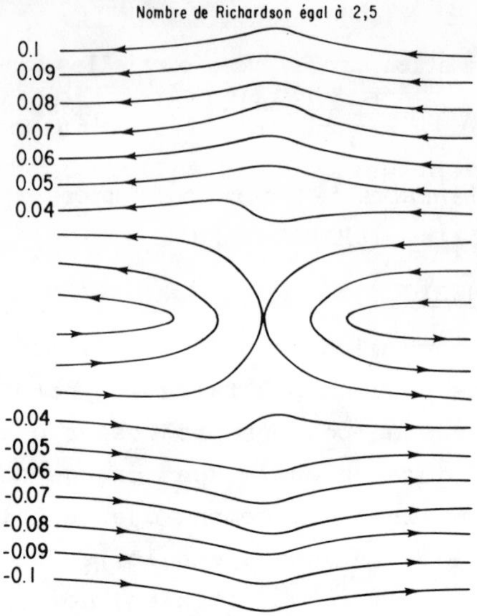

Fig. 2.1 - Ondes de cisaillement avec comme solutions aux limites pour $z_0 = \pm H$ des lignes fluides en forme de courbe de Lorentz

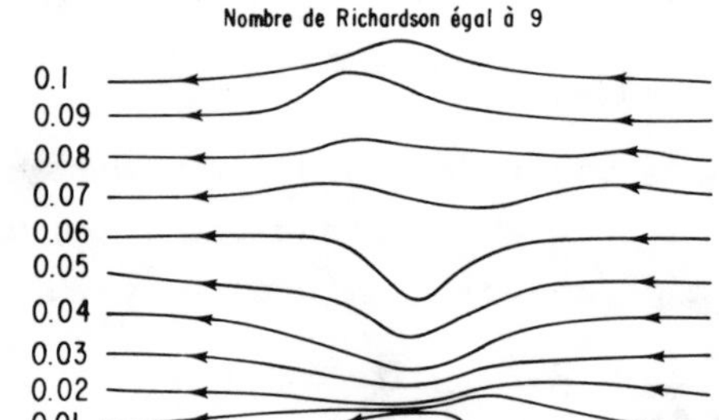

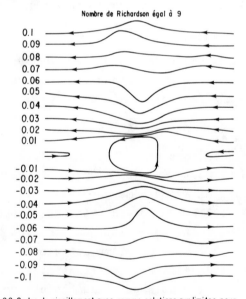

Fig. 2.2 - Ondes de cisaillement avec comme solutions aux limites pour $z_0 = \pm H$ des lignes fluides en forme de courbe de Lorentz

CONCLUSION GENERALE

Lorsque l'on étudie en météorologie des problèmes en amplitude finie, il paraît utile de chercher les solutions analytiques, même qualitatives, ce n'est que grâce à celles-ci que l'on peut avancer dans la connaissance des phénomènes météorologiques. Souvent les méthodes numériques et analytiques se complètent et peuvent être utilisées simultanément; parfois, et c'est le cas dans l'étude des ondes d'inertie dans un courant de Couette, on peut trouver des méthodes analytiques permettant de rendre compte de phénomènes qui échappent aux solutions numériques.

BIBLIOGRAPHIE

BRETHERTON F.P. J. Fluid Mech. 27 (3) 513-539 (1966) : "The propagation of groups of internal gravity waves in a shear flow".
CUNNINGHAM W.J. Dunod Paris (1963) : "Analyse non linéaire".
CHANDRASEKHAR S. Clarendon Press (1961) : "Hydrodynamic and Hydromagnetic stability.
DAVIS H.T. Dover Publication (1963) : "Introduction to non linear differential and integral equations".
GERBIER W. et BERANGER M. J. Roy. Meteorol. Soc. (1961) 87 (371),13-23 "Experimental studies of lee waves in the French Alps".
KUETTNER J. Schweizer Aero Revue (1958) 33, 208-215 : "The rotor flow in the lee of mountains".
LILLY D.K. J.G.R. (1971) : "Observations of mountain induced turbulence"
LONG R.R. Tellus (1953) 15, 42-58 : "Some aspects of the flow of stratified fluids : a theorical investigation".
MILNE W.E. John Wiley § Sons inc. (1953) : "Numerical solution of differential Equations".
MINORSKY N. J.W Edwards Publisher in Mich. (1947) : "Introduction to non linear mechanics".
PANOV J. (1951) "U.R.S.S. formulas for the numerical solution of partial diff. eq. by the method of differences".
QUENEY P. Chicago Press (1947) : "Theory of perturbations in stratified currents with application to air flow over mountains".
VERGEINER I. Quart. J.R. Met. Soc. (1971) 97 pp. 30-60 : "An operational linear lee waves model for arbitrary basic flow and two-dimensional topography".
VERONIS G. Tellus 22, 1 (1971) : "A simple finite difference grid with non constant intervals".
WEILL A. C.R. Acad. Sc. Paris 171, 293-296 (1971) : "Etude analytique des perturbations stationnaires engendrées par une discontinuité simple dans le champ du vent, à l'amont d'un obstacle dans un fluide stratifié".

Lecture Notes in Physics

Bisher erschienen / Already published

Vol. 1: J. C. Erdmann, Wärmeleitung in Kristallen, theoretische Grundlagen und fortgeschrittene experimentelle Methoden. 1969. DM 20,–

Vol. 2: K. Hepp, Théorie de la renormalisation. 1969. DM 18,–

Vol. 3: A. Martin, Scattering Theory: Unitarity, Analyticity and Crossing. 1969. DM 16,–

Vol. 4: G. Ludwig, Deutung des Begriffs physikalische Theorie und axiomatische Grundlegung der Hilbertraumstruktur der Quantenmechanik durch Hauptsätze des Messens. 1970. DM 28,–

Vol. 5: M. Schaaf, The Reduction of the Product of Two Irreducible Unitary Representations of the Proper Orthochronous Quantummechanical Poincaré Group. 1970. DM 16,–

Vol. 6: Group Representations in Mathematics and Physics. Edited by V. Bargmann. 1970. DM 24,–

Vol. 7: R. Balescu, J. L. Lebowitz, I. Prigogine, P. Résibois, Z. W. Salsburg, Lectures in Statistical Physics. 1971. DM 18,–

Vol. 8: Proceedings of the Second International Conference on Numerical Methods in Fluid Dynamics. Edited by M. Holt. 1971. DM 28,–

Vol. 9: D. W. Robinson, The Thermodynamic Pressure in Quantum Statistical Mechanics. 1971. DM 16,–

Vol. 10: J. M. Stewart, Non-Equilibrium Relativistic Kinetic Theory. 1971. DM 16,–

Vol. 11: O. Steinmann, Perturbation Expansions in Axiomatic Field Theory. 1971. DM 16,–

Vol. 12: Statistical Models and Turbulence. Edited by M. Rosenblatt and C. Van Atta. 1972. DM 28,–

Vol. 13: M. Ryan, Hamiltonian Cosmology. 1972. DM 18,–

Vol. 14: Methods of Local and Global Differential Geometry in General Relativity. Edited by D. Farnsworth, J. Fink, J. Porter and A. Thompson. 1972. DM 18,–

Vol. 15: M. Fierz, Vorlesungen zur Entwicklungsgeschichte der Mechanik. 1972. DM 16,–

Vol. 16: H.-O. Georgii, Phasenübergang 1. Art bei Gittergasmodellen. 1972. DM 18,–

Vol. 17: Strong Interaction Physics. Edited by W. Rühl and A. Vancura. 1973. DM 28,–

Vol. 18: Proceedings of the Third International Conference on Numerical Methods in Fluid Mechanics, Vol. I. Edited by H. Cabannes and R. Temam. 1973. DM 18,–

Vol. 19: Proceedings of the Third International Conference on Numerical Methods in Fluid Mechanics, Vol. II. Edited by H. Cabannes and R. Temam. 1973. DM 26,–